I0008963

Applied and Numerical Harmonic Analysis

Series Editor
John J. Benedetto
University of Maryland

Editorial Advisory Board

Akram Aldroubi
NIH, Biomedical Engineering/
Instrumentation

Ingrid Daubechies
Princeton University

Christopher Heil
Georgia Institute of Technology

James McClellan
Georgia Institute of Technology

Michael Unser
NIH, Biomedical Engineering/
Instrumentation

Victor Wickerhauser
Washington University

Douglas Cochran
Arizona State University

Hans G. Feichtinger
University of Vienna

Murat Kunt
Swiss Federal Institute
of Technology, Lausanne

Wim Sweldens
Lucent Technologies
Bell Laboratories

Martin Vetterli
Swiss Federal Institute
of Technology, Lausanne

Wavelet Theory
and Harmonic Analysis
in Applied Sciences

C. E. D'Attellis
E. M. Fernández-Berdaguer
Editors

Springer Science+Business Media, LLC

C. E. D'Attellis
Department of Mathematics
University of Buenos Aires
Paseo Colón 850
1063-Buenos Aires
Argentina

E. M. Fernández-Berdaguer
Instituto de Cálculo
Ciudad Universitaria - Pabellón II
1428-Buenos Aires
Argentina

Library of Congress Cataloging-in-Publication Data

Wavelet theory and harmonic analysis in applied sciences / Carlos E.
    D'Attellis, Elena M. Fernández-Berdaguer, editors.
        p. cm. -- (Applied and numerical harmonic analysis)
    Includes bibliographical references.
    ISBN 978-1-4612-7379-0   ISBN 978-1-4612-2010-7 (eBook)
    DOI 10.1007/978-1-4612-2010-7
    1. Wavelets (Mathematics) 2. Harmonic analysis. I. D'Attellis,
Carlos Enrique.   II. Fernández-Berdaguer, Elena M., 1947-
QA403.3.W374   1997
601 '.5152433--dc21   97-184
                                                                       CIP

Printed on acid-free paper

© 1997 Springer Science+Business Media New York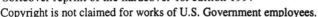
Originally published by Birkhäuser Boston in 1997
Softcover reprint of the hardcover 1st edition 1997
Copyright is not claimed for works of U.S. Government employees.
All rights reserved. No part of this publication may be reproduced, stored in a retrieval system,
or transmitted, in any form or by any means, electronic, mechanical, photocopying, recording,
or otherwise, without prior permission of the copyright owner.

Permission to photocopy for internal or personal use of specific clients is granted by
Springer Science+Business Media, LLC for libraries and other users registered with
the Copyright Clearance Center (CCC), provided that the base fee of $6.00 per copy,
plus $0.20 per page is paid directly to CCC, 222 Rosewood Drive, Danvers, MA
01923, U.S.A. Special requests should be addressed directly to Springer
Science+Business Media, LLC.

ISBN 978-1-4612-7379-0

Typeset by the Editors in LATEX2e.
Cover design by Dutton & Sherman Design, New Haven, CT.

9 8 7 6 5 4 3 2 1

Contents

5   Frames and Riesz bases: a short survey   93

S. J. Favier, R. A. Zalik

6   Fourier Analysis of Petrov-Galerkin Methods Based on Biorthogonal Multiresolution Analyses   119

S. M. Gomes, E. Cortina

9 Characterization of Epileptic EEG Time Series (I): Gabor Transform and Nonlinear Dynamics Methods   179

S. Blanco, S. Kochen, R. Quian Quiroga, L. Riquelme, O. Rosso and P. Salgado

10 Characterization of Epileptic EEG Time Series (II): Wavelet Transform and Information Theory   227

C. D'Attellis, L. Gamero, S. Isaacson, R. Sirne and M. Torres

III Applications in Physical Sciences   263

11 Wavelet Networks for Modelling Nonlinear Processes   265

N. Roqueiro, E. L. Lima

12 Higher order asymptotic boundary conditions for an oxide region in a semiconductor device   301

I. Gamba

Preface

The idea of this book originated in the works presented at the First Latinamerican Conference on Mathematics in Industry and Medicine, held in Buenos Aires, Argentina, from November 27 to December 1, 1995.

A variety of topics were discussed at this meeting. A large percentage of the papers focused on Wavelet and Harmonic Analysis. The theory and applications of this topic shown at the Conference were interesting enough to be published. Based on that we selected some works which make the core of this book. Other papers are contributions written by invited experts in the field to complete the presentation. All the works were written after the Conference.

The purpose of this book is to present recent results as well as theoretical applied aspects of the subject. We have decided not to include a section devoted to the theoretical foundations of wavelet methods for non-specialists. There are excellent introductions already available, for example, Chapter one in Wavelets in Medicine and Biology, edited by A. Aldroubi and M. Unser, 1996, or some of the references cited in the chapter.

The book is addressed to an audience of scientists, engineers, physicists and mathematicians, and contains original articles grouped in three main sections:

1. Theory and Implementation
2. Applications in Biomedical Sciences
3. Applications in Physical Sciences

The first section is addressed to specialists interested in recent theoretical developments in the subject. The other two sections include applications to electroencephalograms, electrocardiograms, geophysics, partial differential equations, identification of nonlinear systems, etc.

We would like to thank Akram Aldroubi for his enthusiastic participation in the Conference and his encouragement in preparing this book. We are also grateful to the authors for their excellent contributions. We thank Abel Lucano for his invaluable editorial help. Finally we would like to thank Wayne Yuhasz and Lauren E. Lavery of Birkhäuser for their patience with international communications which made this book possible.

Carlos Enrique D'Attellis
Elena M. Fernández Berdaguer
Buenos Aires, Argentina

List of Contributors

Akram Aldroubi
National Institute of Health, Biomedical Engineering and Instrumentation Program, Bethesda, Maryland 20892. E-mail: aldroubi@helix.nih.gov

Hugo Aimar
Universidad Nacional del Litoral. INTEC-CONICET, Güemes 3450, (3000)-Santa Fe, Argentina.

Ricardo L. Armentano
University Institute of Biomedical Sciences, Favaloro Foundation. Buenos Aires, Argentina.

Ana Bernardis
Universidad Nacional del Litoral, Güemes 3450, (3000)-Santa Fe, Argentina.

Susana Blanco
Instituto de Cálculo. Facultad de Ciencias Exactas y Naturales, Universidad de Buenos Aires. CONICET. Ciudad Universitaria, Pab. II, (1428)-Buenos Aires, Argentina.

Luis A. Caffarelli
Courant Institute, New York University, New York, NY 10012. E-mail: caffarel@cims.nyu.edu

Elsa Cortina
DIGID. Av Corrientes 1516. (1042)-Buenos Aires, Argentina. E-mail: elsac@is.com.ar

Carlos E. D' Attellis
Universidad Nacional de Buenos Aires. Facultad de Ingeniería, Dpto. de Matemática. Paseo Colón 850. (1063)-Buenos Aires. Argentina. E-mail: ceda@aleph.fi.uba.ar

Marcela A. Fabio
Departamento de Matemática. Facultad de Ciencias Exactas y Naturales, Universidad de Buenos Aires. Ciudad Universitaria, Pab. I, (1428)-Buenos Aires, Argentina.

S. J. Favier
Instituto de Matemática Aplicada, Universidad Nacional de San Luis, 5700 San Luis, Argentina.

Elena Fernández-Berdaguer
Instituto de Cálculo. Facultad de Ciencias Exactas y Naturales, Universidad de Buenos Aires. CONICET. Ciudad Universitaria, Pab. II, (1428)-Buenos Aires, Argentina.
E-mail: efernan@ic.fcen.uba.ar

Irene Gamba
Courant Institute, New York University, New York, NY 10012.
E-mail: gamba@math1.cims.nyu.edu

Lucas G. Gamero
Universidad Nacional de Entre Ríos. Facultad de Ingeniería, Dpto. de Matemática e Informática. C.C. 57, Suc.3. (3100)-Paraná, Entre Ríos. Argentina.

Sônia M. Gomes
IMECC-UNICAMP, Caixa Postal 6065, (13081)- Campinas, SP, Brasil. E-mail: soniag@ime.unicamp.br

Luis Guarracino
Departamento de Geofísica Aplicada, Observatorio Astronómico, Universidad Nacional de La Plata y CIC, La Plata (1900)-Buenos Aires, Argentina

Cristian E. Gutiérrez
Department of Mathematics, Temple University, Philadelphia, PA 19122. E-mail: gutier@euclid.math.temple.edu

Susana I. Isaacson
Universidad Nacional de Buenos Aires. Facultad de Ingeniería, Dpto. de Matemática. Paseo Colón 850. (1063)- Buenos Aires. Argentina.

Silvia Kochen
Centro Municipal de Epilepsia, Hospital Ramos Mejía y CONICET. Buenos Aires, Argentina.

E. L. Lima
Escola de Quimica, Cidade Universitaria, Universidade Federal do Rio de Janeiro, Brasil.

Olivier Meste
Laboratoire D' Informatique, Signaux et Systemes. Université de Nice, Sophia Antipolis. C.N.R.S. E-mail: meste@essi.fr

Rodrigo Quian Quiroga
Institute of Physiology Medical, University of Lübeck, Ratze-burger Alle 160. (23538)-Lübeck, Germany.

Agustín J. Ramírez
University Institute of Biomedical Sciences, Favaloro Foundation. Buenos Aires, Argentina.

Luis Riquelme
Departamento de Fisiología y Biofísica, Facultad de Medicina, Universidad de Buenos Aires.

Hervé Rix
Laboratoire D' Informatique, Signaux et Systemes. Université de Nice, Sophia Antipolis. C.N.R.S.

Marcelo R. Risk
University Institute of Biomedical Sciences, Favaloro Foundation. Buenos Aires, Argentina.

N. Roqueiro
Escola de Quimica, Cidade Universitaria, Universidade Federal do Rio de Janeiro, Brasil. E-mail: nestor@h2o.eq.ufrj.br

Osvaldo A. Rosso
Instituto de Cálculo. Facultad de Ciencias Exactas y Naturales, Universidad de Buenos Aires. CONICET. Ciudad Universitaria, Pab. II, (1428)-Buenos Aires, Argentina.

Pablo Salgado
Centro Municipal de Epilepsia, Hospital Ramos Mejía, Univerisdad de Buenos Aires. Buenos Aires, Argentina.

J. Philip Saul
Department of Cardiology, The Children Hospital, Harvard Medical School, Boston, MA, U.S.A.

Juan E. Santos
Departamento de Geofísica Aplicada, Observatorio Astronómico, Universidad Nacional de La Plata, La Plata, (1900)-Buenos Aires, Argentina and Department of Mathematics, Purdue University, West Lafayette, IN 47907 U.S.A.

Eduardo P. Serrano
Departamento de Matemática. Facultad de Ciencias Exactas y Naturales, Universidad de Buenos Aires. Ciudad Universitaria, Pab. I, (1428)-Buenos Aires, Argentina.

Ricardo O. Sirne
Universidad Nacional de Buenos Aires. Facultad de Ingeniería, Dpto. de Matemática. Paseo Colón 850. (1063)-Buenos Aires. Argentina.

Jamil F. Sobh
Department of Cardiology, The Children Hospital, Harvard Medical School, Boston, MA, U.S.A.

María E. Torres
Universidad Nacional de Entre Ríos. Facultad de Ingeniería, Dpto. de Matemática e Informática. C.C. 57, Suc.3. (3100)-Paraná, Entre Ríos. Argentina.

R. A. Zalik
Department of Mathematics, Auburn University, Auburn, AL 36849-5310, U.S.A. E-mail: zalik@mail.auburn.edu

Part I

Theory and Implementations

Chapter 1

Singular integrals related to the Monge-Ampère equation

Luis A. Caffarelli
Cristian E. Gutiérrez

1 Introduction

The purpose of this note is to describe some results of real analysis related with the Monge-Ampère equation that are proved in [1] and to show its application to the boundedness of certain singular integrals.

The set up is the following. Let $\phi : I\!R^n \to I\!R$ be a convex function which for simplicity is assumed smooth and consider the sets defined as follows: given a point $x \in I\!R^n$ let $\ell(y)$ be a supporting hyperplane of ϕ at the point $(x, \phi(x))$, and given $t > 0$ define the set

$$S(x,t) = S_\phi(x,t) = \{y \in I\!R^n : \phi(y) < \ell(y) + t\}.$$

These sets are obtained by projecting on $I\!R^n$ the points on the graph of ϕ that are below a supporting hyperplane lifted in t. We shall call these sets *sections*. They are convex sets that can give a very degenerate geometry, in the sense that they can very narrow in certain directions as t becomes small. In fact, this is the case when ϕ contains lines in its graph. These degeneracies can be avoided by assuming the following doubling condition. Let μ be the Monge-Ampère measure generated by ϕ, that is, μ is the measure whose density is $\det(D^2\phi(x))$. We assume that there exists a constant $C > 0$ such that

$$\mu(S(x,t)) \leq C\,\mu(\frac{1}{2}\,S(x,t)), \tag{1.1}$$

for every section $S(x,t)$, where $\frac{1}{2}\,S(x,t)$ denotes the $\frac{1}{2}$-dilation of the set $S(x,t)$ with respect to its center of mass. In fact, it is proved in [2] that if μ satisfies the doubling condition (1-1) then the graph of

ϕ does not contain lines, and the sets $S(x,t)$ are of a size that can be controlled by euclidean balls when these sets are re-escaled by using affine transformations. This is true in an essential way because the Monge-Ampère equation

$$M\phi = \det(D^2\phi) = \mu \qquad (1.2)$$

is invariant under affine transformations. This is connected with the following the variant of Fritz John's lemma: *Let S be a convex set in $I\!R^n$ with non-empty interior, and let E be the ellipsoid of minimum volume containing S whose center is the center of mass of S. There exists a constant $\alpha(n)$ depending only on dimension such that*

$$\alpha(n)\, E \subset S \subset E,$$

where $\alpha(n)\, E$ means the $\alpha(n)$-dilation of E with respect to its center.

Since E is an ellipsoid, there is an affine transformation T such that $T(E) = B(0,1)$. In other words, this lemma says that every convex set S with non-empty interior is "similar" to the unit ball at certain scale, i.e., there exists an affine transformation T such that

$$B(0, \alpha(n)) \subset T(S) \subset B(0,1).$$

Given a convex set S with non-empty interior, the set $T(S)$ shall be called a *normalization* of S, and T shall be called an affine transformation that *normalizes S.*

The equation $(1-2)$ is invariant by affine transformations of determinant 1. In fact, if $T : I\!R^n \to I\!R^n$ is an invertible affine transformation, i.e., $Tx = Ax + b$ where A is an $n \times n$ real matrix and $b \in I\!R^n$, and for $\lambda > 0$ we set

$$\psi_\lambda(y) = \frac{1}{\lambda}\phi(T^{-1}y)$$

where ϕ satisfies (1-2), then ψ_λ satisfies the equation

$$M\psi_\lambda(y) = \det(D^2\psi_\lambda)(y) = \frac{1}{\lambda^n}(\det T)^{-2}\mu(T^{-1}y) = \bar{\mu}(y).$$

If μ satisfies (1-1) then by the definition of section it can be shown that the new measure $\bar{\mu}$ is doubling with respect to the sections of ψ_λ, i.e.,

$$\bar{\mu}(S_{\psi_\lambda}(x,t)) \leq C\,\bar{\mu}(\frac{1}{2}\,S_{\psi_\lambda}(x,t))),$$

with the same constant as in (1-1). Moreover, the sections of ϕ and ψ_λ are related in the following way:

$$T(S_\phi(x,t)) = S_{\psi_\lambda}(Tx, t/\lambda).$$

In particular, this formula says that a normalization of a section corresponds to a normalization of the equation $(1-2)$. In other words, if we normalize a section of ϕ we then obtain a section corresponding to another solution of an equation of the form (1-2). These facts and the properties of the sections $S(x,t)$ proved in [2] lead us to consider a general family of convex sets with the following properties.

For each $x \in I\!\!R^n$ we have a one-parameter family of open and bounded convex sets denoted by $S(x,t)$, $t > 0$, $S(x,t) \subset S(x,t')$, for $t \leq t'$, which shall be also called *sections*, and satisfy:

(A) There exist constants K_1, K_2, K_3 and ϵ_1, ϵ_2, all positive, and with the following property:

given two sections $S(x_0, t_0)$, $S(x,t)$ with $t \leq t_0$ such that

$$S(x_0, t_0) \cap S(x,t) \neq \emptyset,$$

and given T an affine transformation that "normalizes" $S(x_0, t_0)$, i.e.,

$$B(0, 1/n) \subset T(S(x_0, t_0)) \subset B(0,1),$$

there exists $z \in B(0, K_3)$ depending on $S(x_0, t_0)$ and $S(x,t)$, such that

$$B\left(z, K_2 \left(\frac{t}{t_0}\right)^{\epsilon_2}\right) \subset T(S(x,t)) \subset B\left(z, K_1 \left(\frac{t}{t_0}\right)^{\epsilon_1}\right),$$

and

$$Tx \in B\left(z, \frac{1}{2}K_2 \left(\frac{t}{t_0}\right)^{\epsilon_2}\right).$$

Here $B(x,t)$ denotes the euclidean ball centered at the point x with radius t.

(B) There exists a constant $\delta > 0$ such that given a section $S(x,t)$ and $z \notin S(x,t)$, if T is an affine transformation that "normalizes" $S(x,t)$ then

$$B(T(z), \epsilon^\delta) \cap T(S(x, (1-\epsilon)t)) = \emptyset,$$

for any $0 < \epsilon < 1$.

(C) $\cap_{t>0} S(x,t) = \{x\}$ and $\cup_{t>0} S(x,t) = \mathbb{R}^n$.

Condition (A) can be phrased in the following way: if two sections intersect then if we normalize the "largest" of the two, the other one looks like a ball with proportional radius at the scale in which the "largest" is normalized.

In addition, we assume that a Borel measure μ finite on compact sets is given, $\mu(\mathbb{R}^n) = +\infty$, and satisfies the following doubling property with respect to the parameter

$$\mu(S(x, 2t)) \leq C\mu(S(x, t)). \tag{1.3}$$

In the case that the sections $S(x,t)$ are coming from the Monge-Ampère equation then condition (1-3) is implied by (1-1).

The main result of [1] is the following variant of the Calderón-Zygmund decomposition.

Theorem 1.1 *Let $S(x,t)$ be an abstract family of sections satisfying properties (A), (B) and (C). Let A be a bounded open subset of $\mathbb{R}^n$, and $0 < \delta < 1$. There exists a countable family of sections $\{S_k = S(x_k, t_k)\}_{k=1}^{\infty}$, $x_k \in A$, and $t_k \leq M$ with the following properties:*

(a) $\frac{\delta}{C_1} \leq \frac{\mu(S_k \cap A)}{\mu(S_k)} \leq \delta$, $C_1 > 0$ *depending only on C in $(1-3)$.*

(b) $A \subset \cup_k S_k$.

(c) $\mu(A) \leq \delta_0 \mu(\cup_1^\infty S_k)$, *where $\delta_0 = \delta_0(\delta, C_2) < 1$, and C_2 is a constant depending only on the parameters in $(1-3)$, (A) and (B), and dimension.*

(d) *If $\tau > 0$ is sufficiently small and $S_k^\tau = S(x_k, (1-\tau)t_k)$ then*

$$\sum_{k=1}^{\infty} \chi_{S(x_k, (1-\tau)t_k)}(x) \leq K \log \frac{1}{\tau},$$

> *where K is a constant depending only on the constants in (A) and (B).*

We remark that by example (b) in §8.8, page 40 of [4], there exists an absolutely continuous doubling measure (with respect to Euclidean balls in $I\!\!R^n$) $\mu = f\,dx$, where f vanishes on a set of positive measure. Therefore, Theorem 1.1 cannot hold for arbitrary bounded measurable sets A.

The properties of the solutions of the linearized Monge-Ampère equation can be naturally described in terms of the sections $S(x,t)$ and a variant of Theorem 1.1 above plays a significant role in this study. The details can be found in [3].

In this note we use some of the tools developed to prove Theorem 1 to show estimates of certain singular integral operators. The organization is as follows. In **2**, we define the Hardy-Littlewood maximal function on sections and we prove that it is of weak-type 1-1. Using this result, we prove in **2** the L^2-boundedness of the singular integrals. This estimate is the starting point for the study of these operators in L^p. The second step is to use Theorem 1 to prove that these operators are of weak-type 1-1. We leave such issue for another occasion.

2 The maximal function

We define the following Hardy-Littlewood maximal function for sections

$$M f(x) = \sup_{\lambda>0} \frac{1}{\mu\left(S(x,\lambda)\right)} \int_{S(x,\lambda)} |f(y)|\,d\mu(y),$$

and we shall show that Mf is of weak-type 1-1 with respect to the measure μ, i.e., there exists a constant $C > 0$ such that

$$\mu\{x : M f(x) > s\} \leq \frac{C}{s} \int_{I\!\!R^n} |f(y)|\,d\mu(y),$$

for all $s > 0$ and any $f \in L^1(I\!\!R^n, d\mu)$. This implies that M is a bounded operator in $L^p(I\!\!R^n, d\mu)$ for $1 < p \leq \infty$.

To prove the weak-type we need the following two lemmas whose proof can be found in [1]. The first is a Besicovitch-type covering lemma for sections.

Lemma 2.1 *Let $A \subset \mathbb{R}^n$ be a bounded set. Suppose that for each $x \in A$ a section $S(x, t)$ is given such that t is bounded by a fix number M. Let us denote by F the family of all these sections. Then there exists a countable subfamily of F, $\{S(x_k, t_k)\}_{k=1}^{\infty}$, with the following properties:*

(i) *$A \subset \cup_{k=1}^{\infty} S(x_k, t_k)$.*

(ii) *$x_k \notin \cup_{j<k} S(x_j, t_j)$, $\forall k \geq 2$.*

(iii) *For $\epsilon > 0$ small (smallness depending only on the constants in A and B) we have that the family*

$$\mathcal{F}_\epsilon = \{S(x_k, (1 - \epsilon)t_k)\}_{k=1}^{\infty}$$

has bounded overlaps, more precisely

$$\sum_{k=1}^{\infty} \chi_{S(x_k,(1-\epsilon)t_k)}(x) \leq K \log \frac{1}{\epsilon},$$

where K is a constant depending only the constants in A and B; χ_E denotes the characteristic function of the set E.

Lemma 2.2 *Let μ be a doubling measure on the sections $S(x, \lambda)$. i.e., μ satisfies $(1 - 3)$. Given $\epsilon > 0$ (small) and $\lambda > 0$ there exists $1 < t \leq 2$, depending on λ and ϵ such that $t - \epsilon \geq 1$ and*

$$\mu(S(x, (t - \epsilon)\lambda)) \geq (1 - \epsilon \log C)\mu(S(x, t\lambda)). \qquad (1.4)$$

Now to the proof of the weak-type 1-1 of M. In order to apply lemma 2.1, we truncate the maximal function by setting

$$M_B f(x) = \sup_{B > \lambda > 0} \frac{1}{\mu\left(S(x, \lambda)\right)} \int_{S(x,\lambda)} |f(y)| \, d\mu(y),$$

where $B > 0$, and consider the set

$$A_s = \{|x| \leq m : M_B f(x) > s\}.$$

Given $x \in A_s$ there exists $S(x, \lambda)$ such that $\lambda < B$ and

$$\frac{1}{\mu\left(S(x, \lambda)\right)} \int_{S(x,\lambda)} |f(y)| \, d\mu(y) > s,$$

and we call $\mathcal{F}$ the family of these sections. Given $\epsilon > 0$, we shall enlarge the family $\mathcal{F}$, i.e., given $S(x, \lambda) \in \mathcal{F}$ by lemma 2.2 there exists $t = t(x, \lambda, \epsilon)$ such that $t - \epsilon \geq 1$, $1 < t \leq 2$. We let $r = t\lambda$ and consider the family $\mathcal{F}' = \{S(\S, \nabla)\}$. Note that

$$S(x, \lambda) \subset S(x, (t - \epsilon)\lambda) \subset S(x, (1 - \frac{\epsilon}{2})r) \subset S(x, r). \qquad (1.5)$$

Also, $S(x, (1 - \frac{\epsilon}{2})r) \subset S(x, 2\lambda)$, and then by the doubling property $\mu\left(S(x, (1 - \frac{\epsilon}{2})r)\right) \leq C_1 \mu\left(S(x, \lambda)\right)$. By lemma 1, we can select a countable subfamily of $\mathcal{F}'$ denoted by $\{S(x_k, r_k)\}_{k=1}^{\infty}$, such that $A_s \subset \cup_k S(x_k, r_k)$ and by doubling we have

$$\mu(A_s) \leq \sum_k \mu\left(S(x_k, r_k)\right) \leq C \sum_k \mu\left(S(x_k, (1 - \frac{\epsilon}{2})r_k)\right). \qquad (1.6)$$

Now note that by (2-2)

$$
\begin{aligned}
s \;&<\; \frac{1}{\mu\left(S(x, \lambda)\right)} \int_{S(x,\lambda)} |f(y)| \, d\mu(y) \\
&\leq\; \frac{\mu\left(S(x, (1 - \frac{\epsilon}{2})r)\right)}{\mu\left(S(x, \lambda)\right)} \frac{1}{\mu\left(S(x, (1 - \frac{\epsilon}{2})r)\right)} \int_{S(x,(1-\frac{\epsilon}{2})r)} |f(y)| \, d\mu(y) \\
&\leq\; \frac{\mu\left(S(x, 2\lambda)\right)}{\mu\left(S(x, \lambda)\right)} \frac{1}{\mu\left(S(x, (1 - \frac{\epsilon}{2})r)\right)} \int_{S(x,(1-\frac{\epsilon}{2})r)} |f(y)| \, d\mu(y) \\
&\leq\; C_1 \frac{1}{\mu\left(S(x, (1 - \frac{\epsilon}{2})r)\right)} \int_{S(x,(1-\frac{\epsilon}{2})r)} |f(y)| \, d\mu(y).
\end{aligned}
$$

This applied to (2-3) yields

$$\mu(A_s) \leq \frac{C}{s} \sum_k \int_{S(x_k,(1-\frac{\epsilon}{2})r_k)} |f(y)| \, d\mu(y) \leq \frac{K \log \frac{1}{\epsilon}}{s} \int_{I\!\!R^n} |f(y)| \, d\mu(y),$$

by lemma 1.

By letting first $m \to \infty$ and then $B \to \infty$, the weak-type 1-1 of M follows.

3 Application to singular integral operators

Let $\rho(x, y) = \phi(y) - \phi(x) - \nabla \phi(x) \cdot (y - x)$, and since we have assumed ϕ smooth $S(x, t) = \{y : \rho(x, y) < t\}$. The properties of the sections imply the following engulfing property: there exists $\theta \geq 1$ such that if $x \in S(y, t)$ then $S(y, t) \subset S(x, \theta t)$. From this property it easy to show that

$$\rho(y, x) \leq \theta \rho(x, y), \tag{1.7}$$

and

$$\rho(x, y) \leq \theta^3 \left(\rho(x, z) + \rho(z, y) \right). \tag{1.8}$$

We actually do not to need to assume that the sections S are coming from a convex function ϕ. It is enough to assume that we have a family of sections satisfying (A), (B), (C), $(1-3)$, and which are given by a function ρ that satisfies (3-1) and (3-2). We recall that $\mu(x) = \det \left(D^2 \phi(x) \right)$, and to simplify the notation we write $S_i(x) = S(x, 2^i)$, $x \in \mathbb{R}^n$, $i \in \mathbb{Z}$.

We shall consider kernels $k(x, y)$ that can be represented in the form

$$k(x, y) = \sum_i k_i(x, y), \tag{1.9}$$

where the k_i's satisfy the following properties:

$$\operatorname{supp} k_i(\cdot, y) \subset S_i(y), \qquad \forall y; \tag{1.10}$$

$$\operatorname{supp} k_i(x, \cdot) \subset S_i(x), \qquad \forall x; \tag{1.11}$$

$$\int_{\mathbb{R}^n} k_i(x, y) \mu(y) \, dy = \int_{\mathbb{R}^n} k_i(x, y) \mu(x) \, dx = 0, \qquad \forall x, y; \tag{1.12}$$

$$\sup_i \int_{\mathbb{R}^n} |k_i(x, y)| \mu(y) \, dy \leq C_1, \qquad \forall x; \tag{1.13}$$

$$\sup_i \int_{\mathbb{R}^n} |k_i(x, y)| \mu(x) \, dx \leq C_1, \qquad \forall y; \tag{1.14}$$

if T is an affine transformation that normalizes the section $S_i(y)$ then k_i satisfies the Lipschitz condition

$$|k_i(T^{-1}u, y) - k_i(T^{-1}v, y)| \leq C_2 \frac{1}{\mu\left(S_i(y)\right)} |u - v|; \tag{1.15}$$

and finally, if T is an affine transformation that normalizes the section $S_i(x)$ then k_i satisfies the Lipschitz condition

$$|k_i(x, T^{-1}u) - k_i(x, T^{-1}v)| \leq C_2 \frac{1}{\mu\left(S_i(x)\right)} |u - v|. \tag{1.16}$$

We shall prove the following theorem.

Theorem 3.1 *Let $k(x, y) = \sum_i k_i(x, y)$ be a kernel as in (3-3) where each k_i satisfies the hypotheses $(3-4)$-$(3-10)$. Then the operator defined by*

$$Hf(x) = \int_{\mathbb{R}^n} k(x, y) f(y) \mu(y) \, dy$$

is bounded in $L^2_\mu(\mathbb{R}^n)$.

Proof:

It uses the method of almost orthogonality developed by Cotlar and our estimates for the maximal function on sections proved in **2**. Given $i \in \mathbb{Z}$, let us define

$$H_i f(x) = \int_{\mathbb{R}^n} k(x, y) f(y) \mu(y) \, dy.$$

Then the adjoint of H_i is given by

$$H_i^* f(y) = \int_{\mathbb{R}^n} k(x, y) f(x) \mu(x) \, dx.$$

A simple calculation shows that

$$H_i \left(H_j^* f \right)(x) = \int_{\mathbb{R}^n} K_{ij}(x, y) f(y) \mu(y) \, dy$$

where

$$K_{ij}(x, y) = \int_{\mathbb{R}^n} k_i(x, z) k_j(y, z) \mu(z) \, dz;$$

and

$$H_i^* \left(H_j f \right)(x) = \int_{\mathbb{R}^n} \tilde{K}_{ij}(x, y) f(y) \mu(y) \, dy$$

where

$$\tilde{K}_{ij}(x, y) = \int_{\mathbb{R}^n} k_i(z, x) k_j(z, y) \mu(z) \, dz.$$

Our goal is to establish that there exist positive constants C and ϵ such that for all $i, j \in \mathbb{Z}$

$$\left\| H_i \left(H_j^* f \right) \right\|_{L^2_\mu} \leq C 2^{-\epsilon |i-j|} \| f \|_{L^2_\mu}$$

and

$$\left\| H_i^* \left(H_j f \right) \right\|_{L^2_\mu} \leq C 2^{-\epsilon |i-j|} \| f \|_{L^2_\mu}.$$

Assuming these estimates, by Cotlar's lemma, see [4], p. 280, the theorem follows. Let us consider the kernel $K_{ij}(x, y)$ and assume

$j \geq i$. By (3-5), $\operatorname{supp}(k_i(x, \cdot)k_j(y, \cdot)) \subset S_i(x) \cap S_j(y)$. If $S_i(x) \cap S_j(y) \neq \emptyset$ and $z \in S_i(x) \cap S_j(y)$ then by (3-1) and (3-2) $\rho(x, y) \leq \theta^3 (2^i + \theta \rho(y, z)) \leq \theta^4(2^i + 2^j) \leq 2\theta^4 2^j$. This implies that $K_{ij}(x, y) = 0$ for $\rho(x, y) > 2\theta^4 2^j$. Let us estimate $K_{ij}(x, y)$ when $\rho(x, y) \leq 2\theta^4 2^j$. By (3-6) and (3-5) we can write

$$
\begin{aligned}
K_{ij}(x, y) &= \int_{\mathbb{R}^n} k_i(x, z)\,(k_j(y, z) - k_j(y, x))\,\mu(z)\,dz \\
&= \int_{S_i(x)} k_i(x, z)\,(k_j(y, z) - k_j(y, x))\,\mu(z)\,dz.
\end{aligned}
$$

Let T_j be the affine transformation that normalizes $S_j(y)$ and in the last integral make the change of variables $z = T_j^{-1}u$. Then

$$
K_{ij}(x, y) = \int_{T_j(S_i(x))} k_i(x, T_j^{-1}u)\left(k_j(y, T_j^{-1}u) - k_j(y, x)\right) \frac{1}{|\det T_j|}\mu(T_j^{-1}u
$$

Hence, by (3-10)

$$
|K_{ij}(x, y)| \leq C_2 \frac{1}{\mu(S_j(y))} \frac{1}{|\det T_j|} \int_{T_j(S_i(x))} |k_i(x, T_j^{-1}u)||u - T_j x|\mu(T_j^{-1}u)\,d
$$

We may assume that $S_i(x) \cap S_j(y) \neq \emptyset$, otherwise $K_{ij}(x, y) = 0$. Hence, by property (A) of the sections and $j \geq i$ we have

$$
T_j(S_i(x)) \subset B\left(z, K_1 \left(\frac{2^i}{2^j}\right)^{\epsilon}\right),
$$

where $|z| \leq K_2$. Therefore

$$
|K_{ij}(x, y)| \leq C_2' \frac{1}{\mu(S_j(y))} \frac{1}{|\det T_j|} 2^{\epsilon(i-j)} \int_{T_j(S_i(x))} |k_i(x, T_j^{-1}u)|\mu(T_j^{-1}u)\,du.
$$

If we make the change of variables $u = T_j T_i^{-1}v$, where T_i is an affine transformation that normalizes $S_i(x)$ we then obtain

$$
|K_{ij}(x, y)| \leq C_2' \frac{1}{\mu(S_j(y))} \frac{1}{|\det T_j|} 2^{\epsilon(i-j)} \int_{T_i(S_i(x))} |k_i(x, T_i^{-1}v)|\mu(T_i^{-1}v)\,dv.
$$

By changing varibles back, $v = T_i u$ and using (3-7) and (3-5) we obtain

$$
|K_{ij}(x, y)| \leq K \frac{1}{\mu(S_j(y))} 2^{\epsilon(i-j)}. \tag{1.17}
$$

Since $\rho(x,y) \leq 2\theta^4 2^j$, it follows that $y \in S(x, 2\theta^4 2^j)$ and by the engulfing property $S(x, 2\theta^4 2^j) \subset S(y, 2\theta^5 2^j)$. Hence, by doubling $\mu\left(S(y, 2^j)\right) \geq C\mu\left(S(x, 2^j)\right)$ and from (3-11) we obtain the estimate

$$|K_{ij}(x,y)| \leq K \frac{1}{\mu\left(S_j(x)\right)} 2^{\epsilon(i-j)},$$

for $j \geq i$ and $\rho(x,y) \leq 2\theta^4 2^j$.

Therefore

$$|K_{ij}(x,y)| \leq K\, 2^{\epsilon(i-j)} \frac{1}{\mu\left(S_j(x)\right)} \chi_{S(x, 2\theta^4 2^j)}(y).$$

Hence,

$$|H_i(H_j^* f)(x)| \leq K\, 2^{\epsilon(i-j)} M_\mu f(x), \qquad \text{for} \quad j \geq i.$$

Since M_μ is bounded in L_μ^2, we obtain the desired estimates. The case when $i > j$ and the operator $H_i^* H_j$ are treated in a similar way.

4   References

[1] L. A. Caffarelli and C. E. Gutiérrez Real Analysis Related to the Monge-Ampère Equation. *Trans. A. M. S.*, vol 348(3): 1075-1092, 1996.

[2] L. A. Caffarelli Some Regularity Properties of Solutions of Monge-Ampère Equation. *Comm. on Pure and App. Math.*, vol XLIV: 965-969, 1991.

[3] L. A. Caffarelli and C. E. Gutiérrez Properties of the Solutions of the Linearized Monge-Ampère Equation. *to appear in Amer. Jour of Math.*

[4] E. M. Stein Harmonic Analysis: Real-Variable Methods, Orthogonality, and Oscillatory Integrals *Princeton Math. Series #43, Princeton U. Press, Princeton, NJ, 1993.*

Chapter 2

Wavelet characterization of functions with conditions on the mean oscillation

Hugo Aimar
Ana Bernardis

1 Introduction

The space BMO of those real functions defined on $I\!R^n$ for which the mean oscillation over cubes is bounded, appears in the pioneer works of J. Moser [13] and John-Nirenberg [9] in the early sixties as a tool for the study of regularity of weak solutions of elliptic and parabolic differential equations. Their main result, known today as John-Nirenberg Theorem, provides a characterization of BMO in terms of the exponential decay of the distribution function on each cube. Although the depth of this result, the space BMO only became well known in harmonic analysis after the celebrated Fefferman-Stein theorem of duality for the Hardy spaces: BMO is the dual of the Hardy space H^1. Since H^1 was already known to be the good substitute of L^1 for many questions in analysis, the space BMO was realized as the natural substitute of L^∞ in the scale of the Lebesgue spaces. Also due to Fefferman and Stein is the Littlewood-Paley type characterization of BMO in terms of the derivatives of the harmonic extension and Carleson measures (see for example [16]). This result has a discrete version: the Lemarié-Meyer characterization of BMO using wavelets, [10].

On the other hand, after the independent works of Campanato [2] and Meyers [12], it became clear that the pointwise Lipschitz α $(0 < \alpha < 1)$ regularity of a function is equivalent to a mean oscillation condition over cubes for that function. Again Lemarié and Meyer [10] for the discrete case and Holschneider and Tchamitchian [6] for the continuous case, gave necesary and sufficient conditions for Lipschitz α in terms of the wavelet coefficients. See also [8].

In 1965, S. Spanne [14], introduced the general setting of $BMO(\rho)$ spaces from which both, BMO and Lipschitz α , are special situations. The aim of this article is to provide the wavelet analysis of the unified approach of Spanne for $BMO(\rho)$ spaces in terms of Carleson type conditions on the wavelet coefficients. As usual when working with $BMO(\rho)$, we do not generaly obtain conditions which are at once necessary and sufficient. Nevertheless in many important cases, for example $\rho(t) = t^\alpha$, we get a characterization. Moreover, we shall notice that the above mentioned conditions on the wavelet coefficients for a compactly supported wavelet are different from those obtained for more general wavelets. It is worth to mention that the wavelet characterization of BMO and Lipschitz α are both recovered from that of $BMO(\rho)$, with proper choices of ρ. The paper is organized as follows: in Section 2 we introduce the wavelet bases of Meyer and Daubechies, in Section 3 we review the basic facts about $\mathcal{MO}$ spaces of functions with conditions on the mean oscillation. In Section 4 we define the Carleson type sequential spaces $\mathcal{C}$. In Section 5 we obtain some necessary conditions for the wavelet coefficients of $\mathcal{MO}$ functions in terms of $\mathcal{C}$ spaces. Section 6 is devoted to obtain norm inequalities of the type $||.||_{\mathcal{MO}} \le C||.||_{\mathcal{C}}$ for finite linear combinations of wavelets. These norm inequalities are proved to be true using the $H^\eta - BMO(\eta)$ duality [17], in Section 7, providing in this way the sufficient condition for $\mathcal{MO}$ in terms of $\mathcal{C}$ sequences.

2   Wavelet Bases

Usually a wavelet is a function ψ of real variable satisfying the following three types of conditions

(i) *smoothness and decay:* $|\partial^l \psi(x)| \le C_{M,l} (1 + |x|)^{-M}$, for all $M > 0$, for all $l \le m_1$ and some $m_1 \in I\!N_0$,

(ii) *vanishing moments:* $\int x^k \psi(x) \, dx = 0$, for all $0 \le k \le m_2$ and some $m_2 \in I\!N_0$,

(iii) *basis:* $\mathcal{B} = \{\psi_{j,k} = 2^{j/2} \psi(2^j x - k) : \quad (j,k) \in \mathbb{Z}^2\}$ is a orthonormal basis of $L^2(I\!R)$.

The set of functions $\mathcal{B}$ is called a wavelet basis for $L^2(I\!R)$. The first example of a basis of wavelets is the well known Haar system, in this case the wavelet is the Haar function $h(x) = \chi_{[0,1/2)} - \chi_{[1/2,1)}$. This function has compact support but has no regularity. The splines of order m generate the classical Battle-Lemarie wavelets with higher regularity. In this work we shall make special use of two wavelet basis: those of Meyer and Daubechies. Meyer gave a wavelet ψ belonging to the class of Schwartz functions $\mathcal{S}$. Daubechies constructed a regular wavelet with compact support. More precisely, given any $N \in I\!N$, there exists a wavelet $\psi \in C^N$ with compact support whose size depends linearly on N.

For a given positive m, we shall write $m[x_0-r, x_0+r] = [x_0-mr, x_0+mr]$. By a dyadic interval we mean one of the form $I_{j,k} = [k2^{-j}, (k+1)2^{-j})$, $j, k \in \mathbb{Z}$, the class of all dyadic intervals shall be denoted by $\mathcal{D}$. For the Daubechies wavelet ψ, for which $\operatorname{supp}\psi \subseteq m[0, 1]$, we have that $\operatorname{supp}\psi_{j,k} \subseteq mI_{j,k}$, for every $j, k \in \mathbb{Z}$.

The basic references for the existence and constructions of all these wavelets are [3], [4] and [11].

3   Functions with conditions on the mean oscillation: $\mathcal{M}\mathcal{O}$ spaces

Even when all the concepts presented here can be defined in the n-dimensional euclidean space, for the sake of simplicity, we shall work with $n = 1$. Let f be a locally integrable real function of a real variable. We shall say that f is a function of bounded mean oscillation, or that $f \in BMO$, if there exist a positive constant C and for each compact interval I a real number C_I such that

$$\frac{1}{|I|} \int_I |f(x) - C_I| \, dx \leq C. \tag{2.1}$$

Let us first notice that condition (3.1) is equivalent to

$$\frac{1}{|I|} \int_I |f(x) - f_I| \, dx \leq C', \tag{2.2}$$

where $f_I = |I|^{-1} \int_I f dx$. The infimum of the constants C' for which (3.2) holds for every interval I is a norm on BMO modulo constant functions, denoted by $||f||_{BMO}$. It is clear from (3.2) that L^{∞} is a subspace of BMO. The function $f(x) = log|x|$ is the basic example of a non-bounded BMO function. The result of John and Nirenberg [9] implies that logaritmic is the maximal growth of BMO functions, since their distributions are of exponential type.

John-Nirenberg: *A function f belongs to BMO if and only if there are positive constants α and β such that for every interval I the inequality*

$$|\{x \in I : |f(x) - f_I| > \lambda\}| \leq \alpha |I| e^{-\beta \lambda} ,$$

holds for every $\lambda > 0$.

Given $\alpha \in (0,1)$ we say that f is a Lipschitz α function, or that $f \in \Lambda_{\alpha}$, if there exists a constant C such that

$$|f(x) - f(y)| \leq C|x - y|^{\alpha} , \tag{2.3}$$

for every x and y in $\mathbb{R}$. The infimum of those constants C is a norm in Λ_{α}, modulo constant functions, denoted by $||f||_{\Lambda_{\alpha}}$. The result of Campanato and Meyers quoted in the introduction is the following

Campanato - Meyers: *Given $\alpha \in (0,1)$, a locally integrable function f satisfies the inequality*

$$\frac{1}{|I|} \int_I |f(x) - f_I| \, dx \leq C|I|^{\alpha} ,$$

for every interval I if and only if f equals almost everywhere a Lipschitz α function.

Given a non-decreasing function $\rho : \mathbb{R}^+ \to \mathbb{R}^+$, we denote by $BMO(\rho)$ the class of all locally integrable functions f, for which there exists a constant C such that

$$\frac{1}{|I|} \int_I |f(x) - f_I| \, dx \leq C\rho(|I|) ,$$

for every interval I. Let $||f||_{BMO(\rho)}$ denote the infimum of those constants C. We can summarize the results of [14] in the next statement. We say that ρ satisfies the Δ_2 condition if $\rho(2t) \leq K\rho(t)$ for some constant K and every $t > 0$. By $\Lambda(\rho)$ we denote the space of functions satisfying an inequality like (3.3) with $\rho(|x - y|)$ instead of $|x - y|^\alpha$. The norm $||f||_{\Lambda(\rho)}$ is defined in the usual way. The previous result by Campanato and Meyers has the following extension due to Spanne [14].

Spanne: *For a given non-decreasing function ρ satisfying the Δ_2 condition, we have that*
(i) $\Lambda(\rho)$ is continuously imbeded into $BMO(\rho)$,
(ii) when $\int_0^1 \frac{\rho(t)}{t} dt < \infty$, $BMO(\rho)$ is contained in $\Lambda(\tilde{\rho})$ with $\tilde{\rho}(t) = \int_0^t \frac{\rho(s)}{s} ds$,
(iii) when $\tilde{\rho}(t) \leq C\rho(t)$, the spaces $BMO(\rho)$ and $\Lambda(\rho)$ coincide,
(iv) when $\frac{\rho(t)}{t}$ is non-increasing and $\int_0^1 \frac{\rho(t)}{t} dt = \infty$, there is a function in $BMO(\rho)$ that is neither bounded nor continuous.

Let us notice that, as is the case for the result of Campanato and Meyers, *(ii)* and *(iii)* hold in the sense of Lebesgue: if $f \in BMO(\rho)$ then f is equals almost everywhere to a $\Lambda(\tilde{\rho})$ function.

Let us finally state the general duality results for $BMO(\rho)$ as proved in [17], actually these results were obtained even in the setting of spaces of homogeneous type. We shall now restrict the class of functions ρ in order to apply the atomic decomposition of the predual of $BMO(\rho)$ as given in [17] (see also [7]). We shall assume now that ρ can be written as $\rho = A + \bar{\rho}$, with $A \geq 0$ and $\bar{\rho}$ a growth function: $\bar{\rho}$ is positive, non-decreasing, $\bar{\rho}(t) \to 0$ when $t \to 0^+$ and has finite upper type, i.e., there exist a finite and positive m and a constant C such that the inequality

$$\bar{\rho}(st) \leq ct^m \bar{\rho}(s),$$

holds for every $t \geq 1$ and every $s > 0$. Let us notice that such ρ satisfies the Δ_2 condition.

Atoms: *Let $\rho(t)$ be a growth function plus a nonnegative constant. A ρ atom, is a function $a(x)$ on $\mathbb{R}$ with support contained in an interval I and satisfying*

$$\int a(x)dx = 0, \tag{2.4}$$

$$||a||_\infty \leq \frac{1}{|I|\rho(|I|)}. \tag{2.5}$$

When $\rho(t) = t^{\frac{1}{p}-1}$, $0 < p \le 1$, a ρ atom is usually called a (p, ∞) atom.

Atomic Hardy spaces: *Let ω be a growth function of positive lower type l, i.e. $\omega(ts) \le Ct^l\omega(s)$, for every $0 < t \le 1$, $s > 0$ and some positive C. If $\rho(t) = \frac{t^{-1}}{\omega^{-1}(t^{-1})}$, we define the atomic space H^ρ, as the linear space of all distributions f in S' which can be represented by $f(\zeta) = \sum_i a_i(\zeta)$, for all ζ in S, where $\{a_i\}_i$ is a sequence of multiples of ρ atoms such that*

$$\sum_i |I_i|\omega(\|a_i\|_\infty) < \infty,$$

where I_i contains the support of a_i and $a_i(\zeta) = \int a_i\zeta$.

Let us notice that, from Lemma 4.7 in [17], we get that also the series $\sum \frac{\|a_i\|_\infty}{\omega^{-1}(|I_i|^{-1})}$ converges. The duality result is the following

Viviani: *Let ω be a function of lower type l, $l > 0$. Assume that $\omega(s)/s$ is non-increasing. Let $\rho(t)$ be the function defined by $\rho(t) = \frac{t^{-1}}{\omega^{-1}(t^{-1})}$. Then $(H^\rho)' \equiv BMO(\rho)$.*

4   Sequential spaces of Carleson type: $\mathcal{C}$ spaces

In [10], Lemarie and Meyer introduce the following Carleson type condition for sequences $\{c_J : J \in \mathcal{D}\}$ of real numbers indexed by the dyadic intervals, the inequality

$$\sum_{\substack{J \in \mathcal{D} \\ J \subseteq I}} |c_J|^2 \le C|I|, \tag{2.6}$$

holds for some constant C and every dyadic interval I. They prove that (4.1), for the wavelet coefficients of a given function f, characterizes the BMO property for f. This result is a discrete version of the Carleson type condition of Fefferman and Stein [5]: let $\psi_t(x) = t\frac{\partial}{\partial t}\varphi_t(x)$, where $\varphi_t(x) = t^{-1}\varphi(x/t)$ and φ is positive, smooth and compactly supported. The result in [5] asserts that, given a function f in BMO the measure μ defined by

$$d\mu(x,t) = |\psi_t * f(x)|^2 \, dx \, \frac{dt}{t},$$

on the Borel sets of $I\!\!R^2_+$ satisfies a Carleson condition

$$\mu(\tilde{I}) \leq C\mu(I),$$

where $\tilde{I} = I \times [0, l]$ and I is a real interval of length l.

In this section we generalize the Carleson condition (4.1) in order to produce spaces of sequences with more general control on the left hand side of (4.1). These spaces are going to describe the natural conditions for the wavelet coefficients of $BMO(\rho)$ functions. Given a non-decreasing function $\rho : I\!\!R^+ \to I\!\!R^+$, we shall say that a real sequence $c = \{c_J : J \in \mathcal{D}\}$ over the dyadic intervals $\mathcal{D}$, belongs to $\mathcal{C}(\rho)$ if there exists a constant A such that

$$\Big(\sum_{\substack{J \in \mathcal{D} \\ J \subseteq I}} |c_J|^2 \Big)^{1/2} \leq A|I|^{1/2}\rho(|I|), \tag{2.7}$$

holds for every $I \in \mathcal{D}$. We will denote $||c||_{\mathcal{C}(\rho)}$ the infimum of those constants A, so that $\mathcal{C}(\rho) = \{c = \{c_J\} : ||c||_{\mathcal{C}(\rho)} < \infty\}$.

5 $\mathcal{C}$ as a necessary condition for the wavelet coefficients of $\mathcal{MO}$ functions

5.1 The Daubechies wavelet case

Theorem A: Let $\{\psi_{j,k} : (j,k) \in \mathbb{Z}^2\}$ a Daubechies wavelet basis. Let ρ be a positive non-decreasing function defined on $I\!\!R^+$ satisfying the Δ_2 condition. If $f \in BMO(\rho)$, then $c = \{c_{j,k} =< f, \psi_{j,k} >: (j,k) \in \mathbb{Z}^2\}$, the wavelet coefficient sequence of f, belongs to $\mathcal{C}(\rho)$, with the usual identification of $\mathcal{D}$ with $\mathbb{Z}^2$, and

$$||c||_{\mathcal{C}(\rho)} \leq C||f||_{BMO(\rho)}.$$

Proof: Let I be a dyadic interval. We write $f(x) = f_1(x) + f_2(x) + f_I$, where $f_1(x) = (f(x) - f_I)\chi_{mI}(x)$ and m is such that supp $\psi \subseteq m[0,1]$. Then, since supp $\psi_{j,k} \subseteq mI_{j,k}$ and ψ has zero integral, from Bessel inequality, we get that

$$
\left(\sum_{I_{j,k} \subseteq I} |c_{j,k}|^2 \right)^{1/2} = \left(\sum_{I_{j,k} \subseteq I} | <f_1, \psi_{j,k}> |^2 \right)^{1/2}
$$

$$
\leq \|f_1\|_2 = \left(\int_{mI} |f(x) - f_I|^2 \, dx \right)^{1/2}
$$

$$
\leq \left(\int_{mI} |f(x) - f_{mI}|^2 \, dx \right)^{1/2} + |mI|^{1/2}|f_{mI} - f_I|
$$

$$
\leq \left(\int_{mI} |f(x) - f_{mI}|^2 \, dx \right)^{1/2}
$$
$$
+ C\|f\|_{BMO(\rho)}|mI|^{1/2}\rho(|mI|).
$$

The second term in the right hand side of this inequality has the desired form since, by the Δ_2 condition, is bounded by $C\|f\|_{BMO(\rho)}|I|^{1/2}\rho(|I|)$. For the first one we only have to notice that the results of Spanne [14] can be applied to show that $BMO(\rho)$ is equivalent to the following bound for the p means of f, with $1 < p < \infty$

$$
BMO(\rho, p) \quad : \quad \left(\frac{1}{|I|} \int_I |f(x) - f_I|^p \, dx \right)^{1/p} \leq C_p \rho(|I|) ,
$$

for every interval I. From which the theorem follows with $p = 2$. $\square$

5.2   The Meyer wavelet case

Theorem B: Let $\{\psi_{j,k} : (j,k) \in \mathbb{Z}^2\}$ be a Meyer wavelet basis. Let ρ be a positive non-decreasing function defined on $\mathbb{R}^+$ satisfying the Δ_2 condition. If $f \in BMO(\rho)$, then $c = \{c_{j,k} : (j,k) \in \mathbb{Z}^2\}$, the wavelet coefficient sequence of f, belongs to $\mathcal{C}(\eta)$, with

$$
\eta(t) = t \int_t^\infty \frac{\rho(s)}{s^2} \, ds,
$$

and

$$
\|c\|_{\mathcal{C}(\eta)} \leq C\|f\|_{BMO(\rho)}.
$$

Proof: Let I be a dyadic interval and, for $\ell \geq 1$, let $I_\ell = 2^\ell I$. Then, we write $f = f_I + \sum_{\ell \geq 0} f_\ell$, where $f_\ell = (f - f_{I_\ell})\chi_{I_\ell \setminus I_{\ell-1}}$. Notice that $|f_{I_{\ell+1}} - f_{I_\ell}| \leq C\rho(|I_{\ell+1}|)$ and $|f_{I_\ell} - f_I| \leq C\sum_{i=1}^{\ell} \rho(|I_i|)$, for all $\ell \geq 1$. Then, by using the equivalence for $BMO(\rho)$, $BMO(\rho, 2)$, we can prove that $\|f_\ell\|_{L^2} \leq C|I_\ell|^{1/2} \sum_{i=0}^{\ell} \rho(|I_i|)$. On the other hand, since $\psi \in S$, we get that

$$\left(\int_{I_\ell \setminus I_{\ell-1}} |\psi_{j,k}(x)|^2 \, dx \right)^{1/2} \leq C(|I_{j,k}||I_\ell|^{-1})^M,$$

for $\ell \geq 2$ and all $M \geq 1$. Now, by Hölder inequality and taking $M = 3/2$, we have that

$$|< f_\ell, \psi_{j,k} >| \leq C\frac{|I_{j,k}|^{3/2}}{|I_\ell|} \sum_{i=0}^{\ell} \rho(|I_i|).$$

So that

$$\sum_{I_{j,k} \subseteq I} \sum_{\ell \geq 2} |< f_\ell, \psi_{j,k} >|^2 \leq C \sum_{I_{j,k} \subseteq I} |I_{j,k}|^3 \sum_{\ell \geq 2} |I_\ell|^{-2} \left(\sum_{i=0}^{\ell} \rho(|I_i|) \right)^2$$

$$\leq C|I| \sum_{\ell \geq 2} 2^{-2\ell} \left(\sum_{i=0}^{\ell} \rho(2^i|I|) \right)^2$$

$$\leq C|I| \sum_{\ell \geq 2} 2^{-2\ell} \left(\int_{|I|}^{2^{\ell+1}|I|} \frac{\rho(t)}{t} \, dt \right)^2$$

$$\leq C|I| \left[|I| \int_{8|I|}^{\infty} \left(\int_{|I|}^{u} \frac{\rho(t)}{t} \, dt \right) \frac{du}{u^2} \right]^2,$$

and, since $\int_{|I|}^{u} = \int_{|I|}^{8|I|} + \int_{8|I|}^{u}$, by Fubini theorem, we get that

$$\sum_{I_{j,k} \subseteq I} \sum_{\ell \geq 2} |< f_\ell, \psi_{j,k} >|^2 \leq C|I|(\rho(|I|) + \eta(|I|))^2$$

$$\leq C|I|\eta^2(|I|).$$

Finally, since for $\ell = 0$ and $\ell = 1$, we get that

$$\sum_{I_{j,k} \subseteq I} |< f_\ell, \psi_{j,k} >|^2 \leq \sum_{j,k} |< f_\ell, \psi_{j,k} >|^2$$

$$= \|f_\ell\|_{L^2}^2$$

$$\leq C|I|\rho^2(|I|)$$

$$\leq C|I|\eta^2(|I|),$$

the theorem follows. □

6 $\mathcal{C}$ as a sufficient condition for $\mathcal{MO}$: The finite case

In this section we prove a boundedness property of $\mathcal{MO}$ norms in terms of the $\mathcal{C}$ norms of the coefficients for finite linear combinations of a wavelet basis. We shall only work with some detail the Daubechies case and we sketch the proof for the Meyer wavelet. Since, usually, the linear span of a wavelet basis $\{\psi_{j,k}\}$ is not dense in $BMO(\rho)$, we will tackle the convergence problem in the next section.

6.1   The Daubechies wavelet case

Theorem C: *Let $\{\psi_{j,k} : (j,k) \in \mathbb{Z}^2\}$ be a Daubechies wavelet basis, ρ a positive non-decreasing function defined on $\mathbb{R}^+$ satisfying the Δ_2 condition and $c = \{c_{j,k} : (j,k) \in \mathbb{Z}^2\}$ a sequence of real numbers vanishing except for a finite subset of $\mathbb{Z}^2$. Then the function $f = \sum_{(j,k) \in \mathbb{Z}^2} c_{j,k} \psi_{j,k}$ belongs to $BMO(\eta)$, where $\eta(t) = t \int_t^\infty \frac{\rho(s)}{s^2} ds$ and satisfies the following norm inequality*

$$\|f\|_{BMO(\eta)} \leq C \|c\|_{\mathcal{C}(\rho)}.$$

Proof: Since $BMO(\eta)$ is translation and dilation invariant and f is a linear combination of a finite number of $\psi_{j,k}$, in order to prove that $f \in BMO(\eta)$, we only need to show that $\psi \in BMO(\eta)$. Since ψ is at least of class C^1 and has compact support, we easily see that $|I|^{-1} \int_I |\psi(x) - \psi_I| \, dx$ is bounded above, both by $\|\psi'\|_\infty |I|$ and $2\|\psi\|_\infty$, and so by $C\eta(|I|)$. Since for $t < 1$, $\eta(t) \geq t \int_1^\infty \frac{\rho(s)}{s^2} \, ds > Ct$ and, for $t \geq 1$, $\eta(t) \geq t\rho(1) \int_t^\infty \frac{1}{s^2} \, ds \geq C$.

Let us now prove the norm inequality. Let I_0 be an interval with center x_0 and length $2r_0$, let $q \in \mathbb{Z}$ such that $2^{-q} \leq r_0 < 2^{-q+1}$. Define the following sets

$$J_1 = \{(j,k) \in \mathbb{Z}^2 : 2^{-j} > 2^{-q}\},$$

$$J_2 = \{(j,k) \in \mathbb{Z}^2 : 2^{-j} \leq 2^{-q} \text{ and } mI_{j,k} \cap I_0 \neq \emptyset\},$$

$$J_3 = \{(j,k) \in \mathbb{Z}^2 : 2^{-j} \leq 2^{-q} \text{ and } mI_{j,k} \cap I_0 = \emptyset\}.$$

Then, we write $f = f_1 + f_2 + f_3$, where $f_i(x) = \sum_{(j,k)\in J_i} c_{j,k}\psi_{j,k}$, $i = 1, 2, 3$. First, let us notice that $f_3(x) = 0$ if $x \in I_0$. On the other hand, if $(j, k) \in J_2$, there is a constant M that only depend of the support of ψ, such that $I_{j,k} \subset MI_0$ and we can choose two dyadic intervals I_1 and I_2 such that $MI_0 \subset I_1 \cup I_2$, then by (4.2) and the properties of ρ we get that

$$
\begin{aligned}
||f_2||_2 &\leq \left(\sum_{I_{j,k}\subset MI_0} |c_{j,k}|^2 \right)^{1/2} \\
&= \left(\sum_{I_{j,k}\subset I_1} |c_{j,k}|^2 + \sum_{I_{j,k}\subset I_2} |c_{j,k}|^2 \right)^{1/2} \\
&\leq ||c||_{\mathcal{C}(\rho)}(|I_1|\rho^2(|I_1|) + |I_2|\rho^2(|I_2|))^{1/2} \\
&\leq C||c||_{\mathcal{C}(\rho)}|I_0|^{1/2}\rho(|I_0|).
\end{aligned}
$$

Now we have that

$$
\begin{aligned}
\int_{I_0} |f(x) - f_1(x_0)|\, dx &\leq \int_{I_0} |f_1(x) - f_1(x_0)|\, dx + \int_{I_0} |f_2(x)|\, dx \\
&\leq |I_0|^{1/2}\left(\int_{I_0} |f_1(x) - f_1(x_0)|^2\, dx \right)^{1/2} \\
&\quad + C||c||_{\mathcal{C}(\rho)}|I_0|\rho(|I_0|).
\end{aligned}
$$

In order to estimate the last integral, let us notice that since the support of ψ is contained in $m[0, 1]$, for each j such that $2^{-j} > 2^{-q}$, only a boundend number of $\{\psi_{j,k}\}$ are non-trivial on I_0. On the other hand, by the regularity of ψ, we get that $|\psi_{j,k}(x) - \psi_{j,k}(x_0)| \leq C2^{3/2j}|x - x_0|$. Now, by taking into account (4.2) again, we see that

$$
\begin{aligned}
\int_{I_0} |f_1(x) - f_1(x_0)|^2\, dx &\leq \int_{I_0} \left(\sum_{(j,k)\in J_1} |c_{j,k}||\psi_{j,k}(x) - \psi_{j,k}(x_0)| \right)^2 dx \\
&\leq C||c||_{\mathcal{C}(\rho)}^2 r_0^2|I_0|\left(\sum_{j<q} 2^j\rho(2^{-j}) \right)^2 \\
&\leq C||c||_{\mathcal{C}(\rho)}^2|I_0|\left(r_0 \int_{r_0}^{\infty} \frac{\rho(s)}{s^2}ds \right)^2 \\
&\leq C||c||_{\mathcal{C}(\rho)}^2|I_0|[\eta(|I_0|)]^2.
\end{aligned}
$$

Finally

$$
|I_0|^{-1} \int_{I_0} |f(x) - f_1(x_0)|\, dx \leq C||c||_{\mathcal{C}(\rho)}\eta(|I_0|),
$$

from which $||f||_{BMO(\eta)} \leq C||c||_{\mathcal{C}(\rho)}$. $\square$

6.2   The Meyer wavelet case

For a Meyer wavelet basis the result of Theorem C remains valid. The main difference with the Daubechies case is now given by the fact that the function f_3 in the decomposition of f is not identically zero on I_0, so that we need an explicit estimate of the $L^2(I_0)$-norm of f_3

$$\|f_3\|_2^2 \leq \sum_{(j,k)\in J_3} \sum_{(j',k')\in J_3} |c_{j,k}||c_{j',k'}| \int_{I_0} |\psi_{j,k}(x)\psi_{j',k'}(x)| \, dx$$

$$\leq \|c\|_{C(\rho)}^2 \sum_{(j,k)\in J_3} \sum_{(j',k')\in J_3} 2^{-\frac{j}{2}}\rho(2^{-j})2^{-\frac{j'}{2}}\rho(2^{-j'})\|\psi_{j,k}\|_2\|\psi_{j',k'}\|_2$$

where, in the above inequality, $\| \cdot \|_2 = \| \cdot \|_{L^2(I_0)}$. Since $\psi \in \mathcal{S}$ we have that, for some M large enough

$$\|\psi_{j,k}\|_{L^2(I_0)}^2 \leq C \int_{I_0} \frac{2^j}{(1 + |2^j x - k|)^{2M}} \, dx$$

$$\leq C\left(\frac{1}{|2^j(x_0 + r_0) - k|^{M-1/2}} + \frac{1}{|2^j(x_0 - r_0) - k|^{M-1/2}}\right),$$

then, by chossing $M > 3/2$

$$\|f_3\|_2 \leq C\|c\|_{C(\rho)}\left(\sum_{j=q}^{\infty} 2^{-j/2}\rho(2^{-j})\right)^{1/2}\left(\sum_{j'=q}^{\infty} 2^{-j'/2}\rho(2^{-j'})\right)^{1/2}$$

$$\leq C\|c\|_{C(\rho)} \int_0^{|I_0|} \frac{\rho(t)}{t^{1/2}} \, dt$$

$$\leq C\|c\|_{C(\rho)}\rho(|I_0|)|I_0|^{1/2},$$

which is the desired estimate.

7 $\mathcal{C}$ as sufficient condition for $\mathcal{MO}$: The general case

After the result of Section 6, we have to prove that given $c \in C(\rho)$ the series $\sum c_{j,k}\psi_{j,k}$ converges in some sense to a function in $BMO(\eta)$, where $\eta(t) = t\int_t^{\infty} \frac{\rho(s)}{s^2}ds$. On the other hand since the type and growth properties of η are those of ρ, in order to use the duality $H^\eta - BMO(\eta)$ stated in Section 3, we shall assume some extra conditions on ρ. These conditions on ρ will allow us to take as a natural convergence for the series, that of the weak $*$ topology of $BMO(\eta)$. The following lemma is easy to prove and enables us to apply the result of Viviani

Lemma: *Let ρ be a non-decreasing function of upper type m, $m \geq 0$ and $\eta(t) = t \int_t^\infty \frac{\rho(s)}{s^2} ds$. Then the function ω defined by $\eta(t) = \frac{t^{-1}}{\omega^{-1}(t^{-1})}$ has lower type $\frac{1}{m+1} > 0$ and $\omega(s)/s$ is non-increasing.*

We are now in position to state and prove the main result of this article. Even when the result is valid also for the Meyer wavelet, we shall only be concerned with the Daubechies case, since their proofs are quite similar.

Theorem D: *Let $\{\psi_{j,k} : (j,k) \in \mathbf{Z}^2\}$ be a Daubechies wavelet basis, let $\rho = A + \bar{\rho}$ be a positive function, where $A \geq 0$ and $\bar{\rho}$ is a growth function such that $\int_1^\infty \frac{\bar{\rho}(s)}{s^2} ds < \infty$. Let $c = \{c_{j,k} : (j,k) \in \mathbf{Z}^2\}$ be a sequence of real numbers that belong to $\mathcal{C}(\rho)$. Then, the series*

$$\sum_{(j,k)\in\mathbf{Z}^2} c_{j,k}\psi_{j,k}$$

converges in the sense of the weak $*$ topology of $BMO(\eta)$ to a function $f \in BMO(\eta)$.

Proof: Let us first prove that $f^N = \sum_{j=-N}^{j=N} \sum_{k \in \mathbf{Z}} c_{j,k}\psi_{j,k} \in BMO(\eta)$, for each $N \in \mathbf{N}$. Let $I_0 = [x_0 - r_0, x_0 + r_0]$ and $q \in \mathbf{Z}$ such that $2^{-q} \leq r_0 < 2^{-q+1}$.

We write $f^N = f_1^N + f_2^N + f_3^N$, where $f_i^N = \sum_{\{(j,k)\in J_i : |j| \leq N\}} c_{j,k}\psi_{j,k}$ and J_i, $i = 1, 2, 3$ are the sets defined in the proof of Theorem C. Now, since for $x \in I_0$, $f_3(x) = 0$, by Hölder inequality we have that

$$|I_0|^{-1} \int_{I_0} |f^N(x) - f_1^N(x_0)| \, dx \ \leq \ (|I_0|^{-1} \int_{I_0} |f_1^N(x) - f_1^N(x_0)|^2 \, dx)^{1/2}$$

$$+ |I_0|^{-1/2} \|f_2^N\|_{L^2(I_0)} \ = \ I + II$$

In order to estimate I, since for each j such that $2^{-j} > 2^{-q}$, only a bounded number of $\{\psi_{j,k}\}$ are non-trivial on I_0, by proceeding as in the finite case we have that $|I| \leq C\|c\|_{\mathcal{C}(\rho)} \eta(|I_0|)$. Notice now that $\|f_2^N\|_{L^2(I_0)} \leq (\sum_{I_{j,k} \subset M I_0} |c_{j,k}|^2)^{1/2}$ and again as in the finite case we get that $II \leq C\|c\|_{\mathcal{C}(\rho)} \rho(|I_0|) \leq C\|c\|_{\mathcal{C}(\rho)} \eta(|I_0|)$. From which we have that $f^N \in BMO(\eta)$.

Now we shall see that f^N converges in the sense of the weak $*$ topology, i.e. we shall prove that $\{f^N\}$ is a Cauchy sequence in this topology. From the above lemma and the result of Viviani it will be enough to show that given $\epsilon > 0$, for each $g \in H^\eta$ we have that $|g(f^N) - g(f^{N'})| < \epsilon$, in other words

$$|\sum_i a_i(f^N) - \sum_i a_i(f^{N'})| < \epsilon,$$

for N and N' large enought and $\{a_i\}_i$ a sequence of multiples of η atoms. Let us take $N' < N$, then

$$|\sum_i a_i(f^N) - \sum_i a_i(f^{N'})| = |\sum_i \int_{I_i} a_i(y)[f^N(y) - f^{N'}(y)] \, dy|$$

$$\leq \sum_i \left| \int_{I_i} a_i(y) \sum_{j=-N}^{j=-N'} \sum_{k \in \mathbb{Z}} c_{j,k} \psi_{j,k}(y) \, dy \right|$$

$$+ \sum_i \left| \int_{I_i} a_i(y) \sum_{j=N'}^{j=N} \sum_{k \in \mathbb{Z}} c_{j,k} \psi_{j,k}(y) \, dy \right|$$

$$= I + II$$

To estimate II, let us assume that $I_i = [x_i - r_i, x_i + r_i]$ and let $q_i \in \mathbb{Z}$ be such that $2^{-q_i} \leq r_i < 2^{-q_i+1}$. We write $f^{N,N'} = \sum_{j=N'}^{j=N} \sum_{k \in \mathbb{Z}} c_{j,k} \psi_{j,k} = f_1^{N,N'} + f_2^{N,N'} + f_3^{N,N'}$, where $f_l^{N,N'} = \sum_{\{(j,k) \in J_l^i: \, N' \leq j \leq N\}} c_{j,k} \psi_{j,k}$, $l = 1, 2, 3$ and J_l^i is the set J_l defined in the proof of Theorem C with I_i instead of I_0. So that, by taking into account that $f_3^{N,N'} = 0$ for $x \in I_i$ and the property (3.4) of the atom a_i we have that

$$II \leq \sum_i \int_{I_i} (|a_i(x) f_2^{N,N'}(x)| + |a_i(x)[f_1^{N,N'}(x) - f_1^{N,N'}(x_i)]|) \, dx$$

$$= \sum_{i=1}^K + \sum_{i=K+1}^\infty = III + IV.$$

By Hölder inequality we have that

$$III \leq \sum_1^K (\|a_i\|_2 \|f_2^{N,N'}\|_{L^2(I_i)} + \|a_i\|_\infty \int_{I_i} |f_1^{N,N'}(x) - f_1^{N,N'}(x_i)| \, dx).$$

Now $\|f_2^{N,N'}\|^2_{L^2(I_i)} \leq \sum_{\{(j,k):I_{j,k} \subset MI_i \text{ and } N' \leq j \leq N\}} |c_{j,k}|^2 < \epsilon$, for N' large enough, since $\sum_{I_{j,k} \subset MI_i} |c_{j,k}|^2$ converges. On the other hand

$$\int_{I_i} |f_1^{N,N'}(x) - f_1^{N,N'}(x_i)| \, dx$$

$$\leq C\|c\|_{\mathcal{C}(\rho)} \int_{I_i} \sum_{\substack{N' \leq j \leq N \\ j < q_i}} |I_{j,k}|^{1/2} \rho(|I_{j,k}|) 2^{3/2j} |x - x_i| \, dx$$

$$\leq C\|c\|_{\mathcal{C}(\rho)} |I_i|^2 \sum_{j=N'}^{N} 2^j \rho(2^{-j})$$

$$\leq C\|c\|_{\mathcal{C}(\rho)} |I_i|^2 \int_{2^{N'}}^{2^N} \frac{\rho(t)}{t^2} \, dt,$$

and, since $\int_1^\infty \frac{\rho(t)}{t^2} dt < \infty$, we have that for N' large enough

$$\|a_i\|_\infty \int_{I_i} |f_1^{N,N'}(x) - f_1^{N,N'}(x_i)| \, dy < \epsilon.$$

Let us estimate IV

$$IV \leq \sum_{i=K+1}^\infty [\|a_i\|_2 \|f_2^{N,N'}\|_{L^2(I_i)}$$

$$+ \|a_i\|_2 \left(\int_{I_i} |f_1^{N,N'}(x) - f_1^{N,N'}(x_i)|^2 \, dy \right)^{1/2}]$$

$$\leq C\|c\|_{\mathcal{C}(\rho)} \sum_{i=K+1}^\infty \|a_i\|_2 [|I_i|^{1/2} \rho(|I_i|) + r_i |I_i|^{1/2} \sum_{j<q_i} 2^j \rho(2^{-j})]$$

$$\leq C\|c\|_{\mathcal{C}(\rho)} \sum_{i=K+1}^\infty \|a_i\|_2 |I_i|^{1/2} \eta(|I_i|)$$

$$\leq C\|c\|_{\mathcal{C}(\rho)} \sum_{i=K+1}^\infty \|a_i\|_\infty |I_i| \eta(|I_i|)$$

$$\leq C\|c\|_{\mathcal{C}(\rho)} \sum_{i=K+1}^\infty \frac{\|a_i\|_\infty}{\omega^{-1}(|I_i|^{-1})},$$

the last sum tends to zero since, as we observed in Section 3, by Lemma 4.7 in [17], the whole series converges. $\square$

The results proved in the Sections 6 and 7 can be generalized to the more general context of spaces of homogeneous type, providing a way to construct $BMO(\eta)$ functions. Let (X, d, μ) be a normal space of homogeneous type of order α, $0 < \alpha \leq 1$, see for example [1] or [17] to recall this concepts. In this spaces, the definitions and the results about the spaces $BMO(\eta)$ and H^η given in Section 7 hold true (see [17]), with a doubling measure μ instead of Lebesgue measure and the function η of upper type m with $0 \leq m < \alpha$.

In [15], Sawyer and Wheeden proved that in the space of homogeneous type it is always possible to construct, for λ positive, a family of balls $\{B_j^k\} = \{B(x_j^k, \lambda^k)\}_{j,k \in \mathbb{Z}}$, such that, if $\tilde{B}_j^k = B(x_j^k, \lambda^{k+1})$, we have

(i) Every ball B_j^k is contained in at least one of the balls $\tilde{B}_j^k$.

(ii) $\sum_j \chi_{B_j^k} \leq M$, for all $k \in \mathbb{Z}$, for some constant M.

(iii) $B_i^k \cap B_j^k = \emptyset$, for $i \neq j$, $k \in \mathbb{Z}$.

Now, we state the following Carleson type condition for the sequence of real numbers $\{\alpha_{j,k} = \alpha_{B_{j,k}} : (j,k) \in \mathbb{Z}^2\}$ associated to the above family of balls: there exists a constant C such that, for every ball $B = B(x, r)$

$$\left(\sum_{B_j^k \subset \bar{B}} |\alpha_{j,k}|^2 \right)^{1/2} \leq C\mu(B)^{1/2}\rho(r),$$

where $\bar{B} = B(x, \lambda^2 r)$. With similar methods to those used in Theorem D we obtain the following result

Let (X, d, μ) be a normal space of homogeneous type of order α and $\{B_j^k : (j,k) \in \mathbb{Z}^2\}$ a family of d-balls in X satisfying (i), (ii) and (iii). Let $\rho = A + \bar{\rho}$ be a positive function, where $A \geq 0$ and $\bar{\rho}$ is a growth function of upper type m, $0 \leq m < \alpha$, such that $\int_1^\infty \frac{\bar{\rho}(s)}{s^2} ds < \infty$. Let $\psi : \mathbb{R}_o^+ \to \mathbb{R}$ a continuosly differentiable function with compact support contained in $(0,1)$ and $\psi_{j,k}(x) = \lambda^{-k/2}\psi(\frac{d(x,x_j^k)}{\lambda^k})$ with $j, k \in \mathbb{Z}$. If $\{\alpha_{j,k} = < f, \psi_{j,k} >_\mu : (j,k) \in \mathbb{Z}^2\}$ is a sequence of real numbers verifying the Carleson condition, then the series $\sum_{j,k} \alpha_{j,k}\psi_{j,k}$ converges in the sense of the weak $*$ topology of $BMO(\eta)$ to a function $f \in BMO(\eta)$.

8   Acknowledgements

This work was partially supported by the Consejo Nacional de Investigaciones Científicas y Técnicas (CONICET) and by the Programación CAI+D, Universidad Nacional del Litoral (UNL).

9   References

[1] Aimar H. Rearrangement and Continuity Properties of $BMO(\phi)$ Functions on Spaces of Homogeneous Type. *Ann. Scuola Norm. Sup. Pisa, Serie IV, Vol XVIII* **3**, 353 - 362, 1991

[2] Campanato S. Proprietà di hölderianità di alcune classi di funzioni. *Ann. Scuola Norm. Sup. Pisa* **17**, 175 - 188, 1963.

[3] Daubechies I. Ten lectures on wavelets. *SIAM CBMS - NSF Regional Conf. Series in Applied Math*, 1992.

[4] Daubechies I. Orthonormal Bases of Compactly Supported Wavelets. *Comm. in Pure and Applied Math, Vol. 41*, 909 - 996, 1988.

[5] Fefferman C. and Stein E. H^p spaces of several variables. *Acta Math.* **129**, 137 - 193, 1971.

[6] Holschneider R. and Tchamitchian F. Pointwise analysis of Riemann's nondifferentiable function. *Centre de Physique Théorique. Marseille.*, 1989.

[7] Janson S. Generalizations of Lipschitz spaces an application to Hardy spaces and bounded mean oscillation. *Duke Math. Journal, Vol 47* **4**, 959 - 982, 1980.

[8] Jaffard S. and Laurencot Ph. Orthonormal Wavelets, Analysis of Operators and Applications to Numerical Analysis. *Wavelets - A Tutorial in Theory and Applications, C. K. Chui (ed.)* Academic Press, Inc., 543 - 601, 1991.

[9] John F. and Nirenberg L. On Functions of Bounded Mean Oscillation. *Comm. on Pure and Applied Math., Vol XIV*, 415 - 426, 1961.

[10] Lemarié P. and Meyer Y. Ondelettes et bases hilbertiennes. *Revista Matemática Iberoamericana, Vol. 2,* **Nos. 1 y 2**, 1 - 18, 1986.

[11] Meyer Y. *Ondelettes et Opérateurs I.* Hermann éditeurs des sciences et des arts. Paris, 1990.

[12] Meyers G. Mean oscillation over cubes and Hölder continuity. *Proc. Amer. Math. Soc.* **15**, 717 - 724, 1964.

[13] Moser J. On Harnack's Theorem for elliptic differential equations. *Comm. Pure and Appl. Math., Vol XIV*, 577 - 591, 1961.

[14] Spanne S. Some function spaces defined using the mean oscillation over cubes. *Ann. Scuola Norm. Sup. Pisa* **19**, 593 - 608, 1965.

[15] Sawyer E. and Wheeden R. Wheighted inequalities for fractional integrals on euclidean and homogeneous spaces. *American Journal of Math.* **114**, 813 - 874, 1962.

[16] Torchinsky A. Real-Variable Methods in Harmonic Analysis. *Academic Press, Inc., Pure and Appl. Math., Vol 123*, 1986.

[17] Viviani B. An Atomic Decomposition of the Predual of $BMO(\rho)$. *Revista Matemática Iberoamericana, Vol 3*, **Nos. 3 y 4**, 401 - 425, 1987.

Chapter 3

Undecimated Wavelet Transform from Orthogonal Spline Wavelets

Eduardo P. Serrano
Marcela A. Fabio

1 Introduction

The *decimated discrete wavelet transform* (DWT) gives us a powerful tool in many signal processing applications. It provides stable *time - scale* representations for any square integrable function as well as a suitable structure of the available information. In connection with this choice, well known families of biorthogonal or orthogonal wavelets are available.

On the other hand, the DWT is not consistent with traslations of the signal. For this reason, it becomes quite inappropriate for the accurate detection of local phenomena or pattern recognition applications. Therefore, techinques based on the *continuous wavelet transform* (CWT) or the *semidiscrete wavelet transform* (SWT) are preferred for these purposes. So, the behaviour of the signal correlates with a continuous - time representation. Local phenomena reflect in characteristic indicators, as zero crossings, vertical edges or relative extremes, while patterns appear independent of the phase.

It is a standard practice to compute these transforms in an appropriate discrete set of parameters. The choice of the grid depends on the mother wavelet, the problem on the hand and the processing strategy.

If the signal is given by sampled data, one usually computes the SWT, by scales, on a discrete time - grid, according to the sampling ratio. Special diadic mother wavelets are employed in this case. The sequence of computed values are called *undecimated discrete wavelet transform* (UWT) of the signal. They lead us to interpolate or approximate the underlying SWT and the above referred indicators can be estimated.

However, these computed values can hardly be considered as *co-efficients* of a wavelet serie. In general we cannot ensure a frame stucture associated with the UWT. This is a remarkable disadvantage because we can not expect as wished, to organize and exploit the numerical information contained in the discrete transform.

These facts suggest very special choices for the diadic wavelet. Then we propose to implement the SWT associated with orthogonal spline wavelets. In this case, the UWT gives us true coeficientes corresponding to an enlarged family of traslated wavelets. Moreover, these elemental functions are naturally disposed in collections, and each collection constitutes an orthonormal basis of the fundamental subspace of spline functions. Thus, the information is organized in a suitable *time - phase - scale* scheme. In other words, we analize the given signal in the context of an *extended multiresolution analysis structure* and efficient techiques for detection and pattern recognition can be combined with the well known advantages of the orthonormal DWT.

According to the above mentioned purpose, we give a brief overview of the SWT and explore its properties in the spline case. Following this, we define the undecimated discrete transform and design an efficient algorithm to compute it from the sampling values of the given signal.

Next, we expose some remarkable properties of the proposed undecimated discrete transform and we propose the extended multiresolution analysis structure, to organize the available numerical information of the signal. Finally, we give some suggestions for signal processing applications.

An extensive literature about spline functions and spline wavelets can be found in the bibliography. We here mention the works of Aldrouibi [2]−[5] Battle [6]−[7], Chui [8]−[10], Daubechies [11]−[12], Lemarie [20], Mallat [21] − [29], Meyer [30] − [35], Shoemberg [39], and Unser [47] − [54].

We also refer to [1], [13], [14], [17], [37], [38] and [55] for further information about related topics.

2   The Undecimated Discrete Wavelet Transform

At the beginning, we give a brief overiew about the semidiscrete wavelet transform (SWT). A function $\psi \in L^1(\mathcal{R}) \cap L^2(\mathcal{R})$, is called *dyadic wavelet* if there exists positive bounds $0 < A \leq B < \infty$, such that the following stability condition holds [9]:

$$A \leq \sum_j |\widehat{\psi}(2^j \omega)|^2 \leq B \quad \text{a.e} \tag{3.1}$$

Clearly, orthogonal wavelets constitute a particular case of dyadic functions with $A = B = 1$ [9], [64].

Given a dyadic wavelet ψ the associated semidiscrete or *dyadic* transform is defined as:

$$W_j^{(s)} L^2(\mathcal{R}) \longrightarrow L^2(\mathcal{R})$$
$$W_j^{(s)} s(y) = 2^j \langle s(x), \psi(2^j(x-y)) \rangle \quad j \in \mathcal{Z}; \; y \in \mathcal{R} \tag{3.2}$$
$$W^{(s)} = \{W_j^{(s)}, \; j \in \mathcal{Z}\} \tag{3.3}$$

Denoting $f^\vee(x) = f(-x)$, we also can write:

$$W_j^{(s)} s(y) = 2^j (s(x) * \psi^\vee(2^j x))(y) \tag{3.4}$$

and taken the Fourier transform, we obtain the formula:

$$(W_j^{(s)} s)^\wedge(\omega) = \widehat{s}(\omega)\widehat{\psi}^*(\omega/2^j) \tag{3.5}$$

that makes explicit the filtering properties of the transform.

A formula for reconstructing any finite - energy function is given by [12]:

$$s(x) = \sum_{j=-\infty}^{\infty} 2^j \int_{-\infty}^{\infty} W_j^{(s)} s(y)\psi^*(2^j(x-y)) \, dy \tag{3.6}$$

where the function ψ^* is a *dyadic dual* of the dyadic wavelet ψ.

Following, let us denote:

$$T^r s(x) = s(x - r), \quad r \in \mathcal{R} \tag{3.7}$$
$$T^{r+1}s = T \, T^r s \tag{3.8}$$

Particullarly, we understand $T^1 = T$, $T^0 = \mathbb{1}$ and $(T^r)^{-1} = T^{-r}$. For simplicity we also write for any sequence in $l^2(\mathcal{Z})$:

$$T^m a(k) = a(k - m) \quad m \in \mathcal{Z} \tag{3.9}$$

A remarkable property of the semidiscrete transform is the consistency respect the translation operator T:

$$W_j^{(s)} T^r s = T^r W_j^{(s)} s \tag{3.10}$$

for any $r \in \mathcal{R}$.

This property makes it highly attractive in signal processing applications. Particularly, for pattern recognition implementations or to detect and characterize singularities and local phenomena. We refer to [48] − [52] for further information about these topics.

However, for computational implementations the translation parameter y must be discretized by restricting it to a discrete set of points $\{y_{jk}\}$.

It is well known that for special choices of the dyadic wavelet ψ and the set $\{y_{jk} = 2^j k, \ k \in \mathcal{Z}\}$, the family:

$$\psi_{jk}(x) = 2^{j/2} \psi(2^j x - k) \quad j, k \in \mathcal{Z} \tag{3.11}$$

provides a *frame* [12], [17], for $L^2(\mathcal{R})$. Then the values:

$$c_j(k) = 2^{-j/2} W_j^{(s)} s(2^{-j} k) \tag{3.12}$$

become coefficientes of a stable representation of the signal:

$$s(x) = \sum_{j=-\infty}^{\infty} 2^{j/2} \sum_{k=-\infty}^{\infty} c_j(k) \psi^*(2^j x - k) \tag{3.13}$$

where ψ^* is the associated dual wavelet. So, the values $W_j^{(s)} s(2^{-j} k)$ provide a *decimated discrete* wavelet transform and they can be arranged into an efficient scheme correlated with a multiresolution analysis structure [12], [22], [23].

On the other hand, decimation destroys the consistency with respect to the translation operator. For this reason the decimated discrete transform is quite inappropriate for the above mentioned applications.

To overcome these problems, an usual procedure is to restrict the translation parameter y to a refined discrete grid. Then, one interpolates the semidiscrete transform from the sampling values.

However, in general, the sampling values $W_j^{(s)}s(y_{jk})$ do not provide a discrete transform in the strict sense. That is, the numerical information is correlated in a complex structure and they cannnot be considered as coefficients of any stable representation of the signal. [22], [25], [38]

Here we will propose an intermediate alternative. Assume that ψ is an orthogonal spline wavelet. Since it is a dyadic wavelet, the semidiscrete transform is well defined. On the other hand, the decimated discrete transform is correlated with an orthonormal basis for $L^2(\mathcal{R})$.

The main idea is to implement an *undecimated discrete* wavelet transform, consistent with integer translations T^m, $m \in \mathcal{Z}$, providing stable representations for any signal given by sampling. In othes words, we intend to combine in a suitable way the desirable properties of the semidiscrete transform with the advantages given by orthogonal spline wavelets [6], [9], [30] and [33].

Denote $V_\nu = \{V_j, j \in \mathcal{Z}\}$ the multiresolution analysis associated with the polynomial spline wavelets of order ν, $\nu \in \mathcal{N}$ and by W_j the orthogonal complement of V_j in V_{j+1}. The fundamental subspace V_0 consists of all functions of $L^2(\mathcal{R})$ which are $\nu - 1$ time continuosly differentiable and equal to a polynonial of degree ν on each interval $[k, k+1)$. As it is well known, there are unique functions $\phi \in V_0$ $\psi \in W_0$ centered on $x_\phi = 0.0$ $x_\psi = 0.5$, respectively, with exponential decay and minimun dispersion [33], such that the family:

$$\phi_{jk}(x) = 2^{j/2}\phi(2^j x - k) \quad j, k \in Z \qquad (3.14)$$

constitutes an orthonormal basis for V_j, and the family:

$$\psi_{jk}(x) = 2^{j/2}\psi(2^j x - k) \quad j, k \in Z \qquad (3.15)$$

constitutes an orthonormal basis for $L^2(\mathcal{R})$. The functions ϕ y ψ are called *orthogonal scale function* and *orthogonal spline wavelet*, respectively.

Assume that ν is an odd positive integer and let $\{W_j^{(s)}s, j \in \mathcal{Z}\}$ the associated semidiscrete transform.

At a first step we state:

Proposition 2.1 *Let $s \in V_0$, then:*

(a) $W_j^{(s)} s \in L^2(\mathcal{R})$.

(b) $W_j^{(s)} s(y)$ *is a spline function of order $2\nu + 1$, with the knot set $\{n \in \mathcal{Z}\}$, for all $j < 0$.*

(c) $W_j^{(s)} s(y)$ *is a spline function of order $2\nu + 1$, with the knot set $\{2^{-(j+1)}n, \, n \in \mathcal{Z}\}$, for all $j \geq 0$.*

Proof:

Denote $\psi^\vee(x) = \psi(-x)$ y $\psi_j(x) = 2^j\psi(2^jx)$. Then $W_j^{(s)} s(y) = s * \psi_j^\vee(y)$.

The first affirmation *(a)* results immediatly from the last expression and the fact $\psi \in L^2(\mathcal{R}) \cap L^1(\mathcal{R})$.

Next, observe that both s and $\psi_j^\vee$, for $j \leq 0$, belong to V_0. Then, we can expand them in the inconditional bases of $\beta_{(\nu)}$-spline functions of order ν. Recall that the generating function is defined as [39]:

$$\beta_{(\nu)}(x) = \underbrace{\chi * \cdots * \chi}_{\nu+1 \ times}(x)$$

Therefore we have:

$$T^n \beta_{(\nu)} * T^k \beta_{(\nu)}(x) = T^{n+k} \beta_{(2\nu+1)}(x)$$

for any pair of integer translations T^n and T^k. Then any convolution in the spline subspace V_0, must be a spline funtion of order $2\nu + 1$ with knots on $\mathcal{Z}$ and we conclude *(b)*.

For $j \geq 0$, observe that $\psi_j^\vee \in V_{j+1}$ and $V_0 \subset V_{j+1}$. Then any function in V_{j+1}, can be expanded in the basis $\{\beta_\nu(2^{(}j+1)x - n), \, n \in \mathcal{Z}\}$. They are spline functions of ν, with knots on $2^{-(j+1)}n, \, n \in \mathcal{Z}$, and we can conclude *(c)* as above. •

And we derive the following:

Corollary 2.1 *The semidiscrete transform $W_j^{(s)} s(y)$ for any signal $s \in V_0$ can be exactly interpolated from the sampling values:*

(a) $W_j^{(s)} s(n)$, $n \in \mathcal{Z}$, *if $j < 0$.*

(b) $W_j^{(s)} s(2^{-(j+1)}n)$, $n \in \mathcal{Z}$, *if $j \geq 0$.* •

Moreover, for these purposes we have on the hand efficient interpolation techniques, [9].

Following, let $s \in \mathcal{V}_0$ and denote:

$$d_j(n) = W_j^{(s)} s(n) \quad j < 0, \; n \in \mathcal{Z} \tag{3.16}$$

Since $s \in \mathcal{V}_0$ we can write:

$$\begin{aligned}
s(x) &= \sum_{j=-\infty}^{-1} \sum_{k=-\infty}^{\infty} c_j(k) \psi_{jk}(x) \\
&= \sum_{j=-\infty}^{-1} 2^{-j/2} \sum_{k=-\infty}^{\infty} d_j(2^{-j}k) \psi_{jk}(x) \tag{3.17}
\end{aligned}$$

Clearly, the sampling values $d_j(n)$, for $j < 0$ and $n \in \mathcal{Z}$ contain the complete information of the signal. The redundancy is needed to preserve the consistency with respect to integer translations of the signal. At this point observe that $T\mathcal{V}_0 = \mathcal{V}_0$.

At last, we define the undecimated discrete transform associated with an orthogonal spline wavelet as:

$$\begin{aligned}
D_j : \mathcal{V}_0 &\longrightarrow l^2(\mathcal{Z}) \; (j < 0) \tag{3.18} \\
D_j s(n) &= d_j(n) \;, n \in \mathcal{Z} \tag{3.19} \\
Ds &= \{D_j s; \, j < 0\} \tag{3.20}
\end{aligned}$$

At the next section we will propose an efficient algorithm to compute the sampling values $d_j(n)$.

We remark that the transform is defined from $\mathcal{V}_0$, the usual representation space for signals given by sampling.

Given N sampling values $\{s(n\Delta x), \, 0 \le n < N\}$, making an appropriate change of variable, it can be supposed that the underlying signal belongs to the referred subspace of spline functions $\mathcal{V}_0$.

We also remark that the values $d_j(n)$ are correlated with the overcomplete family of wavelets $\{2^j \psi(2^j(x-n)) \; j < 0, \; n \in Z\}$. To validate the definition of the undecimated transform, we will demonstrate below that this family provides an efficient and stable representation of the signal.

3   Undecimate Cardinal Algorithm

In this section, we expose an efficient algorithm to compute the undecimated discret transform. We refer to [9], [11], [13], [19], [23] − [37] and [38] about correlated topics.

Given N sampled values $\{s_0(n) = s(n.\Delta x), 0 \leq n < N\}$, we assume $\Delta x = 1$ and the habitual hypothesis that the underlying signal is represented by interpolation in the subspace $\mathcal{V}_0$ of splines functions or order ν, with knots set $\mathcal{Z}$, [9], [25] and [33]. Denote $s_0(x)$ this approximation and let $\mathcal{L}(x)$ the function C-spline of order ν, verifying [14], [39]:

$$\mathcal{L}(n) = \delta_{n,0} \quad \text{for all } n \in \mathcal{Z} \tag{3.21}$$

As it is well known, it generates by translations in $\mathcal{Z}$ an unconditional base for the subspace $\mathcal{V}_0$, [30], [51]. Then, we interpolate:

$$s_0(x) = \sum_n s_0(n)\mathcal{L}(x - n). \tag{3.22}$$

Then, denote by $P_j s_0$ and $Q_j s_0$ the succesive orthogonal projections onto the subspaces V_j and W_j, for each $j < 0$. Such components are spline functions with knots set $\{2^{-j}n\}$ and $\{2^{-(j+1)}n\}$, $n \in \mathcal{Z}$, respectively.

Recall that the wavelets coefficients of the signal are given by:

$$\begin{aligned} c_j(k) &= 2^{j/2}\langle s_0(x); \psi(2^j x - k)\rangle \\ &= 2^{-j/2}\, W_j^{(s)} s(2^{-j}k) \\ &= 2^{-j/2}\, d_j(2^{-j}k) \quad j < 0, \;\; k \in \mathcal{Z}. \end{aligned} \tag{3.23}$$

Since $TV_0 = V_0$, we can analize each translated signal $T^m s_0 \in V_0$, $m \in \mathcal{N}_0$, in the analogous way. So, we also denote:

$$s_j^m(n) = P_j T^m s_0(2^{-j}n) \tag{3.24}$$

$$q_j^m(n) = Q_j T^m s_0(2^{-(j+1)}n) \tag{3.25}$$

$$c_j^m(k) = 2^{j/2}\langle (T^m s_0(x), \psi(2^j x - k)\rangle \tag{3.26}$$

the sampling values of the projections $P_j T^m s_0$ and $Q_j T^m s_0$, and the corresponding wavelet coefficients.

To develop the algorithm we need to give some definitions and demonstrate preliminary lemmas.

First, to compute the coefficients $c_j^m(k)$, directly from the sampling values $s_{j+1}^m(n)$, we define the filter: g as follows:

$$g(2n) = 1/2\,\langle \mathcal{L}(x), \psi(x/2 + n)\rangle \tag{3.27}$$

$$g(2n + 1) = 1/2\,\langle \mathcal{L}(x - 1), \psi(x/2 + n)\rangle \tag{3.28}$$

and the associated operator G as:

$$Ga(k) = \sum_n a(n)g(n-2k) \qquad (3.29)$$

Next, we state:

Lemma 3.1 *Under the previous definitions:*

$$c_j^m(k) = 2^{-j/2}Gs_{j+1}^m(k) \qquad (3.30)$$

for all $j \leq -1$, and $m \in \mathcal{Z}$.

Proof:
Clearly, the result does not depend on the translation T^m, so by simplicity we write off the index m. First, observe that :

$$g(n-2k) = \begin{cases} 1/2\langle\mathcal{L}(x), \psi(x/2+n/2-k)\rangle & \text{if } n \text{ is even} \\ 1/2\langle\mathcal{L}(x-1), \psi(x/2+\lfloor n/2 \rfloor - k)\rangle & \text{if } n \text{ is odd} \end{cases}$$

then, for each $j < 0$:

$$\begin{aligned} c_j(k) &= 2^{j/2}\langle s_{j+1}(x), \psi(2^j x - k)\rangle \\ &= 2^{j/2}\sum_n s_{j+1}(n)\langle\mathcal{L}(2^{j+1}x-n), \psi(2^j x - k)\rangle \quad (3.31) \end{aligned}$$

For n even, making $y = 2^{j+1}x - n$ we obtain:

$$\begin{aligned} 2^{j/2}\langle\mathcal{L}(2^{j+1}x-n), \psi(2^j x - k)\rangle &= 2^{-j/2-1}\langle\mathcal{L}(y), \psi(y/2+n/2-k)\rangle \\ &= 2^{-j/2}g(n-2k) \end{aligned}$$

Analogously, for n odd, making $y = 2^{j+1}x - n + 1$ we also obtain:

$$\begin{aligned} 2^{j/2}\langle\mathcal{L}(2^{j+1}x-n), \psi(2^j x - k)\rangle &= 2^{-j/2-1}\langle\mathcal{L}(y-1), \\ &\quad \psi(y/2+\lfloor n/2 \rfloor - k)\rangle \\ &= 2^{-j/2}g(n-2k) \end{aligned}$$

and substituting in formula 3.31 we conclude the result. $\bullet$

Next, let the sequence of sampling values $r(n) = \psi(n/2)$, $n \in \mathcal{Z}$ and define the reconstructing operator R as:

$$Ra(n) = \sum_k r(n-2k)a(k) \qquad (3.32)$$

Now suppose that, for some $j \leq -1$, we have computed the values $s_{j+1}^m(n)$ and $c_j^m(k)$, and state the reconstruction formulas:

Lemma 3.2 *Under the previous definitions:*

$$q_j^m(n) = 2^{j/2} Rc_j^m(n) \qquad (3.33)$$
$$s_j^m(n) = s_{j+1}^m(2n) - q_j^m(2n) \qquad (3.34)$$

for all $j \leq -1$ and $m \in Z$.

Proof:

We write off again the index m. Since the family
$\{2^{j/2} \psi(2^j x - k), \quad k \in \mathcal{Z}\}$, constitutes an orthonormal base for $\mathcal{W}_j$, we have:

$$Q_j s_0(x) = 2^{j/2} \sum_k c_j(k) \psi(2^j x - k)$$

On the other hand, observe that $Q_j s_0 \subset \mathcal{W}_j \subset \mathcal{V}_{j+1}$, being a spline function with knots set $\{2^{-(j+1)} n; \ n \in \mathcal{Z}\}$. For this reason, we only need to compute the values $Q_j s_0(2^{-(j+1)} n)$:

$$
\begin{aligned}
q_j(n) &= Q_j s_0(2^{-(j+1)} n) \\
&= 2^{j/2} \sum_k c_j(k) \psi(n/2 - k) \\
&= 2^{j/2} \sum_k r(n - 2k) c_j(k) \\
&= 2^{j/2} Rc_j(n)
\end{aligned}
$$

and the first formula holds. Next, we have the orthogonal decomposition:

$$P_{j+1} s_0(x) = P_j s_0(x) \oplus Q_j s_0(x)$$

Since $P_j s \subset \mathcal{V}_j$, being a spline function with knots set $\{2^{-j} n; \ n \in \mathcal{Z}\}$, we can compute:

$$
\begin{aligned}
s_j(n) &= P_j s_0(2^{-j} n) \\
&= P_{j+1} s_0(2^{-j} n) - Q_j s_0(2^{-j} n) \\
&= s_{j+1}(2n) - q_j(2n)
\end{aligned}
$$

and we complete the proof. ●

At this point, we remark that for computing the sampling values $s_j^m(n)$, $q_j^m(2n)$ and the coefficients $c_j^m(k)$ we only need the filters

g and r. Observe that they are computed in a recursive scheme, directly from the sampling values $T^m s_0(n)$, without employing additional filters. Particularly, for $m = 0$, we compute the standard wavelet coefficients and the sampling values of the sucessive projections $P_j s_0$ and $Q_j s_0$. This scheme is something different with respect to the classic Mallat's scheme and presents some numerical advantages. Owing to these comments we will call *cardinal* the proposed algorithm.

Then, the scheme must be adjusted to compute with a minimum of complexity the values $s_j^m(n)$, $q_j^m(2n)$ and $d_j(k)$.

Let us continue characterizing the operator H associated to G, to formalize the computation of the values $s_j^m(n)$ from the previous sequence $s_{j+1}^m(n)$, through the classic Mallat's scheme. We need this formal description of the recursive scheme to prove below the properties of the proposal transform. However, we remark that the action of this operator is accomplished in implicit form by formulas 3.33 and 3.34 .

Let h the sequence defined as:

$$h(0) \quad = \quad 1 - \sum_k g(2k)\psi(k) \tag{3.35}$$

$$h(n) \quad = \quad -\sum_k g(n+2k)\psi(k) \quad \text{for all } n \neq 0 \tag{3.36}$$

and the associated operator H :

$$Ha(n) = \sum_k h(k-2n)a(k) \tag{3.37}$$

and state:

Lemma 3.3 *According to the previous definitions:*

$$s_j^m(n) \quad = \quad Hs_{j+1}^m(n) \tag{3.38}$$

for all $j \leq -1$.

Proof:

From lemma 3.2, we have:

$$s_j(n) \quad = \quad s_{j+1}(2n) - q_j(2n)$$
$$= \quad s_{j+1}(2n) - 2^{j/2}\sum_k c_j(k)\psi(n-k)$$

Next, using lemma 3.1, we develop:

$$
\begin{aligned}
s_j(n) &= s_{j+1}(2n) - \sum_k \left(\sum_m g(m-2k) s_{j+1}(m) \right) \psi(n-k) \\
&= s_{j+1}(2n) - \sum_m \left(\sum_k g(m-2k) \psi(n-k) \right) s_{j+1}(m) \\
&= \left(1 - \sum_k g(2k) \psi(k)\right) s_{j+1}(2n) - \\
&\quad - \sum_{m \neq 2n} \left(\sum_k g(m-2k) \psi(n-k) \right) s_{j+1}(m) \\
&= h(0) s_{j+1}(2n) + \sum_{m \neq 2n} h(m-2n) s_{j+1}(m) \\
&= \sum_m h(m-2n) s_{j+1}(m) \\
&= H s_{j+1}(n)
\end{aligned}
$$

and the proof is complete. •

Using the classic Mallat's scheme from each sequence $T^m s_0$, we can state:

Lemma 3.4 *For all $j < 0$, and denoting $s_0^0 = s_0$, it is verified:*

$$
\begin{aligned}
s_j^m(n) &= H^{-j} T^m s_0^0(n) & (3.39) \\
d_j(n) &= G H^{-(j+1)} T^{(2^{-j}k-n)} s_0^0(k) & (3.40)
\end{aligned}
$$

for arbitrary integer indices m, n and k.

Proof:
From each signal $T^m s_0^0 \in V_0$, using lemmas 3.1 and 3.3, the recursive application of Mallat's algorithm gives us:

$$
\begin{aligned}
s_j^m(n) &= H^{-j} T^m s_0^0(n) \\
c_j^m(k) &= 2^{-j/2} G H^{-(j+1)} T^m s_0^0(k)
\end{aligned}
$$

On the other hand, for each pair of integers $n \in \mathcal{Z}$ and $k \geq 0$, we make $m = 2^{-j}k - n$ and write:

$$
\begin{aligned}
d_j(n) &= 2^j \langle s_0(x), \psi(2^j(x-n)) \rangle \\
&= 2^j \langle s_0(x), \psi(2^j(x+m) - k) \rangle \\
&= 2^j \langle s_0(x-m), \psi(2^j x - k) \rangle
\end{aligned}
$$

$$\begin{aligned} &= 2^j \langle T^m s_0(x), \, \psi(2^j x - k) \rangle \\ &= 2^{j/2} c_j^m(k) \\ &= GH^{-(j+1)} T^{(2^{-j}k-n)} s_0^0(k) \end{aligned}$$

This completes the proof of the lemma. •

The last result leads us to obtain the values $d_j(n)$ by computing the wavelet coefficients of translated versions of the original signal. For this purposes, we rewrite the formula 3.40 as follows. Given $k \in \mathcal{Z}$, we can write the indices n on the range $2^{-j}(k-1) < n \leq 2^{-j}k$, as $n = 2^{-j}k - m$, with $\leq m < 2^{-j}$. Then, the complete sequence $d_j(n)$, $n \in \mathcal{Z}$, can be given by the formula:

$$d_j(2^{-j}k - m) = GH^{-(j+1)} T^m s_0^0(k), \quad k \in \mathcal{Z}, \quad 0 \leq m < 2^{-j} \quad (3.41)$$

Following this, recall that the projection operators G, H do not commute with the translation operator T. Therefore, we need to prove the following formulas:

Lemma 3.5 *For all* $m \in \mathcal{Z}$:

$$\begin{aligned} T^m H &= HT^{2m} & (3.42) \\ TH^m &= H^m T^{(2^m)} & (3.43) \end{aligned}$$

Proof:

It is sufficient to demonstrate that $TH = HT^2$. From formulas 3.33, 3.34 and 3.37 we have:

$$\begin{aligned} Ha(n) &= (\mathbb{1} - RG)a(2n) \\ Ra(n) &= \sum_k a(k)\psi(n/2 - k) \end{aligned}$$

and we prove that $T^2 R = RT$:

$$\begin{aligned} T^2 Ra(n) &= Ra(n - 2) \\ &= \sum_k a(k)\psi(n/2 - 1 - k) \\ &= \sum_k a(k - 1)\psi(n/2 - k) \\ &= RTa(n) \end{aligned}$$

Similarly we can prove that $T^2G = GT$. Now, using these results, we write:

$$
\begin{aligned}
THa(n) &= a(2n-2) - RGa(2n-2) \\
&= T^2a(2n) - T^2RGa(2n) \\
&= T^2a(2n) - RGT^2a(2n) \\
&= (\mathbb{1} - RG)T^2a(2n) \\
&= HT^2a(n)
\end{aligned}
$$

completing the proof. •

At this point, we are in condition to state the recursive scheme of the undecimate cardinal algorithm, to compute the values of the discret transform from the sampling values of the signal. It also gives the successive sequences $s_j^m(n)$, as explicit outputs.

Theorem 3.1 Let $s_0^0(n)$ be the sampling data of the signal, and we denote $m(j) = m - 2^{-(j+1)}$.
Then, for $j \le -1$:

$$
s_j^m(n) = \begin{cases} Hs_{j+1}^m(n) & , \ 0 \le m < 2^{-(j+1)} \\ HTs_{j+1}^{m(j)}(n) & , \ 2^{-(j+1)} \le m < 2^{-j} \end{cases} \tag{3.44}
$$

$$
d_j(2^{-j}k - m) = \begin{cases} Gs_{j+1}^m(k) & , \ 0 \le m < 2^{-(j+1)} \\ GTs_{j+1}^{m(j)}(k) & , \ 2^{-(j+1)} \le m < 2^{-j} \end{cases} \tag{3.45}
$$

for all n, $k \in Z$ and $0 \le m < 2^{-j}$.

Proof:
From lemma 3.3, and using formula 3.24, we have:

$$
\begin{aligned}
s_j^m(n) &= H^{-j}T^m s_0^0(n) \\
&= HH^{-(j+1)}T^m s_0^0(n)
\end{aligned} \tag{3.46}
$$

If $0 \le m < 2^{-(j+1)}$, we have:

$$
s_j^m(n) = Hs_{j+1}^m(n) \tag{3.47}
$$

In the opposite case, if $2^{-(j+1)} \le m < 2^{-j}$, we can write $m = 2^{-(j+1)} + m(j)$. Using the lemma 3.5:

$$
\begin{aligned}
H^{-(j+1)}T^{(2^{-(j+1)})}T^{m(j)} s_0^0(n) &= TH^{-(j+1)}T^{m(j)} s_0^0(n) \\
&= Ts_{j+1}^{m(j)}(n)
\end{aligned}
$$

and replacing this last identity in formula 3.46, we obtain

$$s_j^m(n) = HTs_{j+1}^{m(j)}(n) \tag{3.48}$$

The results 3.47 and 3.48, prove formula 3.44. For the other identity, the proof is quite similar. •.

The recursive scheme in now complete. Let us expose the implementation algorithm. Observing that:

$$
\begin{aligned}
q_j^m(n) &= Q_j T^m s_0(2^{-(j+1)}n) \\
&= 2^{j/2} \sum_k c_j^m(k)\psi(n/2 - k) \\
&= 2^{j/2} Rc_j^m(n) \tag{3.49}
\end{aligned}
$$

recalling that $d_j(2^{-j}k - m) = 2^{j/2}c_j^m(k)$, and using formulas 3.44 and 3.45 we formulate the *undecimate cardinal algorithm* from orthogonal spline wavelets:

Cardinal Algorithm:

- given the initial sequence (sampling values):

$$s_0^0 \in l^2(\mathcal{Z}) \tag{3.50}$$

- for $j = 1, -2 \dots$ and $0 \le m < 2^{-j}$, compute:

$$
d_j(2^{-j}k - m) = \begin{cases} Gs_{j+1}^m(k) & , \ 0 \le m < 2^{-(j+1)} \\ GTs_{j+1}^{m(j)}(k) & , \ 2^{-(j+1)} \le m < 2^{-j} \end{cases} \tag{3.51}
$$

$$q_j^m(n) = Rd_j(2^{-j}. - m)(n) \tag{3.52}$$

$$
s_j^m(n) = \begin{cases} s_{j+1}^m(2n) - q_j^m(2n) & , \ 0 \le m < 2^{-(j+1)} \\ Ts_{j+1}^{m(j)}(2n) - q_j^m(2n) & , \ 2^{-(j+1)} \le m < 2^{-j} \end{cases} \tag{3.53}
$$

By restricting $m = 0$, we justly have the decimate algorithm, to compute the standar wavelet coefficients of the original signal $s_0(x)$.

4   Properties of the Undecimated Discrete Transform

Let us introduce the interpolation operator Λ in the subspace $\mathcal{V}_0$ of spline functions of odd order ν. Given a signal $y \in \mathcal{V}_0$, by the sequence of the sampling values $y_0 = \{y(n),\ n \in \mathcal{Z}\}$, we can interpolate the intermediate values $y(n/2)$ by the action of the *a trous* filter $\mathcal{L}(n/2)$ [9], [39]:

$$
\begin{aligned}
y(n/2) &= \sum_m y(m)\mathcal{L}(n/2 - m) \\
&= \Lambda y_0(n/2)
\end{aligned}
\tag{3.54}
$$

We remark that the filter $\mathcal{L}(n/2)$, is the filter that achieves the two scale equation for the cardinal spline functions [9], that is:

$$
\mathcal{L}(x/2) = \sum_n \mathcal{L}(n/2)\mathcal{L}(x - n)
\tag{3.55}
$$

Suppose to have the corresponding output values of the undecimate cardinal algorithm for any level $j \leq -1$. Firstly, we state the formula to recover the sampling values of the projections $P_{j+1}T^m s_0$.

Proposition 4.1 *Given $j < 0$, and $0 \leq m < 2^{-(j+1)}$, denote $\overline{m} = m + 2^{-(j+1)}$. Then:*

$$
\begin{aligned}
s_{j+1}^m(n) &= 1/2(\Lambda s_j^m(n/2) + q_j^m(n)) + \\
&+ 1/2T^{-1}(\Lambda s_j^{\overline{m}}(n/2) + q_j^{\overline{m}}(n)) \tag{3.56} \\
&= 1/2(\Lambda s_j^m(n/2) + Rd_j(2^{-j}. - m)(n)) + \\
&+ 1/2T^{-1}(\Lambda s_j^{\overline{m}}(n/2) + Rd_j(2^{-j}. - \overline{m})(n)) \tag{3.57}
\end{aligned}
$$

Proof:
Recalling formula 3.53 we write:

$$
s_j^m(n) = \begin{cases} s_{j+1}^m(2n) - q_j^m(2n) & ,\ 0 \leq m < 2^{-(j+1)} \\ Ts_{j+1}^{m(j)}(2n) - q_j^m(2n) & ,\ 2^{-(j+1)} \leq m < 2^{-j} \end{cases}
$$

where $0 \leq m < 2^{-j}$ and $m(j) = m - 2^{-(j+1)}$. Let $0 \leq m < 2^{-(j+1)}$, then:

$$
\begin{aligned}
s_j^m(n) &= s_{j+1}^m(2n) - q_j^m(2n) \\
s_j^{\overline{m}}(n) &= Ts_{j+1}^m(2n) - q_j^{\overline{m}}(2n)
\end{aligned}
$$

Next, applying the interpolation operator Λ and averaging both expressions, we deduce the formula (3.3). The second formula is deduced from the expression 3.52 in an analogous way. •

Below, we state the reconstruction and energy formulas, from the values of the undecimated discrete transform $d_j(n)$.

Theorem 4.1 (*Reconstruction and energy formulas*). *Let $s \in V_0$ be and $d_j(n)$ the values of the undecimated discret transform, then:*

$$\lim_{j \to -\infty} \|s(x) - \sum_{l=j}^{-1} 2^l \sum_n d_l(n)\psi(2^l(x-n))\|_2 \;=\; 0 \qquad (3.58)$$

$$\lim_{j \to -\infty} \sum_{l=j}^{-1} \sum_n |d_l(n)|^2 \;=\; \|s\|_2^2 \quad (3.59)$$

Since we need to prove several previous results, let us postpone the proof of the theorem. The following statement gives us a first version for those formulas:

Proposition 4.2 *Let $s \in V_0$, $j < 0$, and the output sequences s_j^m , $0 \le m < 2^{-j}$ and d_l, for $j \le l < -1$. Then:*

$$s(x) \;=\; 2^j \left(\sum_{m=0}^{2^{-j}-1} T^{-m} P_j T^m s \right)(x) +$$

$$+ \; \sum_{l=j}^{-1} 2^l \sum_n d_l(n)\psi(2^l(x-n)) \qquad (3.60)$$

$$\|s\|_2^2 \;=\; 2^j \sum_{m=0}^{2^{-j}-1} \|P_j T^m s\|^2 + \sum_{l=j}^{-1} \sum_n |d_l(n)|^2 \qquad (3.61)$$

where the first equality and the norm are undestood in sense of $L^2(\mathcal{R})$.

Proof:
   Since $T^m s \in V_0$, for each $0 \le m < 2^{-j}$, we can write:

$$T^m s = P_j T^m s + \sum_{l=j}^{-1} Q_l T^m s$$

and:

$$s \ = \ T^{-m}P_jT^ms + \sum_{l=j}^{-1} T^{-m}Q_lT^ms$$

$$\|s\|^2 \ = \ \|T^ms\|^2$$

$$= \ \|P_jT^ms\|^2 + \sum_{l=j}^{-1} \|Q_lT^ms\|^2$$

adding the 2^{-j} terms and averaging:

$$s \ = \ 2^j \sum_{m=0}^{2^{-j}-1} T^{-m}P_jT^ms + 2^j \sum_{m=0}^{2^{-j}-1}\sum_{l=j}^{-1} T^{-m}Q_lT^ms \quad (3.62)$$

$$\|s\|^2 \ = \ 2^j \sum_{m=0}^{2^{-j}-1} \|P_jT^ms\|^2 + 2^j \sum_{m=0}^{2^{-j}-1}\sum_{l=j}^{-1} \|Q_lT^ms\|^2 \quad (3.63)$$

On the other hand we have:

$$T^{-m}Q_lT^ms(x) \ = \ 2^{l/2}T^{-m}\sum_k c_l^m(k)\psi(2^lx - k)$$

$$= \ T^{-m}\sum_k d_l(2^{-l}k - m)\psi(2^lx - k)$$

$$= \ \sum_k d_l(2^{-l}k - m)\psi(2^l(x + m) - k)$$

$$= \ \sum_k d_l(2^{-l}k - m)\psi(2^l(x - (2^{-l}k - m)))$$

$$(3.64)$$

$$\|Q_lT^ms\|^2 \ = \ \sum_k |c_l^m(k)|^2$$

$$= \ 2^{-l}\sum_k |d_l(2^{-l}k - m)|^2 \quad (3.65)$$

for each $0 \le m < 2^{-j}$ and $j \le l < 0$.

Fixed $j < 0$, observe that, for each $j \le l < 0$ and each $n \in Z$, there are exactly 2^{l-j} pairs of values (k, m), $k \in Z$ and $0 \le m < 2^{-j}$ verifying the equation $n = 2^{-l}k - m$.

In fact, there is a unique integer $m_{n,l}$, $0 \le m_{n,l} < 2^{-l}$ such that $m_{n,l} \equiv -n \pmod{2^{-l}}$. Since $j \le l$, we only have the set of possible

2^{l-j} values m:

$$m_{n,l}; \ m_{n,l} + 2^{-l}; \cdots; \ m_{n,l} + 2^{-l}(2^{l-j} - 1)$$

corresponding with the values $k_m = (n + m)2^l$.

Then, adding in formula 3.62 according to the indices j and m and then grouping the terms for each l, we have:

$$2^j \sum_{m=0}^{2^{-j}-1} \sum_{l=j}^{-1} T^{-m} Q_l T^m s \ = \ 2^j \sum_{m=0}^{2^{-j}-1} \sum_{l=j}^{-1} \sum_k d_l(2^{-l}k - m).$$
$$\cdot \ \psi(2^l(x - (2^{-l}k - m)))$$
$$= \ 2^j \sum_{l=j}^{-1} 2^{l-j} \sum_n d_l(n) \psi(2^l(x - n))$$
$$= \ \sum_{l=j}^{-1} 2^l \sum_n d_l(n) \psi(2^l(x - n))$$

and replacing the second term in formula 3.62, we obtain the equality 3.60.

Similarly, from formula 3.63, we obtain the second equality 3.61, completing the proof. •

In practical signal processing problems, to employ the above formulas may be sufficient, since discrete transform is only computed over a minimum level $J < 0$. However, we wish to prove the limit:

$$\lim_{j \to -\infty} 2^j \sum_{m=0}^{2^{-j}-1} \|P_j T^m s\|^2 = 0 \tag{3.66}$$

Observe that, for each fixed m, the value $\|P_j T^m s\|^2$ decreases to zero when j decreases. On the other hand, we are adding and averaging a growing number of terms as j decreases. A uniform bound for these terms is then required.

Lemma 4.1 Let a multiresolution analysis of spline functions of odd order. Then, given $\epsilon > 0$, there is $N = N(\epsilon) \in \mathcal{N}$, such that for each $j < 0$ and all $f \in V_j$:

$$\frac{1}{2\pi} \int_{|\omega| \ge 2^j N\pi} |\hat{f}(\omega)|^2 \, d\omega \ \le \ \epsilon \|f\|^2 \tag{3.67}$$

Proof:

It is sufficient to demonstrate the result for $j = 0$. If $f(x) \in V_j$, then $f(2^{-j}x) \in V_0$ and the extension is obtained by a variable change. We require for the orthogonal scaled function $\phi \in V_0$. From its properties follows that, given $\epsilon > 0$, there is $N = N(\epsilon)$ such that:

$$\sum_{|n| \geq N} |\widehat{\phi}(\omega + 2n\pi)|^2 \leq \epsilon, \quad \omega \in [-\pi, \pi] \quad (a.e)$$

On the other hand, this function generates by translations an orthonormal basis for V_0. Then we can write:

$$f(x) = \sum_m a(m)\phi(x - m)$$
$$\widehat{f}(\omega) = \left(\sum_m a(m)e^{-\omega m\omega}\right)\widehat{\phi}(\omega)$$
$$= F(\omega)\widehat{\phi}(\omega)$$

where $F(\omega)$ is a 2π- periodic function in $L^2(0, 2\pi)$ and

$$\frac{1}{2\pi}\int_{-\pi}^{\pi} |F(\omega)|^2 d\omega = \sum_m |a(m)|^2 = \|f\|^2$$

Then, for the integer N above defined:

$$\frac{1}{2\pi}\int_{|\omega| \geq N\pi} |\widehat{f}(\omega)|^2 \, d\omega = \frac{1}{2\pi}\int_{-\pi}^{\pi} |F(\omega)|^2 \sum_{|n| \geq N} |\widehat{\phi}(\omega + 2n\pi)|^2 \, d\omega$$
$$\leq \frac{\epsilon}{2\pi}\int_{-\pi}^{\pi} |F(\omega)|^2 \, d\omega$$
$$\leq \epsilon\|f\|^2$$

and the proof is complete . •

Next, we state:

Lemma 4.2 *Let $s \in V_0$. Given $\delta > 0$ there is $J = J(\delta; s) < 0$ such that for all $m \in \mathcal{Z}$ and $j \leq J$:*

$$\|P_j T^m s\|^2 < \delta\|s\|^2 \tag{3.68}$$

Proof:

Assume $\|s\| = 1$. Let $\delta > 0$ and $\epsilon = \epsilon(\delta) > 0$ to be precisely defined below. Let $N = N(\epsilon)$ be, the value verifying the condition of lemma 4.1. For each $j < 0$ and $m \in \mathcal{Z}$ we have the orthogonal decomposition expressed in the Fourier transform domain:

$$(T^m s)^\wedge(\omega) = (P_j T^m s)^\wedge(\omega) + \sum_{l=j}^{-1}(Q_l T^m s)^\wedge(\omega)$$

For simplicity, denote $A_j(\omega)$ and $B_j(\omega)$ the respective terms in the second member. Next, observe that $(P_j T^m s) \in \mathcal{V}_j$. Then we can write:

$$
\begin{aligned}
|\hat{s}(\omega)|^2 &= |(T^m s)^\wedge)(\omega)|^2 \\
&= |A_j(\omega)|^2 + |B_j(\omega)|^2 + A_j(\omega)\overline{B_j}(\omega) + B_j(\omega)\overline{A_j}(\omega)
\end{aligned}
$$
$$(3.69)$$

Following, choose α such that:

$$\frac{1}{2\pi}\int_{|\omega|\leq\alpha}|\hat{s}(\omega)|^2\,d\omega \leq \epsilon$$

and integrate expression 3.69. Then:

$$\frac{1}{2\pi}\int_{|\omega|\leq\alpha}(|A_j(\omega)|^2 \ + \ |B_j(\omega)|^2 +$$
$$+ \ A_j(\omega)\overline{B_j}(\omega) + B_j(\omega)\overline{A_j}(\omega))\,d\omega \ \leq \epsilon$$

Unfolding the integral and observing that $|B_j(\omega)|^2$ is not negative:

$$\frac{1}{2\pi}\int_{|\omega|\leq\alpha}|A_j(\omega)|^2\,d\omega \ -$$
$$- \ |\frac{1}{2\pi}\int_{|\omega|\leq\alpha}A_j(\omega)\overline{B_j}(\omega) + +B_j(\omega)\overline{A_j}(\omega))\,d\omega| \leq$$
$$\leq \ \frac{1}{2\pi}\int_{|\omega|\leq\alpha}|A_j(\omega)|^2\,d\omega +$$
$$+ \ \frac{1}{2\pi}\int_{|\omega|\leq\alpha}A_j(\omega)\overline{B_j}(\omega) + B_j(\omega)\overline{A_j}(\omega))\,d\omega \leq$$
$$\leq \ \epsilon$$
$$(3.70)$$

Let J the grater integer $j < 0$ such that $2^j N\pi \leq \alpha$. Then for all $j \leq J$:

$$[-2^j N\pi, \ 2^j N\pi] \subset [-\alpha, \alpha]$$

Since $(P_j T^m s) \in \mathcal{V}_j$, applying lemma 4.1, we have:

$$\frac{1}{2\pi} \int_{|\omega| \leq \alpha} |A_j(\omega)|^2 \, d\omega \; \geq \; \frac{1}{2\pi} \int_{|\omega| \leq 2^j N\pi} |A_j(\omega)|^2 \, d\omega$$

$$\geq \; (1 - \epsilon) \| P_j T^m s \|^2 \qquad (3.71)$$

On the other hand, given the orthogonality of the decomposition

$$\frac{1}{2\pi} \int_{-\infty}^{\infty} A_j(\omega)\overline{B_j}(\omega) + B_j(\omega)\overline{A_j}(\omega)) \, d\omega \; = \; 0$$

so that:

$$|\frac{1}{2\pi} \int_{|\omega| \leq \alpha} A_j(\omega)\overline{B_j}(\omega) + B_j(\omega)\overline{A_j}(\omega)) \, d\omega \, | \; =$$

$$= \; |\frac{1}{2\pi} \int_{|\omega| \geq \alpha} A_j(\omega)\overline{B_j}(\omega) + B_j(\omega)\overline{A_j}(\omega)) \, d\omega \, | \; \leq$$

$$\leq \; 2(\frac{1}{2\pi} \int_{|\omega| \geq \alpha} |A_j(\omega)|^2 \, d\omega)^{1/2} \, (\frac{1}{2\pi} \int_{|\omega| \geq \alpha} |B_j(\omega)|^2 \, d\omega \,)^{1/2} \; \leq$$

$$\leq \; 2\epsilon^{1/2} \| P_j T^m s \| \, \| \sum_{l=j}^{-1} Q_l T^m s \| \; \leq$$

$$\leq \; 2\epsilon^{1/2} \| P_j T^m s \| \qquad (3.72)$$

We remark that the last inequality follows the fact that the norm of the summing of the components $Q_l T^m s$, is bounded by the norm of the signal. Now, using inequalities 3.71 and 3.72 in 3.70 we conclude:

$$(1 - \epsilon)\| P_j T^m s \|^2 - 2\epsilon^{1/2} \| P_j T^m s \| \; \leq \; \epsilon$$

Therefore:

$$\| P_j T^m s \| \; \leq \; \frac{\epsilon^{1/2} + (2\epsilon - \epsilon^2)^{1/2}}{1 - \epsilon}$$

Let ϵ verifying:

$$\frac{7\epsilon}{(1 - \epsilon)^2} \; \leq \; \delta$$

it follows:

$$\| P_j T^m s \|^2 \; \leq \; \delta$$

for all $j < J$ and $m \in \mathcal{Z}$, as it was required. $\bullet$

Now, we can demonstrate the theorem above stated:

Proof: (Theorem 4.1)   Given $\delta > 0$, we take $J > 0$, satisfying the condition of lemma 4.2, the energy formula 3.59 is proved as follows. Assuming that $\|s\| = 1$, from 3.61 we deduce:

$$1 - \sum_{l=J}^{-1} \sum_{n} |d_l(n)|^2 \ = \ 2^J \sum_{m=0}^{2^{-J}-1} \|P_J T^m s\|^2$$

$$\leq \ \delta$$

The first formula is easely deduced from 3.60 :

$$\|s(x) - \sum_{l=j}^{-1} 2^l \sum_{n} d_l(n)\psi(2^l(x-n))\| \ = \ \|2^J \sum_{m=0}^{2^{-J}-1} T^{-m} P_J T^m s\|$$

$$\leq \ 2^J \sum_{m=0}^{2^{-J}-1} \|P_J T^m s\|$$

$$\leq \ \delta^{1/2}$$

This completes the demonstration. •

Finally, we can resume the properties of the undecimated discret wavelet transform D defined in the preceding section. The norm of the operator is defined as:

$$\|D\| = \sup_{s \in V_0} \frac{\left(\sum_{j<0} \|D_j s\|_{l^2(\mathcal{Z})}^2\right)^{1/2}}{\|s\|_{L^2(\mathcal{R})}} \qquad (3.73)$$

and we state:

Corollary 4.1 *The operator D, corresponding with the undecimated discrete wavelet transform, verifies:*

$$(a) \quad \|D\| = 1 \qquad\qquad (3.74)$$
$$(b) \quad DT^m = T^m D \qquad\qquad (3.75)$$

Proof:
   The first affirmation flows from Theorem 4.1.
   On the other hand, we easily deduce:

$$D_j T^m s(n) \ = \ 2^j \langle s(x-m), \psi(2^j(x-n)) \rangle$$
$$= \ 2^j \langle s(x), \psi(2^j(x-(n-m))) \rangle$$
$$= \ T^m D_j s(n)$$

for all $j < 0$.•

We remark that the range of D is $l^2(\mathcal{Z}_{\leq 0} \times \mathcal{Z})$ and the image of s by D is the double sequence, $d_j(n)$, with $j \in \mathcal{Z}_{\leq 0}$ and $n \in \mathcal{Z}$. Moreover, the operator D is injective, according to (a), but it is not surjective. To illustrate the last affirmation, it can easily be checked that the sequence $e_j(n) = \delta_{j,-1}\delta_{n,0}$ for $j < 0$ does not belong to the image of D.

In general, given a sequence $d(n)$, it is a non trivial problem to decide if it corresponds with the undecimated discrete transform $d_j(n)$ of some signal, for some index j. We refer to [25] for a deeper treatmet of this question.

On the other hand, we remark that values of the undecimated transform of any given signal in $\mathcal{V}_0$ esentially consist in the standard wavelet coefficients of all translated signals $T^m s$. They provide redundant information, but making possible to detect and characterize local phenomena or recognize structures, despite any translation of the signal in its sampling net.

Moreover, the above demonstrated results tell us that redundancy can be controlled. In the next section we propose a hierarchical scheme to organize the information provided by the undecimated discrete transform.

5   Extended Multiresolution Analysis for $\mathcal{V}_0$

At this point we propose an appropriate scheme to organize the numerical information given by the values $d_j(n)$. Since these values are directly correlated with the wavelet coefficients of all possible translations of the signal in $\mathcal{V}_0$:

$$d_j(2^{-j}k - m) = 2^{j/2}\langle T^m s, \psi_{jk}\rangle \tag{3.76}$$

they can be disposed according to an extended multirresolution structure. It essentially consists of the family of translated subspaces $\{T^{-n}\mathcal{V}_j\}$, $0 \leq n < 2^{-j}$ and $j \leq 0$. These subspaces can be organized in collections corresponding with a library of orthonormal bases of translated wavelets, for the fundamental subspace $\mathcal{V}_0$.

For each $j \leq 0$ denote $\mathcal{Z}_j$ the set of integers $\{0 \leq n < 2^{-j}\}$. Let $l \in \mathcal{Z}_j$ and define the subspace:

$$\mathcal{V}_j^{(l)} = \{f(x) : T^l f(x) \in \mathcal{V}_j\} \tag{3.77}$$

Clearly $\mathcal{V}_j^{(l)}$ consists of all spline functions in $\mathcal{V}_0$, with knots on the lattice $\{n \equiv -l \pmod{2^{-j}}\}$. It is closed for the operators T^p, with $p \equiv 0 \pmod{2^{-j}}$. Now we state:

Proposition 5.1 *Let* $0 \leq l < q < 2^{-j}$ *for some* $j < 0$. *Then:*

$$\mathcal{V}_j^{(l)} \cap \mathcal{V}_j^{(q)} = \{0\} \tag{3.78}$$

Proof:

Suppose $f \in \mathcal{V}_j^{(l)} \cap \mathcal{V}_j^{(q)}$. Since $l \neq q$ the sets $\{n \equiv -l \pmod{2^{-j}}\}$ and $\{n \equiv -q \pmod{2^{-j}}\}$ are disjointed. Then, the spline f must be differentiable in all $x \in \mathcal{R}$, that is a polynomial in $\mathcal{V}_0$. We conclude that $f \equiv 0$. •

Next, given $m \in \mathcal{N}_0$ we define the sequence:

$$m_j \equiv m \pmod{2^{-j}} \quad \text{for each } j \leq 0 \tag{3.79}$$

Clearly $m_0 = 0$ and we can write:

$$m = \sum_{k \geq 0} e_k 2^k \quad ; \ e_k = 0, 1 \tag{3.80}$$

Then for each $j \leq -1$ we have:

$$m_j = \sum_{k=0}^{-(j+1)} e_k 2^k \quad \text{if } m \geq 2^{-j} \tag{3.81}$$

$$m_j = m \quad \text{if } m < 2^{-j} \tag{3.82}$$

Moreover, if $q < j$:

$$m_q = \begin{cases} m_j & \text{if } m_q \in \mathcal{Z}_j \\ m_j + 2^{-j} S & \text{if } m_q \in \mathcal{Z}_q / \mathcal{Z}_j \end{cases} \tag{3.83}$$

where $\mathcal{Z}_q / \mathcal{Z}_j$ denotes the difference set and S is the integer:

$$S = \sum_{k=0}^{-(q+1)+j} e_{k-j} 2^k$$

Therefore,

$$m_q \equiv m_j \pmod{2^{-j}} \tag{3.84}$$

Now we can state:

Proposition 5.2 *Given $q \leq j \leq 0$ and $m \in \mathcal{N}_0$, we have:*

$$V_q^{(m_j)} \subset V_j^{(m_j)} \tag{3.85}$$

$$V_q^{(m_q)} \subset V_j^{(m_j)} \tag{3.86}$$

Proof:

If $f(x) \in V_q^{(m_j)}$ then

$$f(x - m_j) \in V_q \subset V_j$$

and the first result holds. On the other hand, if $f(x) \in V_q^{(m_q)}$ then

$$f(x - m_q) \in V_q \subset V_j$$

Therefore,

$$f(x - m_q) = f(x - m_j + (m_j - m_q)) \in V_j$$

and since $m_j - m_q \equiv 0 \pmod{2^{-j}}$, we finally have $f(x - m_j) \in V_j$.
•

For the particular case $q = j - 1$ we conclude:

Corollary 5.1

$$V_{j-1}^{(m_j)} \subset V_j^{(m_j)} \tag{3.87}$$

$$V_{j-1}^{(m_j+2^{-j})} \subset V_j^{(m_j)} \tag{3.88}$$

Proof:

Let $l = m_j + 2^{-j}$. Since $m_j < 2^{-j}$, it follows $l \in \mathcal{Z}_{j-1}/\mathcal{Z}_j$. Then $l_j = m_j$, $l_{j-1} = l$ and the result holds applying 3.87 and 3.88 for the integer l . •

These results lead us to define the *extended multiresolution analyis stucture* for V_0 as follows:

$$V_E = \{V_j^{(l)}; \ l \in \mathcal{Z}_j, \ j \leq 0\} \tag{3.89}$$

and following the propositions already demonstrated, we summarize the properties of these closed subspaces:

Theorem 5.1 *The family V_E defined by 3.89 verifies:*

(a) $\quad \cdots V_j^{(m_j)} \subset V_{j+1}^{(m_{j+1})} \subset \cdots \subset V_0^{(0)}$, *for all* $m \in \mathcal{N}_0$ (3.90)

(b) $\quad f(x) \in V_j \Longleftrightarrow f(x+m) \in V_j^{(m_j)}$, $m \in \mathcal{N}_0$ (3.91)

(c) $\quad f(x) \in V_j^{(m_j)} \Longleftrightarrow f(2x - m_j) \in V_{j+1} \ m \in \mathcal{N}_0, \ j \leq -1$

(3.92)

(d) $\quad V_j^{(l)} \cap V_j^{(q)} = \{0\}$, $l, q \in \mathcal{Z}_j \ p \neq q$. (3.93)

(e) *The subspace $V_0^{(0)}$ has an orthonormal basis generated by the translations in $\mathcal{Z}$ of a unique function.* •

Note that $V_0^{(0)}$ denotes here the fundamental subspace V_0. We can see that given $m \in \mathcal{N}_0$ the subfamily:

$$V_E^{(m)} = \{V_j^{(m_j)}; \ m \equiv m_j \in \mathcal{Z}_j, \ j \leq 0\} \qquad (3.94)$$

essentially has the same structure of the standard multiresulution analysis for $L^2(\mathcal{R})$, [22].

Note that each subfamily $V_E^{(m)}$ corresponds with a branch of the tree. If $m = 0$ we have the branch indexed by 0, corresponding with the standard multiresolution analysis for V_0. If $m \geq 1$, the branch is indexed by the associated sequence m_j.

It can also be seen that from the knot indexed by $(j+1, l)$, for $l \in \mathcal{Z}_{j+1}$, descend two knots indexed by (j, l) and $(j, l + 2^{-(j+1)})$ respectively. Then, given the integers $m1 \equiv m2$ (mod $2^{-(j+1)}$, with $m1_{j+1} = m2_{j+1} = l$) and $m1_j = m2_j + 2^{-(j+1)}$, the corresponding branch bifurcates at the knot $calV_{j+1}^{(l)}$.

We remark that the proposition (d), in theorem 5.1, *is not true* in general. Here, the fact that V_0 consists of spline functions cannot be obviated. On the other hand, this hypothesis provides a sufficient condition. At the present, we cannot formulate analogous results beyond the spline case.

Following this we construct the correlated library of orthonormal bases for V_0. Given $j < 0$ and $l \in \mathcal{Z}_j$, we define the translated scale functions and wavelets in V_E:

$$\phi_{jk}^{(l)}(x) \ = \ 2^{j/2}\phi(2^j(x+l) - k) \qquad (3.95)$$

$$\psi_{jk}^{(l)}(x) \ = \ 2^{j/2}\psi(2^j(x+l) - k) \qquad (3.96)$$

Clearly they are spline functions with knots on the lattices $\{n \equiv -l \pmod{2^{-j}}\}$ and $\{n \equiv -l \pmod{2^{-(j+1)}}\}$ respectively. That is:

$$\phi_{jk}^{(l)} \in \mathcal{V}_j^{(l)} \tag{3.97}$$

$$\psi_{jk}^{(l)} \in \mathcal{V}_{j+1}^{(p)}; \quad p \equiv l \pmod{2^{-(j+1)}} \tag{3.98}$$

and each family $\{\phi_{jk}^{(l)}, k \in \mathcal{Z}\}$ provides an orthonormal basis for the subspace $\mathcal{V}_j^{(l)}$. On the other hand, we remark that if $p \in \mathcal{Z}_{j+1}$, we would find two families of wavelets in the subspace $\mathcal{V}_{j+1}^{(p)}$, corresponding with the supraindices p and $p + 2^{-(j+1)}$ respectively.

Next, given $m \in \mathcal{N}_0$ and the associated sequence $\{m_j\}$, for each $j < 0$, we define:

$$\mathcal{W}_j^{(m_j)} = span\{\psi_{jk}^{(m_j)}\}; \quad k \in \mathcal{Z} \tag{3.99}$$

Clearly, each $\mathcal{W}_j^{(m_j)}$ is isomorphous with $\mathcal{W}_j$ and the wavelets $\{\psi_{jk}^{(m_j)}\}$ give us an orthonormal basis for this subspace. Now we state:

Proposition 5.3

$$\mathcal{V}_{j+1}^{(m_{j+1})} = \mathcal{V}_j^{(m_j)} \oplus \mathcal{W}_j^{(m_j)} \tag{3.100}$$

Proof:
The inclusion $\mathcal{V}_{j+1}^{(m_{j+1})} \supset \mathcal{W}_j^{(m_j)}$ is trivial. Let $f(x) \in \mathcal{V}_{j+1}^{(m_{j+1})}$. Since $m_j \equiv m_{j+1} \pmod{2^{-(j+1)}}$, we also have $T^{m_j-m_{j+1}} f(x) \in \mathcal{V}_{j+1}^{(m_{j+1})}$, then $T^{m_j} f(x) \in \mathcal{V}_{j+1}$ and we can write:

$$T^{m_j} f(x) = \sum_n a(n)\phi(2^j x - n) + \sum_k b(k)\psi(2^j x - k)$$

Therefore, we can represent the given function in $\mathcal{V}_{j+1}^{(m_{j+1})}$ in the form:

$$f(x) = \sum_n a(n)\phi(2^j(x + m_j) - n) + \sum_k b(k)\psi(2^j(x + m_j) - k)$$

On the other hand, since the operator T is an isometry, we conclude the ortogonality of the decomposition. •

Observe that, according to proposition 5.3, the subspace $\mathcal{V}_{j+1}^{(p)}$, $p \in \mathcal{Z}_{j+1}$, can be decomposed in two ways:

$$
\begin{aligned}
\mathcal{V}_{j+1}^{(p)} &= \mathcal{V}_j^{(p)} \oplus \mathcal{W}_j^{(p)} \\
&= \mathcal{V}_j^{(p+2^{-(j+1)})} \oplus \mathcal{W}_j^{(p+2^{-(j+1)})}
\end{aligned}
\qquad (3.101)
$$

At last, we define the library of bases. For each $m \in \mathcal{N}_0$ we choose the wavelet basis for $V_E^{(m)}$:

$$
B_m = \{\psi_{jk}^{(m_j)} \; ; j < 0; \; k \in \mathcal{Z}\} \qquad (3.102)
$$

and the full collection $\{B_m, \; m \in \mathcal{N}_0\}$ is the library of orthonormal bases for the subspace $\mathcal{V}_0$, associated with the undecimated discrete wavelet transform.

At last, given $m \in \mathcal{N}_0$ y $s \in \mathcal{V}_0$, we compute:

$$
\begin{aligned}
\langle s(x), \psi_{jk}^{(m_j)}(x) \rangle &= \langle T^{m_j} s(x), \psi_{jk}(x) \rangle \\
&= 2^{-j/2} d_j (2^{-j} k - m_j)
\end{aligned}
\qquad (3.103)
$$

and

$$
\begin{aligned}
\langle T^m s(x), \psi_{jk}(x) \rangle &= \langle s(x), \psi_{jk}^{(m_j)}(x + 2^{-j}\lfloor m2^j \rfloor) \rangle \\
&= 2^{-j/2} d_j (2^{-j}(k - \lfloor m2^j \rfloor) - m_j) \quad (3.104)
\end{aligned}
$$

We conclude that each sequence $\{d_j(n), \; n \equiv -m_j \pmod{2^{-j}}\}$ corresponds with the standard wavelet spectrum of the signal $T^{m_j} s$, in the level j. The structure of the spectrum of any signal $T^p s$, with $m_j \equiv p \pmod{2^{-j}}$, is quite similar.

In summary, the extended multiresolution analysis scheme, provides a suitable way to organize the numerical information given by the values $d_j(n)$. Moreover, we can characterize the wavelet spectrum of any integer translation for a given signal in $\mathcal{V}_0$. This ability leads to implement efficient techniques in pattern recognition applications.

On the other hand, we can design estrategies to choose appropiate bases to represent the given signal, according to the problem on the hand.

Particularly, one can exploit the difference of information given by different representations to detect and characterize local phenomena [41].

6   Decomposition of a Signal in $\mathcal{V}_0$

Let $\mathcal{V}_0$ the regular subspace consisting of spline functions of odd order $\nu \geq 3$ and $s \in \mathcal{V}_0$. We now consider further properties of the semidiscrete spline transform and the correlated undecimated discrete transform, already defined. We also assume that $s \in L^1(\mathcal{R})$, being this a reasonable supposition for any experimental signal, given by finite sampling values [22] and [41]. Therefore, $W_j^{(s)} \in L^1(\mathcal{R})$ and $d_j \in l^1(\mathcal{Z})$, for each $j \leq -1$ [39].

Firstly, we examine the location of the discrete representation of the signal in the time domain. Since:

$$
\begin{aligned}
d_j(n) &= W_j^{(s)} s(n) \\
&= 2^j \langle s(x), \psi(2^j(x-n)) \rangle
\end{aligned}
\tag{3.105}
$$

and the spline wavelet ψ is centered on $x_0 = 0.5$, the value $d_j(n)$ resume the local information of the signal s at a neighbour of the center $x_{jn} = 2^{-(j+1)}+n$, corresponding with the wavelet $\psi(2^j(x-n))$. Therefore, we will locate the values $d_j(n)$ at these points.

Following this, we wish to characterize the filtering properties of the discrete transform. For each $j < 0$, denote

$$
\begin{aligned}
\tilde{P}_j &= 2^j \sum_{m=0}^{2^{-j}-1} T^{-m} P_j T^m \\
\tilde{Q}_j &= 2^j \sum_{m=0}^{2^{-j}-1} T^{-m} Q_j T^m
\end{aligned}
\tag{3.106}
$$

Now, fixed j and recalling formula 3.62, we can decompose the given signal as:

$$
s(x) = \tilde{P}_j s(x) + \sum_{l=j}^{-1} \tilde{Q}_l s(x)
\tag{3.107}
$$

and

$$
\tilde{P}_j s(x) = \tilde{P}_{j-1} s(x) + \tilde{Q}_{j-1} s(x)
\tag{3.108}
$$

Observe that the components $\tilde{P}_j s$ and $\tilde{Q}_l s$, with $j \leq l \leq -1$, average the projections $T^{-m} P_j T^m s$ and $T^{-m} Q_l T^m s$ onto the respective subspaces $\mathcal{V}_j^{(m)}$ and $\mathcal{W}_l^{(m)}$, respectively. Such decomposition is not orthogonal. Moreover, since P_j and Q_l do not exchange with the translation operator T the operators $\tilde{P}_j$ and $\tilde{Q}_l$ are not projectors:

$$\tilde{P}_j^2 \neq \tilde{P}_j$$
$$\tilde{Q}_l^2 \neq \tilde{Q}_l \qquad (3.109)$$

We can understand the above decomposition as follows. First, recall that the spline wavelet Fourier transform $\hat{\psi}(2^{-j}\omega)$ is well concentrated on the two-side band:

$$\Omega_j = 2^j \pi \leq |\omega| \leq 2^{(j+1)}\pi$$

In addition there is a number $c \cong \sqrt{2}/2$ such that [41]:

$$c \leq |\hat{\psi}(2^{-j}\omega)| \leq 1 \qquad \omega \in \Omega_j \qquad (3.110)$$

Therefore, from this properties and the identity:

$$(W_j^{(s)}s)^\wedge(\omega) = \hat{s}(\omega)\hat{\psi}^*(2^{-j}\omega) \qquad (3.111)$$

we can see that $(W_j^{(s)}s)^\wedge$ is well concentrated on Ω_j. Moreover, we also deduce the local formula:

$$\hat{s}(\omega) = \frac{(W_j^{(s)}s)^\wedge(\omega)}{|\hat{\psi}(2^{-j}\omega)|} e^{-i\omega 2^{-(j+1)}} \qquad \omega \in \Omega_j \qquad (3.112)$$

So that, we conclude that $(W_j^{(s)}s)^\wedge$ achieves the complete spectral information $\hat{s}(\omega)$, for $\omega \in \Omega_j$.

Now, denote $\mathcal{L}^{(2\nu+1)}$ the corresponding cardinal spline function of order $2\nu + 1$. We remark that its Fourier transform $\hat{\mathcal{L}}^{(2\nu+1)}$, is real and positive on $[-\pi, \pi]$. Moreover, it can be considered an almost ideal low-pass filter on this range, [3]. Then we can write:

$$\hat{\mathcal{L}}^{(2\nu+1)}(\omega) = \chi_{[-\pi,\pi]}(\omega) + \epsilon^{(2\nu+1)}(\omega) \qquad (3.113)$$

where the term $\epsilon^{(2\nu+1)}(\omega)$ quickly vanishes as ν increases, [3], [41].

Next, we characterize the action of the operator $\tilde{Q}_j$, $j < 0$. Recalling that $W_j^{(s)}s$ is a spline function of order $2\nu + 1$ with knot set $\mathcal{Z}$ we can write:

$$W_j^{(s)}s(x) = \sum_n d_j(n)\mathcal{L}^{(2\nu+1)}(x - n) \qquad (3.114)$$

$$\tilde{Q}_j s(x) = 2^j \sum_n d_j(n)\psi(2^j(x - n)) \qquad (3.115)$$

and transforming:

$$(W_j^{(s)}s)^\wedge(\omega) = \left(\sum_n d_j(n)\, e^{-\imath\omega n}\right) \widehat{\mathcal{L}}^{(2\nu+1)}(\omega) \qquad (3.116)$$

$$(\tilde{Q}_j s)^\wedge(\omega) = \left(\sum_n d_j(n)\, e^{-\imath\omega n}\right) \widehat{\psi}(2^{-j}\omega) \qquad (3.117)$$

These formulas suggest similarity between both representations of the signal. We can exploit it by writing:

$$(\tilde{Q}_j s)^\wedge(\omega)\, \widehat{\mathcal{L}}^{(2\nu+1)}(\omega) = (W_j^{(s)}s)^\wedge(\omega)\widehat{\psi}(2^{-j}\omega) \qquad (3.118)$$

Restrict $\omega \in \Omega_j$. Then, from the the properties of the function $\widehat{\mathcal{L}}^{(2\nu+1)}$ and using 3.112 we finally derive the remarkable formulas:

$$\begin{aligned}
(\tilde{Q}_j s)^\wedge(\omega) &= (W_j^{(s)}s)^\wedge(\omega)\frac{\widehat{\psi}(2^{-j}\omega)}{\widehat{\mathcal{L}}^{(2\nu+1)}(\omega)} \\
&= \widehat{s}(\omega)\frac{|\widehat{\psi}(2^{-j}\omega)|^2}{\widehat{\mathcal{L}}^{(2\nu+1)}(\omega)} \qquad (3.119)
\end{aligned}$$

for all $\omega \in \Omega_j$.

We can see that $(\tilde{Q}_j s)^\wedge$ and $(W_j^{(s)}s)^\wedge$ can be considered as alternative approximations of $\widehat{s}(\omega)$ on Ω_j. In other words, the operator $\tilde{Q}_j$ acts as an almost band-pass filter. Therefore, for practical purposes, one can employ the approximation:

$$(\tilde{Q}_j s)^\wedge(\omega) \approx \widehat{s}(\omega)\,\chi_{\Omega_j}(\omega) \qquad (3.120)$$

and from the formulas:

$$\begin{aligned}
\widehat{s}(\omega) &= \sum_{j<0} (\tilde{Q}_j s)^\wedge(\omega) \\
&\approx \sum_{j<0} \widehat{s}(\omega)\,\chi_{\Omega_j}(\omega) \qquad (3.121)
\end{aligned}$$

understanding the first equation in the sense given by the theorem 4.1, we can conclude that the components $\tilde{Q}_j s$ are *nearly* orthogonals.

On the other hand, we also conclude that the undecimated discrete transform values $d_j(n)$ resume the complete spectral information of the signal at the range Ω_j. Particularly, opposite to the standar

wavelet coefficients, they also provide the phase information. Therefore the UWT becomes an efficient tool to detect and characterize
local phenomena by scales.

Finally, let us walk in the opposite way. For simplicity, take $j = -1$. As we already remarked the spline wavelet Fourier transform $\widehat{\psi}(\omega/2)$ is almost concentrated at $\pi/2 \leq |\omega| \leq \pi$. On the other hand, we noted that it does not vanish on $0 < |\omega| < 2\pi$. Therefore the filter:

$$\frac{|\widehat{\psi}(\omega/2)|^2}{\widehat{\mathcal{L}}^{(2\nu+1)}(\omega)}$$

is bounded and does not vanish on $0 < |\omega| < 2\pi$. This suggests that the formula 3.119 could be inverted beyond Ω_{-1}, particularly on the range $0 < |\omega| \leq \pi$. Justly, if $s_0 \in \mathcal{V}_0$, its Fourier transform is totally defined from the values $\widehat{s}(\omega)$ for $-\pi \leq \omega \leq \pi$. In other words, can we reconstruct the signal $s(x)$ from the values $d_{-1}(n)$? That is, does $(\tilde{Q}_{-1}s)^{\wedge}(\omega)$ carry the complete information of the signal? Observe that this conjecture appears as contradictory with respect to the above results.

However, let us give a simple illustration, providing by the undec-
imated Haar transform. In this case, for $j = -1$, we have $d_{-1}(n) = (1/2)(s(n+1) - s(n))$ and we easily construct the reconstruction scheme:

$$
\begin{align}
s(0) \quad & : \quad \text{initial condition} & (3.122)\\
s(n+1) \; &= \; 2d_{-1}(n) + s(n), \quad n \geq 0 & (3.123)\\
s(n-1) \; &= \; -2d_{-1}(n-1) + s(n), \quad n \leq 0 & (3.124)
\end{align}
$$

Note that we required the initial condition $s(0)$. Of course, we cannot guarantee the stability of the formula for it suggests the completeness of the Haar family $\{\psi_H(\frac{1}{2}(x-n)), \; n \in \mathcal{Z}\}$ in the subspace $\mathcal{V}_0$.

Beyond the Haar case, we can demonstrate the following result, in the spline multiresolution analysis context.

Proposition 6.1 *Let $\mathcal{V}_0$ a multiresolution analysis consisting of spline functions of odd order $\nu < \infty$. Then the family:*

$$\psi(\frac{1}{2}(x-n)), \quad n \in \mathcal{Z}$$

is complete in $\mathcal{V}_0$.

Proof:

Let $s \in V_0$. Then $P_{-1}s \in V_{-1}^{(0)}$ and $P_{-1}Ts \in V_{-1}^{(1)}$, as we have seen it previously.

Then, by Proposition 5.1 we can assure that both s and Ts cannot belong to $V_{-1}^{(0)}$, unless $||s|| = 0$.

Let $s \in V_0$ be such that:

$$\langle s(x),\ \psi(\tfrac{1}{2}(x - n)\,)\rangle = 0, \quad \text{for all } n \in \mathcal{Z}$$

This is equivalent to say that they simultaneously verify:

$$\langle s(x),\ \psi(\tfrac{1}{2}x - n)\rangle = 0, \quad \text{for all } n \in \mathcal{Z}$$

and

$$\langle Ts(x),\ \psi(\tfrac{1}{2}x - n)\rangle = 0, \quad \text{for all } n \in \mathcal{Z}$$

but this implies that s and Ts are elements of $V_{-1}^{(0)}$, then $||s|| = 0$, and the result follows. •

We can conclude that for $j = -1$ the values $d_{-1}(n)$ contain all the information of the signal in V_0. This result has theorical interest. However, the refined family of wavelets $\{\psi(\tfrac{1}{2}(x - n)\,),\quad n \in \mathcal{Z}\}$, is not a frame for V_0. At present, we do not know how to exploit this property. Particularly, we cannot give a reconstruction algorithm besides the Haar case.

This appears to be a hard problem. To illustrate, take $\epsilon > 0$ and $s \in V_j$ for an appropriate $j \ll -1$. Then we have:

$$||\tilde{Q}_{-1}||^2 = \sum_n d_{-1}^2 \leq \epsilon ||s||^2$$

and observe that although the values corresponding with $j = -1$ contain all the information of the signal, they must be almost zero and carry on to the inestability of the reconstruction.

We want to underline that these results, commentaries and observations are strongly based on the fact that the undecimated transform is defined from orthogonal spline wavelets.

Furthermore, it is important to remark that the result of proposition 6.1 *is not valid* for all wavelet families. For instance, in Littlewood-Paley case, associated to ideal cardinal sinc function [3], the values

$d_{-1}(n)$, do not contain more than the strict spectral information in range $\pi/2 \leq |\omega| \leq \pi$. We also remark that this case is not included in the hypothesys of the above mentioned proposition.

To finish, we give some suggestions for the use of the proposed UWT in signal processing applications. We refer to [41] for more details about these implementations.

Firstly, the discrete transform leads us to interpolate the spline function SWT by using appropriate cardinal filters [5] – [9]. Furthermore, well known techiques can be implemented as the zero-crossing or modulus maxima representations [25], [27], or the multiscale edge detection methods [16], [29].

Beyond these standard alternatives, we have on the hand another techniques from the undecimated discrete transform.

One can implement an efficient pattern recognition scheme by correlating the respective transforms, corresponding with the pattern and the signal. Then we can detect and recognize common structures living along the signal, by scales, [41].

On the other hand, we can design techiques to detect local phenomena, by analizing the derivatives of the given signal $s \in \mathcal{V}_0$.

For these purposes, we employ an appropriate different scheme from the values of the discrete transform. Then we can estimate locally k^{th} order modulus of continuity for the signal s, by scales [41]. This technique appears as an efficient alternative to the above refered multiscale edge detection.

Moreover, the extended multiresolution scheme and the correlated library of basis, give us the chance to implement flexible and efficient strategies to analize the signal, according to the problem on the hand. Particularly, they lead us to compress the available information, using the well known minima entropy criterion.

7   Conclusion

We have proposed the use of the undecimated discrete wavelet transform associated with orthonormal spline wavelets. They lead us to combine the advantages given by the orthogonal spline wavelet analysis with the property of consistency with respect to the translations of the signal.

For these purposes, we expose an efficient algorithm to compute the discrete transform and construct an extended multiresolution structure to organize the available information.

Summarizing, we outline remarkable properties of the undecimated discrete wavelet transform. They can be exploited in signal processing applications, according to the problem on the hand. Moreover, they suggest efficient tools and strategies to these purposes. We hope that these results ecourage further research.

8   References

[1] J. M. Ahlberg, E. N. Nilson and J. L. Walsh, *The Theory of Splines and Their Applications*, Academic Press, New York, 1967.

[2] A. Aldroubi and M. Unser, Families of Multiresolution and Wavelet Spaces with Optimal Properties, Numer. Funct. Anal. and Optimz. **14**, 1993.

[3] A. Aldroubi, M. Unser and M. Eden, Cardinal Spline Filters: Stability and Convergence to the Ideal Sinc Interpolator, Signal Process. **28**, 1992.

[4] A. Aldroubi and M. Unser, Families of Wavelet Transforms in Connection with the Shannon's Sampling Theory and the Gabor Transform, In *Wavelets: A Tutorial in Theory and Applications*, Academic Press Inc., San Diego, 1992.

[5] A. Aldroubi and M. Unser, Sampling Procedures in Function Spaces and Asymptotic Equivalence with Shannon's Sampling Theory, Numer. Funct. Anal. and Optimz. **15**, 1994.

[6] G. Battle, A block spin construction of ondelettes. Part I: Lemarié functions, Comm. Math. Phys. **110**, 1987.

[7] G. Battle, Cardinal Spline Interpolation and the Block Spin Construction of Wavelets, In *Wavelets: A Tutorial in Theory and Applications*, Academic Press Inc., San Diego, 1992.

[8] C. K. Chui, *Multivariate Splines*, SIAM, Philadelphia, 1991.

[9] C. K. Chui, *An Introduction to Wavelets*, Academic Press, Boston, 1992.

[10] C. K Chui *et al.*, *Wavelets: A Tutorial in Theory and Applications*, Academic Press Inc., San Diego, 1992.

[11] I. Daubechies, Orthonormal bases of compactly supported wavelets, Comm. Pure Appl. Math. **41**, 1988.

[12] I. Daubechies, *Ten Lectures on Wavelets*, SIAM, Philadelphia, 1992.

[13] P. Dutilleux, An Implementation of the Algorithm *A Trous* to Compute the Wavelet Transform, In *Proceeding of the International Conference on Wavelets, Time - Frequency Methods and Phase Space - Marseille - France*, Berlin: Springer-Verlag, 1987.

[14] P. J. Davies, *Interpolation & Approximation*, Dover Pub. Inc., New York, 1975.

[15] A. Grossmann, R. Kronland-Martinet, J. Morlet, Reading and Understanding Cotinuous Wavelet Transforms, In *Proceeding of the International Conference on Wavelets, Time - Frequency Methods and Phase Space - Marseille - France*, Berlin: Springer-Verlag, 1987.

[16] A. Grossmann, Wavelet Transforms and Edge Detection, Preprint, CNRS, Marseille, 1986.

[17] C. E. Heil and D. F. Walnut, Continuous and discrete wavelet transforms, SIAM Rev. **31**, 1989.

[18] J. R. Higgins, Five Short Stories about the Cardinal Series, Bull. Am. Math. Soc., **12**, N° 1, 1985.

[19] M. Holschneider, R. Kronland-Martinet, J.Morlet, Ph. Tchamitchian, A Real Time Algorithm for Signal Analysis with the Help of the Wavelet Transforms, In *Proceeding of the International Conference on Wavelets, Time - Frequency Methods and Phase Space - Marseille - France*, Berlin: Springer-Verlag, 1987.

[20] P. G. Lemarié, Une nouvelle Base d'ondelettes de $L^2(\mathcal{R}^n)$. J. Math. Pure et Appl. **67**, 1988.

[21] S. Mallat, A theory of multiresolution signal decomposition: the wavelet representation, IEEE Trans. Pattern Anal. Machine Intell. **11**, 1989.

[22] S. Mallat, Multiresolution approximations and wavelet orthonormal bases of $L^2(\mathcal{R})$, Trans. Amer. Math. Soc. **315**, 1989.

[23] S. Mallat, *Multiresolution Representations and Wavelets*, Grasp. Lab. **153**, Univ. of Pennsylvania, Philadelphia, 1988.

[24] S. Mallat, Multifrequency Channel Decompositions of Images and Wavelet Models, IEEE Trans. Acoust., Speech and Signal Process., **41**, N° 12, 1989.

[25] S. Mallat, Zero Crossing of Wavelet Transform, IEEE Trans. on Information Theory **37**, 1991.

[26] S. Mallat and Z. Zhang, *Complete Signal Representation with Multiscale Edges*, Technical Report N^o **483**, New York Univ., 1989.

[27] S. Mallat and W. L. Hwang, *Singularity Detection and Processing with Wavelets*, Technical Report N^o **549**, New York Univ., 1991.

[28] S. Mallat and W. L. Hwang, Singularity Detection and Processing with Wavelets, IEEE Trans. on Information Theory, **38**, N° 2, 1992.

[29] S. Mallat and Z. Zhang, *Characterization of Signals from Multiscale Edges*, Technical Report N^o **592**, New York Univ., 1991.

[30] Y. Meyer, *Ondelettes, fonctions splines et analyses graduees*, Rend. Sem. Mat. Univers. Politecn. Torino **45**, Torino, 1987.

[31] Y. Meyer, *Wavelets and Operators*, Cahiers de Mathematiques de la decision **8704**, CEREMADE, Universite de Paris Dauphine, Paris, 1989.

[32] Y. Meyer, *Ondelettes, Filters Miroirs en Quadrature et Traitment Numerique de l'Imagen*, Cahiers de Mathematiques de la decision **8914**, CEREMADE, Universite de Paris Dauphine, Paris, 1989.

[33] Y. Meyer, *Ondelettes et Operateurs*, Tomes I, II, III. Hermann, Paris, 1990.

[34] Y. Meyer, *Wavelets: Algorithms & Applications*, SIAM, Philadelphia, 1993.

[35] Y. Meyer, In *Book Reviews*, Bull. Amer. Math. Soc., **28**, 1993.

[36] M. J. Powell, *Approximation Theory and Methods*, Cambridge University Press, Cambridge, 1981.

[37] O. Rioul and P. Duhamel, Fast Algorithms for Discrete and Contiuous Wavelet Transforms, IEEE Trans. on Information Theory, **38**, N° 2, 1992.

[38] M. J. Shensa, The Discrete Wavelet Transform: Wedding the *A Trous* and Mallat Algorithms, IEEE Trans. Acoust., Speech and Signal Process. **40**, N° 10, 1992.

[39] I. J. Schoenberg, *Cardinal Spline Interpolation*, SIAM, Philadelphia, 1993.

[40] E. Serrano y D. Melas, *Algoritmo rápido para el análisis de señales mediante ondelettes spline*, Publicaciones Previas IAM - Conicet **191**, 1992.

[41] E. Serrano, *Tesis Doctoral*, Dep. Mat., FCEyN, Universidad de Buenos Aires, 1996.

[42] E. Serrano y M. Fabio, Variaciones en la Transformada Wavelet, In *Actas del Segundo Congreso Dr. A. Monteiro* Universidad del Sur, Bahia Blanca, 1993.

[43] E. Serrano and M. Fabio, Applications of the Wavelet Transform to Acoustic Signal Processing IEEE Trans. Acoust., Speech and Signal Process., **44**, N° 5, 1996.

[44] E. Serrano, Some Remarks about Compactly Supported Spline Wavelets, Appl. Comput. Harmonic Analysis., **3**, N° 1, 1996. 1996.

[45] Special Issue on Wavelet Transform Multiresolution Signal Analysis, IEEE Trans. on Information Theory, **38**, N° 2, 1992.

[46] *Special Issue on Wavelets and Signal Processing*, IEEE Trans. Acoust., Speech and Signal Processing, **41**, N° 12, 1993.

[47] M. Unser, A. Aldroubi and M. Eden, Fast B-Spline Transforms for Continuous Image Representation and Interpolation, IEEE Trans. Pattern Anal. Machine Intell. **13**, 1991.

[48] M. Unser, A. Aldroubi and M. Eden, On The Asymptotic Convergence of B-Spline to Gabor Functions, IEEE Trans. on Information Theory, **38**, N° 2, 1992.

[49] M. Unser and M. Eden, FIR Approximations of Inverse Filters and Perfect Reconstruction Filter Banks, Signal Process. **36**, 1992.

[50] M. Unser, A. Aldroubi and M. Eden, The L_2 Polynomial Spline Pyramid', IEEE Trans. Pattern Anal. Machine Intell. **15**, 1993.

[51] M. Unser, A. Aldroubi and M. Eden, A family of polynomial spline wavelet transforms, Signal Process. **30**, 1993.

[52] M. Unser, A. Aldroubi and M. Eden, B-Spline Signal Proccesing: Part I- Theory, IEEE Trans. Acoust., Speech and Signal Process. **41**, N° 12, 1993.

[53] M. Unser, A. Aldroubi and M. Eden, B-Spline Signal Proccesing: Part II- Efficient Design and Applications, IEEE Trans. Acoust., Speech and Signal Process. **41**, N° 12, 1993.

[54] M. Unser, A. Aldroubi, Polynomial Splines and Wavelets-A Signal Processing Perpective In *Wavelets: A Tutorial in Theory and Applications*, San Diego: Ac. Press. Inc., 1992.

[55] G. G. Walter, Traslation and Dilation Invariance in Orthogonal Wavelets, Appl. Comput. Harmonic Analysis., 1994.

Chapter 4

Oblique Multiwavelet Bases

Akram Aldroubi

1 Introduction

The goal of this chapter is to introduce the new concept of oblique wavelet and multiwavelet bases described in [1]. These wavelet bases contain, as special cases, the orthogonal, semiorthogonal, and biorthogonal theory of wavelets and multiwavelets [6, 9, 12, 13, 22]. The main advantage of oblique wavelets is that they give more flexibility in choosing wavelet bases, without compromising the fast filter bank implementation algorithms.

1.1 An oblique wavelet based on the Haar multiresolution

This first example introduces the concept of oblique wavelets. Our starting point is a piecewise constant function f_0 which belongs to the Haar multiresolution

$$V_0(\chi_{[0,1]}) = \left\{ \sum_{k \in \mathcal{Z}} c_0(k) \chi_{[0,1]}(x-k); \ c_0 \in l_2 \right\}, \qquad (4.1)$$

where $\chi_{[a,b]}$ is the characteristic function on the interval $[a, b]$ (i.e., $\chi_{[a,b]}(x) = 1, \forall x \in [a, b]$, and $\chi_{[a,b]}(x) = 0, \forall x \notin [a, b]$), and where l_2 is the space of square summable sequences. We wish to approximate the function

$$f_0(x) = \sum_{k \in \mathcal{Z}} c_0(k) \chi_{[0,1]}(x-k), \qquad (4.2)$$

in V_0 by a coarser function f_{V_1} in V_1.

One way to do this is to downsample f_0, and then hold the sample values for interval lengths of size two as shown in Fig. 1

$$f_{V_1}(x) = \sum_{k \in \mathcal{Z}} c_0(2k) \chi_{[0,2]}(x-2k). \qquad (4.3)$$

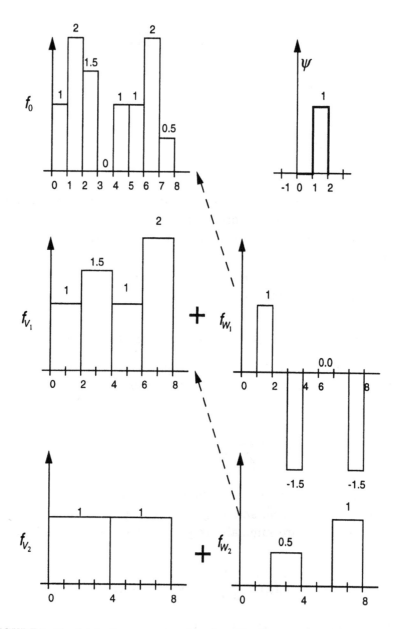

FIGURE 1. To obtain the approximation f_{V_1}, The function f_0 is ignored in the intervals $[2k - 1, 2k]$, while its values on the intervals $[2k, 2k + 1]$ are extended. The error $f_{W_1} = f_0 - f_{V_1}$ is clearly a linear combination of the rectangular pulse ψ and its shifts on the grid of even integers $2\mathcal{Z}$.

The difference between the function f_{V_1} and the starting function f_0 is given by

$$f_{W_1}(x) = \sum_{k \in Z} \left(c_0(2k+1) - c_0(2k) \right) \chi_{[1,2]}(x - 2k), \qquad (4.4)$$

where the error function f_{W_1} belongs to the space W_1, defined by

$$W_1 = \left\{ \sum_{k \in Z} d_1(k) \chi_{[1,2]}(x - 2k);\ d_1 \in l_2 \right\}. \qquad (4.5)$$

It is not difficult to see that our original function f_0 can be written as

$$f_0(x) = f_{V_1}(x) + f_{W_1}(x)$$
$$= \sum_{k \in Z} c_0(2k) \chi_{[0,2]}(x - 2k) + \sum_{k \in Z} \left(c_0(2k+1) - c_0(2k) \right) \chi_{[1,2]}(x - 2k).$$

Clearly, $V_1 \subset V_0$ and $W_1 \subset V_0$. Thus, we can apply the same decomposition algorithm to f_{V_1} and f_{W_1}, now both viewed as elements of V_0. By inspection, we see that this procedure leaves f_{V_1} and f_{W_1} invariant.

Moreover, we have that the intersection $V_1 \cap W_1 = \{0\}$. To see this (see Figure 2), we simply note that if $w_1 \in W_1$, then $w_1(2k + 1/2) = 0$ for any $k \in Z$. On the other hand, if $w_1 \in V_1$, then $w_1(2k + 1/2) = c_1(k)$. Thus, $c_1(k) = 0$ for all $k \in Z$, and we conclude that $w_1(x) = 0$. The previous argument implies that $f_{V_1} = P_{V_1//W_1} f_0$ is the oblique projection of f_0 in V_1 in a direction parallel to W_1, and $f_{W_1} = P_{W_1//V_1} f_0$ is the oblique projection of f_0 in W_1 in a direction parallel to V_1 (see Figure 3). Moreover we have

$$f_0 = P_{V_1//W_1} f_0 + P_{W_1//V_1} f_0. \qquad (4.6)$$

It is easy to see that the set $\left\{ \phi_{1,k}(x) = 2^{-1/2} \chi_{[0,2]}(x - 2k);\ k \in Z \right\}$ is an orthogonal basis of V_1, and that $\left\{ \psi_{1,k}(x) = \chi_{[1,2]}(x - 2k);\ k \in Z \right\}$ is an orthogonal basis of W_1. Moreover, it can be shown that the set

$$\{\phi_{1,k}(x),\ \psi_{1,k}(x);\ k \in Z\}$$

forms a Riesz basis for the space V_0 of piecewise constants with integer knot-points [1].

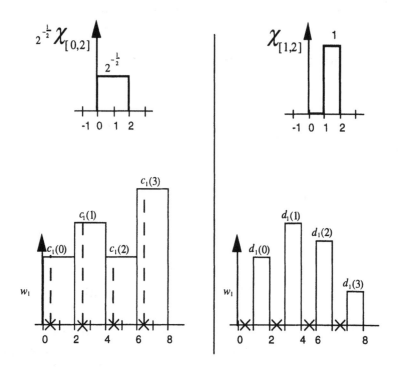

FIGURE 2. A function w_1 that belongs to V_1 must be a linear combination of $\{\chi_{[0,2]}(x - 2k);\ k \in \mathcal{Z}\}$ (left panel), while a function w_1 that belongs W_1 must be a linear combination of $\{\chi_{[1,2]}(x - 2k);\ k \in \mathcal{Z}\}$ (right panel). Thus, the only possible function that belongs to $V_1 \cap W_1$ is the zero function $w_1 = 0$.

If we repeat the decomposition procedure, we obtain a series of coarser approximation of f_0 and a series of error functions as shown in Fig. 1. The decomposition algorithm consists of the repetitive application of the perfect reconstruction filter algorithm shown in Figure 4.

Therefore, for any function $f \in L_2$ (Here, L_2 is the space of square integrable functions), we can choose a sufficiently small value J_2 so that the orthogonal projection $f_{J_2} = P_{V_{J_2}} f$ of f into the space of piecewise constants V_{J_2} is arbitrarily close to f. Then, we can repeatedly apply the decomposition algorithm (4.6) to obtain

$$f_{J_2} = P_{V_{J_1} // W_{J_1}} f_{J_2} + \sum_{j=J_2+1}^{J_1} P_{W_j // V_j} f_{J_2}. \qquad (4.7)$$

It follows that the set $\left\{ \phi_{J_1,k}, \psi_{j,k}(x) = 2^{\frac{-j+1}{2}} \chi_{[1,2]}(2^{-j+1}x - 2k); \right.$

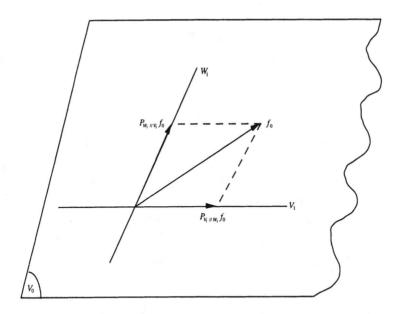

FIGURE 3. Schematic representation of the oblique projection of f_0 on V_1 in a direction parallel to W_1.

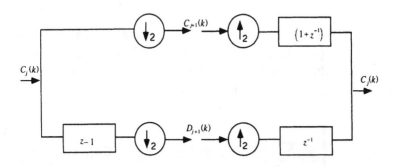

FIGURE 4. Perfect reconstruction filter-bank for computing a Haar oblique wavelet transform of order zero. The transfer functions of the filters are obtained by setting $z = e^{-i2\pi f}$.

$k \in \mathcal{Z}, j \leq J_1\}$ is dense in L_2. However, this set is not a Riesz basis of L_2, but simply *an inductive limit Riesz basis* (see Theorem 3.1): any finite number of members in this set are linearly independent, and finite linear combinations of this set are dense in L_2. What is interesting is that we still have a fast and stable computational algorithm for calculating the expansion in terms of this countable basis.

We call the set $\{\phi_{J_1,k}, \psi_{j,k}; \; k \in \mathcal{Z}, j \leq J_1\}$ an *inductive-limit Riesz basis* of L_2, and we call the spaces W_j that are generated by the Riesz bases

$$\{\psi_{j,k}; \; k \in \mathcal{Z}\}$$

oblique wavelet spaces. Unlike other wavelets, the function $\psi(x) = 2^{\frac{1}{2}}\chi_{[1/2,1]}$ does not have an average of zero. Thus, it is not an orthogonal, semiorthogonal, or biorthogonal wavelet [3, 18]. In contrast with biorthogonal wavelets, this Haar oblique wavelet has neither an associated dual wavelet spaces, nor a dual wavelet basis. What is remarkable is that this transform is faster than the classical Haar transform. Moreover, the support of the wavelet (in this case the function $\chi_{[\frac{1}{2},1]}$) is half the support of the orthogonal Haar wavelet.

2   Multiscaling Functions and Multiwavelet Bases

2.1   Multiscaling functions

If the multiresolution spaces $\{V_j\}_{j \in \mathcal{Z}}$

$$\cdots V_2 \subset V_1 \subset V_0 \subset V_{-1} \subset V_{-2} \subset \cdots$$

are generated by the translations and dilations of $r > 1$ functions $\phi^1(x), ..., \phi^r(x)$, then the vector function $\Phi = (\phi^1(x), ..., \phi^r(x))^T$ is called a *multiscaling function* (Here, $(\bullet)^T$ denotes the matrix transpose operator). Thus, we have that

$$V_j = \left\{ \sum_{i=1}^{r} \sum_{k \in \mathcal{Z}} c_j^i(k) 2^{-j/2} \phi^i \left(\frac{x}{2^j} - k \right) ; c_j^i \in l_2, \; i = 1, \cdots, r \right\}$$

$$= \left\{ \sum_{k \in \mathcal{Z}} \mathbf{C}_j^T(k) \Phi_j(x - 2^j k); \mathbf{C}_j(k) \in l_2^r \right\},$$

where $\mathbf{C}_j(k)$ is the vector $\mathbf{C}_j = (c_j^1(k), c_j^2(k), \ldots, c_j^r(k))^T$, $l_2^r = l_2 \times \cdots \times l_2$, and the vector $\Phi_j(x)$ is defined to be

$$\Phi_j(x) = \left(\phi_j^1(x), \phi_j^2(x), \cdots, \phi_j^r(x) \right)^T$$

$$= 2^{-j/2} \left(\phi^1 \left(\frac{x}{2^j} \right), \phi^2 \left(\frac{x}{2^j} \right), \cdots, \phi^r \left(\frac{x}{2^j} \right) \right)^T$$

In particular, a function $f_0 \in V_0$ is given by

$$
\begin{aligned}
f_0 &= \sum_{k \in \mathbb{Z}} c_0^1(k) \phi^1(x-k) + \ldots + c_0^r(k) \phi^r(x-k) \\
&= \sum_{k \in \mathbb{Z}} \mathbf{C}_0^T \Phi(x-k).
\end{aligned}
$$

A prototypical example of a multiscaling function is constructed from the Hermite cubic splines that are used in finite element methods [13]. This multiscaling function consists of the two functions $\phi^1(x)$ and $\phi^2(x)$ shown Figure 5. Each of these two functions is concocted of piecewise polynomial splines of order 3 that are joined at the integer knot points in such a way as to be continuous and to have a continuous first derivative. Thus, $\phi^1(x)$ and $\phi^2(x)$ are $C^1(\mathcal{R})$. They are also constructed to have minimal compact support: Support(ϕ^i) = $[0, 2]$, for $i = 1, 2$, and they have the property that $\phi^1(0) = 1$, $\left(\frac{d}{dx} \phi^1 \right)(0) = 0$, and $\phi^2(0) = 0$, $\left(\frac{d}{dx} \phi^2 \right)(0) = 1$.

2.2   Multiwavelets

If the spaces $\{W_j\}_{j \in \mathbb{Z}}$ that complement the multiresolution spaces $\{V_j\}_{j \in \mathbb{Z}}$ (i.e., $V_{j+1} + W_{j+1} = V_j$) are generated by a set of r functions $\psi^1(x), \psi^2(x), \cdots, \psi^r(x)$, then the vector $\mathbf{\Psi}(x) = (\psi^1(x), \psi^2(x), \cdots, \psi^r(x))^T$ is called a *multiwavelet*, as long as the set

$$\left\{ \psi_{j,k}^i(x) = 2^{-\frac{i}{2}} \psi_j^1(x - 2^j k), \cdots, 2^{-\frac{i}{2}} \psi_j^r(x - 2^j k); \ k \in \mathbb{Z} \right\}$$

forms a Riesz basis of

$$
\begin{aligned}
W_j &= \left\{ \sum_{i=1}^r \sum_{k \in \mathbb{Z}} d_j^i(k) 2^{-j/2} \psi^i \left(\frac{x}{2^j} - k \right); d_j^i \in l_2, \ i = 1, \cdots, r. \right\} \\
&= \left\{ \sum_{k \in \mathbb{Z}} \mathbf{D}_j^T(k) \mathbf{\Psi}_j(x - 2^j k); \mathbf{D}_j(k) \in l_2^r \right\},
\end{aligned}
$$

where $\mathbf{\Psi}_j(x)$ is defined to be

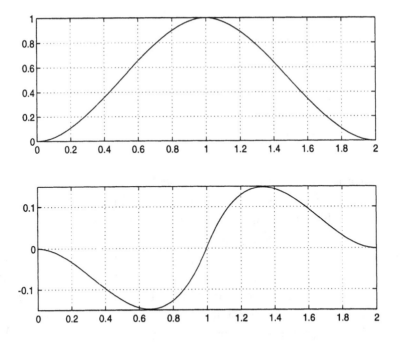

FIGURE 5. The Hermite cubic spline multiscaling function $\Phi(t) = (\phi^1(t), \phi^2(t))$.

$$\Psi_j(x) := 2^{-j/2}\Psi\left(\frac{x}{2^j}\right). \tag{4.8}$$

Note that we do not require orthogonality between the spaces V_j and W_j. Therefore, the spaces W_j are not necessarily orthogonal to each other.

2.3   Classification of wavelet bases

According to the angles between the spaces $\{W_j\}_{j\in Z}$ and the choice of bases for these spaces, wavelets and multiwavelets can be classified into several categories:

- *Orthogonal* wavelet and multiwavelet bases [7, 9, 15, 16, 20]: For this case, there are two conditions that must be satisfied:

  1. The wavelet spaces must be orthogonal to each other:

  $$W_j \perp W_l, \ \forall j \neq l.$$

2. The set $\left\{\psi_{j,k}^i;\ k \in \mathcal{Z},\ i = 1, \cdots, r.\right\}$ must form an orthogonal basis of W_j.

- *Semiorthogonal* wavelet and multiwavelet bases, [2, 4, 8, 19, 21]:
  For these bases, the only requirement is that $W_j \perp W_l,\ \forall j \neq l$.
  In this case, the set $\left\{\psi_{j,k}^i;\ k \in \mathcal{Z},\ i = 1, \cdots, r.\right\}$ is not necessarily an orthogonal basis of W_j.

- *Biorthogonal* wavelet bases [5]: This case consists of a pair of wavelet bases $\left\{\psi_{j,k}^i;\ k \in \mathcal{Z},\ i = 1, \cdots, r.\right\}$ and $\left\{\tilde{\psi}_{j,k}^i;\ k \in \mathcal{Z},\ i = 1, \cdots, r.\right\}$ generating a pair of wavelet spaces W_j and $\tilde{W}_j$, and satisfying the biorthogonality condition

$$\left\langle \psi_{j,k}^i, \tilde{\psi}_{m,n}^l \right\rangle_{L_2} = \delta_0(j - m)\delta_0(k - n)\delta_0(i - l), \qquad (4.9)$$

  where $\delta_0(k)$ is the impulse sequence located at 0, i.e., $\delta_0(0) = 1$, and $\delta_0(k) = 0,\ k \neq 0$.

- *Oblique* wavelet and multiwavelet bases [1]: In this case, we require only that the set $\left\{\psi_{j,k}^i;\ k \in \mathcal{Z},\ i = 1, \cdots, r.\right\}$ be a Riesz basis of W_j. We do not require orthogonality between the wavelet spaces, as in the orthogonal and semiorthogonal cases; and unlike the biorthogonal case, we do not require the existence of the associated wavelet spaces $\tilde{W}_j$ or the existence of a biorthogonal wavelet.

The first orthogonal multiwavelet bases were introduced by Donovan, Geronimo, Hardin, and Massopust [10]. These bases have the remarkable properties of being orthogonal, regular, compactly supported, and symmetrical. It was not possible to construct wavelet bases ($r = 1$) satisfying these four properties before [7]. Other orthogonal multiwavelet bases have also been found [20]. Semiorthogonal constructions of multiwavelets were first introduced in [11, 12]. The first biorthogonal multiwavelet constructions of which we are aware has been introduced in [17]. These multiwavelets are special cases of oblique multiwavelets [1]. In fact, the theory underlying oblique multiwavelet bases encompasses the theories of orthogonal, semiorthogonal, and biorthogonal multiwavelets.

2.4   Two-scale equations

Since $V_1 \subset V_0$, $\Phi\left(\frac{x}{2}\right)$ must be a linear combination of the basis of V_0. Therefore, we must have the two-scale relation

$$\Phi\left(\frac{x}{2}\right) = 2\sum_{k\in\mathbb{Z}} \mathbf{H}_1(k)\Phi(x-k), \tag{4.10}$$

where $\mathbf{H}_1(k)$ is an $r \times r$ matrix-sequence called the *generating sequence*. As an example, the matrix-sequence generating the Hermite cubic splines is given by [13]

$$\mathbf{H}_1(0) = \begin{bmatrix} 1/2 & 3/4 \\ -1/8 & -1/8 \end{bmatrix}, \quad \mathbf{H}_1(1) = \begin{bmatrix} 1 & 0 \\ 0 & 1/2 \end{bmatrix}, \quad \mathbf{H}_1(2) = \begin{bmatrix} 1/2 & -3/4 \\ 1/8 & -1/8 \end{bmatrix}. \tag{4.11}$$

Another example is the Cohen-Daubechies-Plonka 2×2 generating matrix-sequence $\mathbf{H}_1(k)$ which can be characterized by its Fourier transform $\hat{\mathbf{H}}_1(z)|_{z=e^{-i2\pi f}}$ as [6]

$$\hat{\mathbf{H}}_1(f) = \frac{1}{4}\begin{bmatrix} z^1 + 2z^2 + z^3 & (z^1 - 2z^3 + z^5)/2 \\ 1/32 & z \end{bmatrix}. \tag{4.12}$$

The associated multiscaling function $\Phi(x) = (\phi^1(x), \phi^2(x))^T$ consists of the pair of functions depicted in Figure 2.4.

In a similar fashion, since $W_1 \subset V_0$, $\Psi\left(\frac{x}{2}\right)$ must be a linear combination of the basis of V_0:

$$\Psi\left(\frac{x}{2}\right) = 2\sum_{k\in\mathbb{Z}} (\delta_1 * \mathbf{G}_1)(k)\Phi(x-k), \tag{4.13}$$

where $\mathbf{G}_1(k)$ is a matrix-sequence; $\delta_i(k)$ is the unit impulse sequence located at $k = i$, and the *generalized convolution* $(a * \mathbf{B})(k)$ between a scalar-sequence $a(k)$ and a matrix-sequence $\mathbf{B}(k)$ is the matrix-sequence $\mathbf{C}(k)$ defined by

$$\mathbf{C}_{i,j}(k) := \sum_{l\in\mathbb{Z}} a(l)\mathbf{B}_{i,j}(k-l).$$

From this definition, it follows that the unit impulse $\delta_1(k)$ in (4.13) is used to shift the matrix-sequence $\mathbf{G}_1(k)$, i.e., $(\delta_1 * \mathbf{G}_1)(k) =$

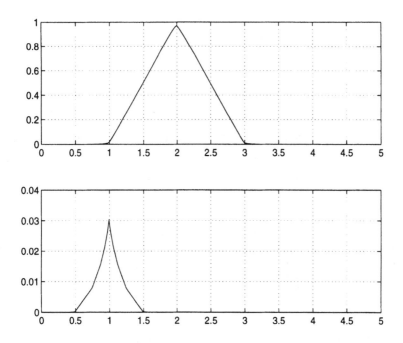

FIGURE 6. The Cohen-Daubechies-Plonka multiscaling function $\Phi(t) = (\phi^1(t), \phi^2(t))$ corresponding to the generating matrix-sequence given by (4.12).

$\mathbf{G}_1(k - 1)$. In the context of this paper, we will also use the symbol "$*$" to denote another generalized convolution between two sequences of matrices. Specifically, the convolution $\mathbf{C}(k) = (\mathbf{A} * \mathbf{B})(k)$ between the $m \times r$ matrix-sequence $\{\mathbf{A}(k)\}_{k \in \mathcal{Z}}$ and the $r \times n$ matrix-sequence $\{\mathbf{B}(k)_{k \in \mathcal{Z}}\}$ is the $m \times n$ matrix-sequence, defined in terms of the convolution between the entries of $\mathbf{A}$ and $\mathbf{B}$ as

$$\mathbf{C}_{i,j}(k) := \sum_{l=1}^{r} \sum_{h \in \mathcal{Z}} \mathbf{A}_{i,l}(h) \mathbf{B}_{l,j}(k - h). \qquad (4.14)$$

3 Construction of Oblique Wavelet and Multiwavelet Bases

The main result for finding oblique multiwavelets relies on a construction in the Fourier domain. Specifically, we use the $r \times r$ matrix-function $\widehat{\mathbf{H}}_1(f)$ (which is usually known) to construct a $2r \times 2r$ invertible matrix function

$$\widehat{\mathbf{S}}_1(f) = \begin{bmatrix} \widehat{\mathbf{H}}_1^T(f) & \widehat{\mathbf{G}}_1^T(f) \\ \widehat{\mathbf{H}}_1^T(f - \tfrac{1}{2}) & -\widehat{\mathbf{G}}_1^T(f - \tfrac{1}{2}) \end{bmatrix}. \qquad (4.15)$$

The main goal is to choose the $r \times r$ matrix-function $\widehat{\mathbf{G}}_1^T(f)$ appropriately so that $\widehat{(\mathbf{S}_1(f))}^{-1}$ exits. Then, the inverse Fourier transform $\mathbf{G}_1(k)$ of $\widehat{\mathbf{G}}_1^T(f)$ gives rise to an oblique multiwavelet (or wavelet if $r = 1$) by the formula (4.13) [1]. The main result can be stated as follows [1]:

Theorem 3.1 *If $\widehat{\mathbf{S}}_1^{-1}(f)$ exists for almost all f, and if the matrix-norms of $\widehat{\mathbf{S}}_1(f)$ and $\widehat{\mathbf{S}}_1^{-1}(f)$ are uniformly bounded, i.e., if there exist two constants $m > 0$ and $M > 0$ independent of f such that $\left\|\widehat{\mathbf{S}}_1(f)\right\| \leq M$ and $\left\|\widehat{\mathbf{S}}_1^{-1}(f)\right\| \leq m$, then*

$$\Psi_1(x) = 2^{-\frac{1}{2}} \Psi\left(\frac{x}{2}\right) = 2^{\frac{1}{2}} \sum_{k \in \mathbb{Z}} (\delta_1 * \mathbf{G}_1)(k) \Phi(x - k) \qquad (4.16)$$

is an oblique multiwavelet. It is associated with the multiresolution generated by the multiscaling function $\Phi(x)$, whose two-scale sequence is $\mathbf{H}_1(k)$. The set

$$\left\{ \psi_{j,k}^i(x); \ k \in \mathbb{Z}, \ i = 1, \cdots, r. \right\}$$

is a Riesz basis of W_j; For any J_1, the set

$$\left\{ \phi_{J_1,k}^i, \psi_{j,k}^i; \ k \in \mathbb{Z}, \ j \leq J_1, i = 1, \cdots, r. \right\}$$

is an inductive limit Riesz basis since for any $J_1 \in \mathbb{Z}$ and $J_2 \in \mathbb{Z}$, the set

$$\left\{ \phi_{J_1,k}^i(x), \psi_{j,k}^i(x); \ k \in \mathbb{Z}, \ J_2 + 1 \leq j \leq J_1, \ i = 1, \cdots, r. \right\}$$

is a Riesz basis of V_{J_2}. Moreover, for any $g \in V_{J_2}$, we have the decomposition

$$g = P_{V_{J_1}//W_{J_1}} g + \sum_{j=J_2+1}^{J_1} P_{W_j//V_j} g. \qquad (4.17)$$

The example in Section 1.1 can be directly derived from Theorem 3.1. For this case, $r = 1$, and $H_1(k) = 2^{-1}(\delta_0(k) + \delta_1(k))$. For $G_1(k)$, we choose $G_1(k) = 2^{-\frac{1}{2}}\delta_0(k)$. The 2×2 matrix function $\widehat{\mathbf{S}}_1(f)$ is given by

$$\widehat{\mathbf{S}}_1(f) = \begin{bmatrix} \frac{1+z^{-1}}{2} & 2^{-\frac{1}{2}} \\ \frac{1-z^{-1}}{2} & -2^{-\frac{1}{2}} \end{bmatrix}, \qquad (4.18)$$

where $z = e^{i2\pi f}$. We have that $\det(\widehat{\mathbf{S}}_1(f)) = -2^{-\frac{1}{2}}$ for all f; thus, $\widehat{\mathbf{S}}_1(f)$ has an inverse. Moreover, it is easy to see that the conditions of Theorem 3.1 on the norms of $\widehat{\mathbf{S}}_1(f)$ and its inverse are satisfied. Therefore, Equation (4.16) gives us the oblique wavelet $\psi_1(x/2) = \chi_{[1,2]}(x)$. The filter bank implementing the decomposition and reconstruction algorithms is depicted in Figure 4. If, instead of the previous choice for G_1, we set $G_1(k) = 2^{-1}(\delta_{-1}(k) - \delta_0(k))$, we obtain the well-known orthogonal Haar wavelet. It is easy to check that, in this case, $\widehat{\mathbf{S}}_1(f)$ is given by

$$\widehat{\mathbf{S}}_1(f) = 2^{-1} \begin{bmatrix} 1 + z^{-1} & z - 1 \\ 1 - z^{-1} & z + 1 \end{bmatrix} \qquad (4.19)$$

and that $\det(\widehat{\mathbf{S}}_1(f)) = 1$ for all f. As expected, the wavelet function given by (4.16) is precisely the orthogonal Haar wavelet. Another choice for $\widehat{G}_1(f)$ is to take the trigonometric polynomial given by $\widehat{G}_1(f) = (z - 1)^n$. For $n = 0, 1$, we get the previous two cases. For $n = 2$ we get that $\det(\widehat{\mathbf{S}}_1(f)) = -3 - e^{-i2\pi f}$, which is nonzero for all f. Hence, we again obtain an oblique wavelet, but now with two vanishing moments.

3.1 Oblique multiwavelet bases based on the Hermite cubic splines MRA

The Hermite cubic spline multiscaling function has the 2×2 generating sequence given by (4.11). We select sequences $\mathbf{G}_1^n(k)$ defined by the Fourier transforms $\widehat{\mathbf{G}}_1^n(f) = (z - 1)^n \mathbf{I}_2|_{z=e^{i2\pi f}}$, where $\mathbf{I}_2$ is the 2×2 identity matrix. For $n = 0$, we easily check that the determinant of $\widehat{\mathbf{S}}_1^0(f)$ is non-zero for all f (the superscript 0 relates to the index $n = 0$). We conclude that the inverse $\widehat{\mathbf{S}}_1^0(f)$ exists, and

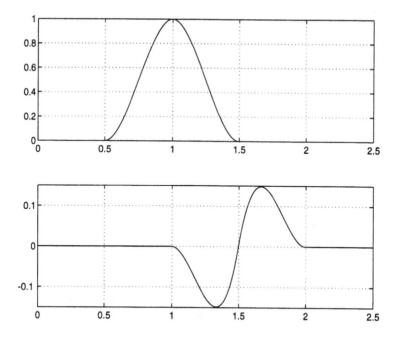

FIGURE 7. The Hermite cubic spline oblique multiwavelet $\Psi(x) = (\psi^1(x), \psi^2(x))$ corresponding to $\widehat{\mathbf{G}}_1^0(f) = \mathbf{I}_2$ $(n = 0)$.

its entries are ratios of trigonometric polynomials. It follows that the conditions of Theorem 3.1 are satisfied. Therefore, the choice $\widehat{\mathbf{G}}_1^0(f) = \mathbf{I}_2$ gives rise to the oblique multiwavelet shown in Figure 7. For this case, the oblique multiwavelet is exactly the same as the multiscaling function except for a contraction by a factor of 2. Thus, the wavelet functions have their support in the interval $[1, 2]$. It can also be shown in Figure 8.

4   Fast Filter-Bank Algorithms

Oblique wavelet and multiwavelet transforms can be implemented using perfect reconstruction filter banks [1]. Any function $g \in V_0$ can be decomposed into a low resolution approximation $g_J \in V_J$ and the sum of the error terms in the spaces $\{W_j\}_{J \leq j \leq 1}$

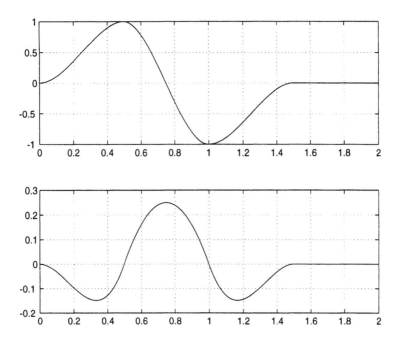

FIGURE 8. The Hermite cubic spline oblique multiwavelet $\Psi(x) = (\psi^1(x), \psi^2(x))$ corresponding to $\widehat{\mathbf{G}}_1^1(f) = (z - 1)\mathbf{I}_2|_{z=e^{i2\pi f}}$ $(n = 1)$.

$$g = \uparrow_{2^J} \left[\mathbf{C}_J^T \right] * \Phi_J + \sum_{j=1}^{J} \uparrow_{2^j} \left[\mathbf{D}_j^T \right] * \Psi_j, \qquad (4.20)$$

where $\uparrow_2$ is the upsampling operator defined by

$$\begin{aligned} \uparrow_2 [b] (2k) &= b(k) \\ \uparrow_2 [b] (2k+1) &= 0 \end{aligned} \qquad \forall k \in \mathcal{Z}. \qquad (4.21)$$

Since $g(x)$ belongs to a multiresolution space V_0, e.g., $g = \mathbf{C}_0^T * \Phi$, the procedure for finding the coefficients $\mathbf{C}_J(k)$ and $\{\mathbf{D}_j(k)\}_{j=J,\dots,1}$ from the coefficients $\mathbf{C}_0(k)$ can be obtained by a fast vector-filter-bank algorithm depicted in the left part of Figure 9. Specifically, we repetitively apply a series of filtering and downsampling given by

$$\begin{aligned} \mathbf{C}_{j+1} &= 2^{\frac{1}{2}} \downarrow_2 [\mathbf{H}_2 * \mathbf{C}_j], \\ \mathbf{D}_{j+1} &= 2^{\frac{1}{2}} \downarrow_2 [\delta_{-1} * \mathbf{G}_2 * \mathbf{C}_j]. \end{aligned}$$

88   Akram Aldroubi

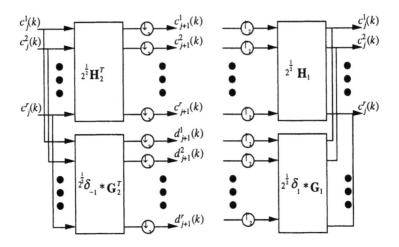

FIGURE 9. Perfect reconstruction filter-bank for computing the oblique multi-wavelet transform. The coefficients $\mathbf{C}_j$ are decomposed into two sequences $\mathbf{C}_{j+1}$ and $\mathbf{D}_{j+1}$ (left filter-bank pair). The two sequences $\mathbf{C}_{j+1}$ and $\mathbf{D}_{j+1}$ can then be combined by the right pair of filters to reconstruct $\mathbf{C}_j$.

k	h_{11}	h_{12}	h_{21}	h_{22}	g_{11}	g_{12}	g_{21}	g_{22}
-1	1/4	3/8	-1/16	-1/16	1	0	0	1
0	1/2	0	0	1/4	-1/16	0	0	1
1	1/4	-3/8	1/16	-1/16	0	0	0	0

TABLE 4.1. Reconstruction filters: $\mathbf{H}_1$, $\mathbf{G}_1$: $[\mathbf{H}_1]_{i,j} = h_{i,j}$, $[\mathbf{G}_1]_{i,j} = g_{i,j}$.

where $\downarrow_2$ is the downsampling (or decimation) operator defined by

$$\downarrow_2 [b](k) = b(2k) \quad \forall k \in \mathcal{Z}, \tag{4.22}$$

and where the filters $\mathbf{H}_2$, and $\mathbf{G}_2$ are the filters defined in the previous section and are associated with the oblique wavelets.

Similarly, the procedure for finding the coefficients $\mathbf{C}_0(k)$ from the knowledge of $\mathbf{C}_J(k)$ and $\{\mathbf{D}_j(k)\}_{j=J,\dots,1}$ can be obtained by the filtering algorithm depicted on the right side of Figure 9. This is done by the repetitive application of a series of upsampling and filtering given by

k	h_{11}	h_{12}	h_{21}	h_{22}	g_{11}	g_{12}	g_{21}	g_{22}
-14	0	0	0.0000023	-0.0000019	0	0	0	0
-13	0	0	0	0	0	0	0	0
-12	0.0000047	-0.0000039	0.0000232	-0.0000189	0	0	0	0
-11	0	0	0	0	-0.0000024	0.0000019	-0.0000058	0.0000047
-10	0.0000468	-0.0000382	0.0002293	-0.0001872	0	0	0	0
-9	0	0	0	0	-0.0000234	0.0000191	-0.0000573	0.0000468
-8	0.0004634	-0.0003786	0.0022701	-0.0018536	0	0	0	0
-7	0	0	0	0	-0.0002317	0.0001892	-0.0005675	0.0004634
-6	0.0045871	-0.0037454	0.0224721	-0.0183484	0	0	0	0
-5	0	0	0	0	-0.0022936	0.0018727	-0.0056181	0.0045871
-4	0.0454077	-0.0370752	0.2224513	-0.1816307	0	0	0	0
-3	0	0	0	0	-0.0227038	0.0185376	-0.0556128	0.0454077
-2	0.4494897	-0.3670068	2.2020410	-1.797959	0	0	0	0
-1	0	0	0	0	-0.2247449	0.1835034	-0.5505103	0.4494897
0	0.4494897	0.3670068	-2.2020410	-1.797959	0.5	0	0	0.5
1	0	0	0	0	-0.2247449	-0.1835034	0.5505103	0.4494897
2	0.0454077	0.0370752	-0.2224513	-0.1816307	0	0	0	0
3	0	0	0	0	-0.0227038	-0.0185376	0.0556128	0.0454077
4	0.0045870	0.0037454	-0.0224721	-0.0183484	0	0	0	0
5	0	0	0	0	-0.0022936	-0.018727	0.0056180	0.0045871
6	0.0004634	0.0003786	-0.0022701	-0.0018536	0	0	0	0
7	0	0	0	0	-0.0002317	-0.0001892	0.0005675	0.0004634
8	0.0000468	0.0000382	-0.00002293	-0.0001872	0	0	0	0
9	0	0	0	0	-0.0000234	-0.0000191	0.0000573	0.0000468
10	0.0000047	0.0000039	-0.00002317	-0.000019	0	0	0	0
11	0	0	0	0	-0.0000024	-0.0000019	0.0000058	0.0000047
12	0	0	-0.0000023	-0.0000019	0	0	0	0

TABLE 4.2. Analysis filters: H_2, G_2: $[H_2]_{i,j} = h_{i,j}$, $[G_2]_{i,j} = g_{i,j}$.

$$\mathbf{C}_j^T = 2^{\frac{1}{2}} \uparrow_2 \left[\mathbf{C}_{j+1}^T \right] * \mathbf{H}_1 + 2^{\frac{1}{2}} \uparrow_2 \left[\mathbf{D}_{j+1}^T \right] * \delta_1 * \mathbf{G}_1. \qquad (4.23)$$

where the filters $\mathbf{H}_1$, $\mathbf{G}_1$, are the filters defined in the previous section and are associated with the oblique wavelets. The decomposition and reconstruction filter-banks constitute together the perfect reconstruction filter-bank structure. For the case of the Hermite cubic spline filters, and for the choice $n = 0$ in the construction of Section 3.1, the filters $\mathbf{H}_1$, $\mathbf{G}_1$, $\mathbf{H}_2$, and $\mathbf{G}_2$ are given in Table 4.1, and Table 4.2.

4.1 Initialization

In practice, the starting point is not the function $g = \mathbf{C}_0^T * \Phi$, but instead a discrete sequence $s(k)$ which is obtained from the samples of some analog function $s(t)$. The first step is then to find the coefficients $C_0(k)$ that correspond to the sequence $s(k)$. One way to do this, is to assume that $s(k)$ has been obtained by sampling the function $s(t) = (C_0^T * \Phi)(t)$ at the sampling points $t = k/r$ (r is the number scaling functions). This assumption allows us, in general, to find the coefficients $C_0(k)$ by a simple filtering of the data $s(k)$[22]. Another method that uses projectors is described in [23].

5   References

[1] A. Aldroubi. Oblique and hierarchical multiwavelet bases. *Journal of Appl. and Comput. Harmonic Analysis*, to appear.

[2] A. Aldroubi and M. Unser. Families of multiresolution and wavelet spaces with optimal properties. *Numer. Funct. Anal. and Optimiz.*, 14(5):417–446, 1993.

[3] C. K. Chui and X. L. Shi. Bessel sequences and affime frames. *Proc. Amer. Math. Soc.*, 1:29–49, 1993.

[4] C. K. Chui and J. Z. Wang. On compactly supported spline wavelets and a duality principle. *Trans. Amer. Math. Soc.*, 330:903–915, 1992.

[5] A. Cohen, I. Daubechies, and J.C. Fauveau. Biorthogonal bases of compactly supported wavelets. *Comm. Pure Appl. Math.*, 45:485–560, 1992.

[6] A. Cohen, I. Daubechies, and G. Plonka. Regularity of refinable function vectors. *preprint*, 1995.

[7] I. Daubechies. Orthonormal bases of compactly supported wavelets. *Comm. Pure Appl. Math.*, 41:909–996, 1988.

[8] C. de Boor, R.A. DeVore, and A. Ron. On the construction of multivariate pre-wavelets. *Constr. Approx.*, 9(2):123–166, 1993.

[9] G.C. Donovan, J.S. Geronimo, D.P. Hardin, and P.R. Massopust. Construction of orthogonal wavelets using fractal interpolation functions. *preprint*, 1994.

[10] J.S. Geronimo, D.P. Hardin, and P.R. Massopust. Fractal functions and wavelet expansions based on several scaling function. *J. Approx. Theory*, 1994.

[11] T.N.T. Goodman and W.S. Lee. Wavelets of multiplicity r. *Trans. Amer. Math. Soc.*, 342:307–324, 1994.

[12] T.N.T. Goodman, S.L. Tang, and W.S. Lee. Wavelet wandering subspaces. *Trans. Amer. Math. Soc.*, 338(2):639–654, 1993.

[13] C. Heil, G. Strang, and V. Strela. Approximation order by translates of refinable functions. *preprint*, 1995.

[14] A.F. Laine and M. Unser, editors. *Mathematical Imaging: Wavelet Applications in Signal and Image Processing*, volume 2569. SPIE– The International Society for Optical Engineering, Bellingham, Washington USA, 1995.

[15] P.G. Lemarié and Y. Meyer. Ondelettes et bases hilbertiennes. *Rev. Math. Iberoamericana*, 2, 1986.

[16] S. Mallat. Multiresolution approximations and wavelet orthonormal bases of $L^2(\mathcal{R})$. *Trans. Amer. Math. Soc.*, 315(1):69–97, 1989.

[17] J. A. Marasovich, editor. *Biorthogonal multiwavelets*. Dissertation, Vanderbilt, 1996.

[18] Y. Meyer. *Opérateurs de Calderon-Zygmund*. Hermann, Paris, France, 1990.

[19] C.A. Michelli. Using the refinement equation for the construction of pre-wavelet. *Numerical Algorithms*, 1:75–116, 1991.

[20] G. Strang and V. Strela. Orthogonal muliwavelets with vanishing moments. *J. Optica Eng.*, 33:2104–2107, 1994.

[21] M. Unser, A. Aldroubi, and M. Eden. A family of polynomial spline wavelet transforms. *Signal Processing*, 30:141–162, 1993.

[22] X.G. Xia, , J.S. Geronimo, D.P. Hardin, and B.W.Sutter. Design of prefilters for discrete multiwavelet transforms. *IEEE Trans. Signal Process.*, 44(1):25–35, 1996.

[23] M.J. Vrhel and A. Aldroubi, Pre-filtering for initialization of multi-wavelet transforms, in *Proc. IEEE ICASSP,* April 21–24, 1997, Munich, Germany, to appear.

Chapter 5

Frames and Riesz bases: a short survey

S. J. Favier
R. A. Zalik

1 Introduction

Orthogonal bases provide a classical method to represent an element of a Hilbert space in terms of simpler ones. A Riesz basis is the image of an orthogonal basis under an invertible continuous linear mapping. In practical applications, when using orthogonal bases one introduces small truncation errors. As we shall see in the sequel, perturbing a Riesz bases (in particular, an orthogonal basis) by a small amount, yields another Riesz basis. This is one of the reasons that motivate the study of such bases.

Frames have many of the properties of bases, but lack a very important one: uniqueness. Frames need not be linearly independent. This turns out to be useful in image and signal processing applications, since this redundancy yields robustness, i. e. less sensitivity to truncation or transmission errors. As with Riesz bases, perturbing a frame by a small amount, also yields a frame. Chui and Shi's oversampling theorems, stated below, provide methods to generate wavelet frames from wavelet Riesz bases. For bases in general, the study of their stability is an old problem [37, 38, 44].

2 Frames and Riesz bases in Hilbert Spaces.

In the sequel, $\mathcal{Z}, \mathcal{Z}^+, \mathcal{R}$ and $\mathcal{C}$ will respectively denote the integers, the strictly positive integers, the real numbers, and the complex numbers; d and N will always be elements of $\mathcal{Z}^+$. $\mathcal{H}$ will denote a (separable) Hilbert space with inner product $\langle \cdot, \cdot \rangle$ and norm $\| \cdot \| := \langle \cdot, \cdot \rangle^{1/2}$. We shall call two inner products in $\mathcal{H}$ *equivalent*, if they generate equivalent norms. Two sequences $\{f_n, n \in \mathcal{Z}^+\}$ and $\{g_n, n \in \mathcal{Z}^+\}$ in

$\mathcal{H}$ are called *biorthogonal* if $\langle f_n, g_k \rangle = \delta_{n,k}$, where δ is Krönecker's delta.

A sequence $\{f_n, n \in Z^+\} \subset \mathcal{H}$ is called a *Riesz basis* if it is the image of an orthonormal basis under a bounded invertible linear operator $U : \mathcal{H} \rightarrow \mathcal{H}$ (cf., e. g. [44]). We begin with a theorem that states a number of important properties of such bases.

Theorem 2.1 *([44]) Let $\{f_n, n \in Z^+\} \subset \mathcal{H}$. Then the following propositions are equivalent:*

(a) *$\{f_n, n \in Z^+\}$ is a Riesz basis for $\mathcal{H}$.*

(b) *There is an equivalent inner product on $l\mathcal{H}$, with respect to which $\{f_n, n \in Z^+\}$ becomes an orthonormal basis for $\mathcal{H}$.*

(c) *$\{f_n, n \in Z^+\}$ is complete in $\mathcal{H}$, and there exists positive constants A and B such that for any arbitrary positive integer n and arbitrary scalars $c_1 \ldots c_n$,*

$$A \sum_{i=1}^{n} |c_i|^2 \leq \left\| \sum_{i=1}^{n} c_i f_i \right\|^2 \leq B \sum_{i=1}^{n} |c_i|^2$$

(d) *$\{f_n, n \in Z^+\}$ is complete in $\mathcal{H}$, and its Gram matrix*

$$(\langle f_i, f_j \rangle)_{i,j=1}^{\infty}$$

generates a bounded invertible operator in ℓ^2.

(e) *$\{f_n, n \in Z^+\}$ is complete in $\mathcal{H}$ and possesses a complete biorthogonal sequence $\{g_n, n \in Z^+\}$ such that*

$$\sum_{i=1}^{n} |\langle f, f_i \rangle|^2 < \infty \qquad \text{and} \qquad \sum_{i=1}^{n} |\langle f, g_i \rangle|^2 < \infty$$

for every $f \in \mathcal{H}$.

A sequence $\{f_n, n \in Z^+\} \subset \mathcal{H}$ is called a *frame* if there are constants A and B such that for every $f \in \mathcal{H}$

$$A\|f\|^2 \leq \sum_{n \in Z^+} |\langle f, f_n \rangle|^2 \leq B\|f\|^2.$$

The constants A and B are called *bounds* of the frame. If $A = B$, the frame is called *tight*. The supremum of all such A and the infimum of all such B are called *best bounds*. If only the right-hand inequality is satisfied for all $f \in \mathcal{H}$, then $\{f_n, n \in Z^+\}$ is called a *Bessel sequence* with bound B. Excellent introductions to the theory of frames and Bessel sequences can be found in [6, 19, 20, 31, 44].

The redundancy inherent in frame decompositions can be used for noise reduction. In fact, the following results from an elementary calculation.

Theorem 2.2 *([2]) Let $\{\phi_n\}$ be a tight frame for H with frame bounds $A = B > 1$, and suppose each $\|\phi_n\| = 1$. Let $\{e_n\}$ be an orthonormal basis (ONB) of H, and let $\{w_n\}$ be a sequence of uncorrelated random variables with 0 mean and such that $\sum Var(w_n) = 1$. Consider the representation of a signal in H in terms of the ONB and the frame, i.e.,*

$$\forall f \in H, \qquad f = \sum \langle f, e_n \rangle e_n,$$

and

$$\forall f \in H, f = \frac{1}{A} \sum \langle f, \phi_n \rangle \phi_n,$$

respectively. If the coefficients $\{\langle f, e_n \rangle\}$ and $\{\langle f, \phi_n \rangle\}$ are both perturbed componentwise by the "random noise" $\{\epsilon w_n\}$, then the expected reconstruction error due to the noise is less in the frame expansion than in the ONB expansion. In fact, there is a gain of $(1 - \frac{1}{A^2})$ in the frame case over the ONB case.

One of the properties that will be used frequently is the following: $\{f_n, n \in Z^+\}$ is a Bessel sequence with bound M if and only if, for every *finite* sequence of scalars $\{c_k\}$,

$$\|\sum_k c_k f_k\|^2 \leq M \sum_k |c_k|^2. \tag{5.1}$$

(cf. eg [44, p.155, Theorem 3]). As remarked by Chui and Shi in [16, Lemma 4], it is a straightforward consequence of this statement that $\{f_n, n \in Z^+\}$ is a Bessel sequence with bound M if and only if (5.1) is satisfied for every sequence $\{c_k\}$ in ℓ^2.

For a given sequence $\{f_n, n \in Z^+\} \in \mathcal{H}$, let $S : \mathcal{H} \to \mathcal{H}$ be defined by

$$S(f) := \sum_{i=1}^{\infty} \langle f, f_n \rangle f_n.$$

A number of important properties of frames are described forthwith:

Theorem 2.3 ([6]) *Let* $\{f_n : n \in Z^+\} \subset H$.

(a) *The following are equivalent.*

(i) $\{f_n, n \in Z^+\}$ *is a frame for* $\mathcal{H}$ *with frame bounds* A *and* B.

(ii) $S : \mathcal{H} \to \mathcal{H}$ *is a topological isomorphism with norm bounds* $\|S\| \leq B$ *and* $\|S^{-1}\| \leq A^{-1}$.

(b) *In the case of either condition in part a, we have that*

$$AI \leq S \leq BI, \qquad B^{-1}I \leq S^{-1} \leq A^{-1}I,$$

$\{S^{-1}f_n\}$ *is a dual frame for* H *with frame bounds* B^{-1} *and* A^{-1}, *and, for all* $f \in H$,

$$f = \sum \langle f, S^{-1}f_n \rangle f_n \tag{5.2}$$

and

$$f = \sum \langle f, f_n \rangle S^{-1}f_n. \tag{5.3}$$

$\{S^{-1}x_n\}$ is called the dual frame of $\{f_n\}$, (2) is the frame decomposition of f, and (3) is the dual frame decomposition of f. (I is the identity map, and $S \leq BI$ means that $\langle (BI - S)f, f \rangle \geq 0$ for each $f \in \mathcal{H}$.)

A frame is called *exact*, if upon the removal of any single element of the sequence, it ceases to be a frame. A sequence is called *minimal*, if no element of the sequence is in the closure of the linear span of the other elements.

Theorem 2.4 ([6]) *Let* $\{f_n : n \in Z^+\} \subset H$ *be a frame for* H *with frame bounds* A *and* B.

(a) *For each* $c \equiv \{c_n\} \in \ell^2(Z^+)$, $f_c \equiv \sum c_n x_n$ *converges in* $\mathcal{H}$ *and* $\|f_c\|^2 \leq B\|c\|^2_{\ell^2(Z^+)}$.

(b) *For each* $f \in \mathcal{H}$, *there is* $c_f \equiv \{c_n\} \in \ell^2(Z^+)$ *such that* $f = \sum c_n f_n \in \mathcal{H}$. *In fact, we can take* $c_n = \langle f, S^{-1}f_n \rangle$ *for each* n.

(c) *If* $f \in H$ *and* $b_f \equiv \{b_n\} \subset C$ *have the property that* $f = \sum b_n f_n \in \mathcal{H}$ *and* $b_n \neq \langle f, S^{-1}f_n \rangle$ *for some* n, *then* $\{b_n\}$ *is not of the form* $\{\langle f, S^{-1}f_n \rangle\}$ *for any* $f \in \mathcal{H}$, *and*

$$\sum |b_n|^2 = \sum |\langle f, S^{-1}f_n \rangle|^2 + \sum |\langle f, S^{-1}f_n \rangle - b_n|^2.$$

(d) Let $F_m = \{f_n : f_n \neq f_m\}$, where m is fixed. Then F_m is a frame if $\langle f_m, S^{-1} f_m \rangle \neq 1$, and F_m is incomplete if $\langle c_m, S^{-1} f_m \rangle = 1$. Further, if $\langle f_m, S^{-1} f_m \rangle = 1$ then $\langle f_m, S^{-1} f_n \rangle = 0$ for all $n \neq m$.

(e) If $\{f_n\}$ is an exact frame then $\{f_n\}$, $\{S^{-1} f_n\}$ are biorthonormal, $\{S^{-1} f_n\}$ is the unique sequence in $\mathcal{H}$ which is biorthonormal to $\{f_n\}$, and F_m (defined in part d) is incomplete.

(f) $\{f_n\}$ is an exact frame if and only if it is minimal.

Theorem 2.5 ([1]) Let $\{\phi_n\}_{n \in \mathbb{Z}}$ be a frame of $\mathcal{H}$ with frame bounds $0 < A \leq B$. If T is a bounded linear operator from $\mathcal{H}$ into $\mathcal{H}$, then $\{\theta_n = T\phi_n\}_{n \in \mathbb{Z}}$ is a frame for $\mathcal{H}$ if and only if there exists a positive constant γ such that the adjoint operator T^* satisfies

$$\|T^* v\|_{\mathcal{K}}^2 \geq \gamma \|v\|_{\mathcal{K}}^2.$$

Theorem 2.6 ([1]) Let $\mathcal{H}_1 \subset \mathcal{H}$ be a closed subspace, and let $\{\phi_n\}_{n \in \mathbb{Z}}$ be a frame of $\mathcal{H}$. If T is a bounded linear operator from $\mathcal{H}$ into $\mathcal{H}_1$, then $\{\theta_n = T\phi_n\}_{n \in \mathbb{Z}}$ is a frame of $\mathcal{H}_1$ if and only if the restriction T_r^* of T^* to $\mathcal{H}_1$ is coercive, i.e., there exists a positive constant γ such that

$$\|T^* v\|_{\mathcal{H}}^2 \geq \gamma \|v\|_{\mathcal{H}}^2, \forall v \in \mathcal{H}_1.$$

In the next theorem $\{\dot{\phi}_n\}_{n=1}^{\infty}$ denotes the dual frame of $\{\phi_n\}_{n=1}^{\infty}$ in $\mathcal{H}$.

Theorem 2.7 ([1]) If $\mathcal{H}_1 \subset \mathcal{H}$ is a closed subspace of $\mathcal{H}$ and if P is the orthogonal projection on $\mathcal{H}_1$, then $\{\dot{\theta}_n = P\phi_n\}_{n \in \mathbb{Z}}$ and $\{\dot{\theta}_n = P\dot{\phi}_n\}_{n \in \mathbb{Z}}$ are dual frames of $\mathcal{H}_1$. Moreover, the frame bounds A and B of $\{\phi_n\}_{n \in \mathbb{Z}}$ are also frame bounds for $\{\theta_n\}_{n \in \mathbb{Z}}$.

Theorem 2.8 ([1]) Let $\{\phi_n\}_{n \in \mathbb{Z}}$ be a frame of $\mathcal{H}$, and U be a bounded linear operator from ℓ_2 into itself. Then $\{\theta_n = U\phi_n\}_{n \in \mathbb{Z}}$ is a frame of $\mathcal{H}$ if and only if there exists a constant $\gamma > 0$ such that for all $x \in X = Range((p(\cdot, \phi))$ we have

$$\|Ux\|_{\ell^2}^2 \leq \gamma \|x\|_{\ell^2}^2, \forall x \in X.$$

Definition 2.1 We say $\{x_n\} \subset \mathcal{H}$ generates a pseudo frame decomposition for $\mathcal{H}$ if there exists $\{x_n^*\} \subset \mathcal{H}$ such that

$$\forall x, y \in \mathcal{H}, \langle x, y \rangle = \sum_n \langle x, x_n^* \rangle \langle x_n, y \rangle \text{ in } \mathcal{H}. \tag{5.4}$$

Theorem 2.9 *([41]) Let $\{x_n\}$ and $\{x_n^*\}$ be both Bessel sequences such that (4) holds. Then both $\{x_n\}$ and $\{x_n^*\}$ are frames for $\mathcal{H}$.*

Theorem 2.10 *([41]) Let $\{x_n\}$ be a frame of $\mathcal{H}$. Let $\{x_n^*\}$ be a Bessel sequence. Then $\{x_n\}$ and $\{x_n^*\}$ forms a pseudo frame decomposition if and only if*

$$x_m = \sum_n \langle x_m, x_n^* \rangle x_n = \sum_n \langle x_m, x_n \rangle x_n^*, \qquad \forall m \in \mathcal{Z}. \qquad (5.5)$$

Let $\{x_m\}$ be a frame in $\mathcal{H}$. Assume $\{x_n^*\}$ is Bessel. Define $U_1 : \mathcal{H} \to \ell^2(\mathcal{Z})$ and $V_1 : \mathcal{H} \to \ell^2(\mathcal{Z})$ as follows

$$\forall x \in \mathcal{H}, U_1 x = \{\langle x, x_n \rangle\}, \qquad (5.6)$$

and

$$\forall x \in \mathcal{H}, V_1 x = \{\langle x, x_n^* \rangle\}.$$

Theorem 2.11 *([41]) Let $\{x_n\}$ be a frame. Let $\{x_n^*\}$ be a Bessel sequence, and assume (5) holds. Then,*

$$V_1 * U_1 = I.$$

Conversely, if $\{x_n\}$ is a frame, and $\{x_n^*\}$ is Bessel such that $V_1^* U_1 = I$, then (5) holds.

Theorem 2.12 *([41]) Assume $\{x_n\}$ is a frame, and $\{x_n^*\}$ is a Bessel sequence. Let U_1 and V_1 be defined as above. Then*

$$x_n = U_1^* e_n, \qquad x_n^* = V_1^* e_n,$$

where $\{e_n\}$ is the standard orthonormal basis for $\ell^2(\mathcal{Z})$. And, $\{x_n^*\}$ is Bessel if and only if V_1^* is a bounded linear operator.

Theorem 2.13 *([41]) Let $\{x_n^*\}$ be a frame. Let U_1 be defined by (6). Generate bounded $V_1^* : \ell^2(\mathcal{Z}) \to \mathcal{H}$ in terms of $V_1^* U_1 = I$, and so define $x_n^* = V_1^* e_n$. Then, we have*

$$\forall x \in \mathcal{H}, \qquad x = \sum_n \langle x, x_n^* \rangle x_n,$$

and $\{x_n^*\}$ is a frame too.

The relationship between frames and Riesz bases is discussed in the next theorem.

Theorem 2.14 ([6]) *Let* $\{f_n : n \in Z^+\} \subset \mathcal{H}$. *The following are equivalent.*

(a) $\{f_n\}$ *is an exact frame for* $\mathcal{H}$;

(b) $\{f_n\}$ *is a bounded unconditional basis of* $\mathcal{H}$;

(c) $\{f_n\}$ *is a Riesz basis of* $\mathcal{H}$.

Thus, orthonormal bases are exact frames, but not every frame is a Riesz basis.

Frames were introduced by Duffin and Schaeffer [22] to study an irregular sampling problem (for other applications, see, e.g.[21]). In the same paper they introduced the *frame algorithm*, which makes it possible to reconstruct uniquely and stably any element $f \in \mathcal{H}$ from the sequence of coefficients $\{\langle f, f_k \rangle, k \in Z^+\}$, and the frame bounds A and B.

Theorem 2.15 ([22],[28]) *(Frame algorithm) Let* $\{e_n, n \in Z^+\}$ *be a frame for* $\mathcal{H}$ *with frame bounds* $A, B > 0$. *Then every* $f \in \mathcal{H}$ *can be reconstructed from the coefficients* $\langle f, e_n \rangle$, $n \in Z$ *by* $f_0 = 0$ *and for* $n \geq 1$

$$f_n = f_{n-1} + \frac{2}{A+B} S(f - f_{n-1})$$

Then $f = \lim_{n \to \infty} f_n$ *with the error estimate*

$$\|f - f_n\| \leq (\frac{B-A}{B+A})^n \|f\|$$

The frame algorithm has the disadvantage that the frame bounds A and B are required for its implementation. This difficulty has recently been resolved by Gröchenig [28]:

Theorem 2.16 ([28]) *(conjugate gradient acceleration method). Let* $\{f_n, n \in Z^+\} \subset \mathcal{H}$ *be a frame with frame bounds* $A, B > 0$ *and* $\rho = (B - A)/(B + A)$, *and let* S *be the associated frame operator. Then the sequence* g_n, *defined by* $g_0 = 0$, $g_1 = Sg$, $\lambda_1 = 2$, *and*

$$g_n = \lambda_n(g_{n-1} - g_{n-2} + S(f - g_{n-1})) + g_{n-2},$$

$$\lambda_n = (1 - \frac{\rho^2}{4}\lambda_{n-1})^{-1} \text{ for } n \geq 2$$

converges to f in $\mathcal{H}$ and satisfies the error estimate

$$\|f - g_n\|_\mathcal{H} \le \frac{2\sigma^n}{1 + \sigma^{2n}}\|f\|_\mathcal{H},$$

where

$$\sigma = \frac{\rho}{1 + \sqrt{1 - \rho^2}} = \frac{B - A}{B + A + 2\sqrt{AB}} = \frac{\sqrt{B} - \sqrt{A}}{\sqrt{B} + \sqrt{A}}.$$

Although Gröchenig's algorithm can be implemented without knowing the frame bounds, these are still needed to obtain error estimates.

If we say that $\{f_n, n \in Z^+\}$ is a Riesz basis with bounds A and B, we mean that A and B are its frame bounds.

3   Frame and Basis Perturbations

Theorem 3.1 *[26] Let $\{f_n, n \in Z^+\}$ be a frame in $\mathcal{H}$ with bounds A and B, let $\{g_n, n \in Z^+\}$ be a Bessel sequence in $\mathcal{H}$ with bound M, and let λ he a complex number such that $|\lambda| < (A/M)^{1/2}$. Then $\{f_n + \lambda g_n, n \in Z^+\}$ is a frame in $\mathcal{H}$ with frame bounds $[(A)^{1/2} - |\lambda|(M)^{1/2}]^2$ and $[(B)^{1/2} + |\lambda|(M)^{1/2}]^2$.*

The proof of this theorem is a straightforward consequence of the triangle inequality for sequences in ℓ^2.

Setting $\lambda = -1$ and replacing g_n by $f_n - g_n$ in Theorem 3.1, we obtain a result of Christensen ([13, Corollary 2.7] or, more explicitly, [15, Corollary 6]). It was also obtained as a corollary of a more general statement on Banach frames (c.f. e.g., [15, Theorem 1]).

Theorem 3.2 *([26]) Let $\{f_n, n \in Z^+\}$ be a frame in $\mathcal{H}$ with bounds A and B. Assume $\{g_n, n \in Z^+\} \subset \mathcal{H}$ is such that $\{f_n - g_n, n \in Z^+\}$ is a Bessel sequence with bound $M < A$. Then $\{g_n, n \in Z^+\}$ is a frame with bounds $[1 - (M/A)^{1/2}]^2 A$ and $[1 + (M/B)^{1/2}]^2 B$.*

Since $|\langle f, f_n - g_n\rangle| \le \|f\|\,\|f_n - g_n\|$, Theorem 3.2 immediately yields another result of Christensen (cf.[11]) and also [10, Proposition 2.4]):

Theorem 3.3 *([26]) Let $\{f_n, n \in Z^+\}$ be a frame in $\mathcal{H}$ with bounds A and B, and assume that*

$$M := \sum_{n \in Z^+} \|f_n - g_n\|^2 < A.$$

Then $\{g_n, n \in Z^+\}$ is a frame in $\mathcal{H}$ with bounds $A[1 - (M/A)^{1/2}]^2$ and $B[1 + (M/B)^{1/2}]^2$.

We also have:

Theorem 3.4 *([26]) Let $\{f_n\}$ be a Riesz basis in $\mathcal{H}$ with bounds A and B. Assume $\{g_n, n \in Z^+\} \subset \mathcal{H}$ is such that $\{f_n - g_n, n \in Z^+\}$ is a Bessel sequence with bound $M < A$. Then $\{g_n, n \in Z^+\}$ is a Riesz basis with bounds $[1 - (M/A)^{1/2}]^2 A$ and $[1 + (M/B)^{1/2}]^2 B$. Conversely, if $\{f_n, n \in Z^+\}$ and $\{g_n, n \in Z^+\}$ are Riesz bases in $\mathcal{H}$ with bounds A_1, B_1 and A_2, B_2 respectively, and U is a linear homeomorphism such that $Uf_n = g_n, n \in Z^+$, then $\{f_n - g_n, n \in Z^+\}$ is a Bessel sequence with bound $M := \min\{B_1 \|I - U\|^2, B_2 \|I - U^{-1}\|^2\}$.*

Applying the Cauchy-Schwarz inequality and Theorem 3.4 we obtain:

Theorem 3.5 *([26]) Let $\{f_n, n \in Z^+\}$ be a Riesz basis in $\mathcal{H}$ with bounds A and B, and assume that*

$$M := \sum_{n \in Z^+} \|f_n - g_n\|^2 < A.$$

Then $\{g_n, n \in Z^+\}$ is a Riesz basis in $\mathcal{H}$ with bounds $A[1 - (M/A)^{1/2}]^2$ and $B[1 + (M/B)^{1/2}]^2$.

The condition $M < A$ in the preceding four theorems cannot be improved. This follows from the following:

Example 1. Let $\{f_n, n \in Z^+\}$ be an orthonormal basis in $\mathcal{H}$, and let

$$g_n := \begin{cases} 0, & \text{if } n = 1 \\ f_n, & \text{otherwise} \end{cases}$$

Then $\{f_n, n \in Z^+\}$ is a frame with bounds $A = B = 1$, $\{f_n - g_n, n \in Z^+\}$ is a Bessel sequence with bound 1, and

$$\sum_{n \in Z^+} \|f_n - g_n\|^2 = \|f_1\|^2 = 1.$$

However, $\{g_n, n \in Z^+\}$ is not dense in $\mathcal{H}$, and therefore it can be neither a frame nor a Riesz basis.

A sequence $\{f_i\}_{i=1}^{\infty}$ is called a *frame sequence* if $\{f_i\}_{i=1}^{\infty}$ is a frame for $\overline{span}\{f_i\}_{i=1}^{\infty}$.

Theorem 3.6 *([14]) Let $\{f_i\}_{i=1}^{\infty}$ be a frame with frame bounds A, B. Let $\{g_i\}_{i=1}^{\infty} \subseteq \mathcal{H}$ and assume that there exist constants $\lambda_1, \lambda_2, \mu \geq 0$ such that $\max(\lambda_1 + \frac{\mu}{\sqrt{A}}, \lambda_2) < 1$ and*

$$\left\|\sum_{i=1}^{n} c_i(f_i - g_i)\right\| \leq \lambda_1 \left\|\sum_{i=1}^{n} c_i f_i\right\| + \lambda_2 \left\|\sum_{i=1}^{n} c_i g_i\right\| + \mu \left(\sum_{i=1}^{n} |c_i|^2\right)^{\frac{1}{2}}, \quad (5.7)$$

for all $c_1, \ldots, c_n (n \in N)$. Then $\{g_i\}_{i=1}^{\infty}$ is a frame with bounds

$$A\left(1 - \frac{\lambda_1 + \lambda_2 + \frac{\mu}{\sqrt{A}}}{1 + \lambda_2}\right)^2, \quad B\left(1 + \frac{\lambda_1 + \lambda_2 + \frac{\mu}{\sqrt{B}}}{1 - \lambda_2}\right)^2.$$

Easy examples show the optimality of the bounds on $\lambda_1, \lambda_2, \mu$ in the sense that the conclusion fails if (7) is only satisfied with $\lambda_1 = 1$ (or $\lambda_2 = 1$ or $\mu = \sqrt{A}$). It is also easy to see that the theorem breaks down if $\{g_i\}_{i=1}^{\infty}$ is not in $\overline{span}\{f_i\}_{j=1}^{\infty}$:

Example 2. Let $\{e_i\}_{i=1}^{\infty}$ be an orthonormal basis for $\mathcal{H}$. Then $\{f_i\}_{i=1}^{\infty} := \{e_1, e_2, 0, 0, 0, \ldots\}$ is a frame sequence. Given $\epsilon > 0$ we define

$$\{g_i\}_{i=1}^{\infty} = \{e_1, e_2, \frac{\epsilon}{3} e_3, \frac{\epsilon}{4} e_4, \ldots, \frac{\epsilon}{n} e_n, \ldots\}$$

By choosing ϵ small enough we can make $\{g_i\}_{i=1}^{\infty}$ so close to $\{f_i\}_{i=1}^{\infty}$ as we want, in the sense that

$$\left\|\sum_{i=1}^{n} c_i(f_i - g_i)\right\| \leq \frac{\epsilon}{3} \left(\sum_{i=1}^{n} |c_i|^2\right)^{\frac{1}{2}}, \quad \forall\{c_i\}_{i=1}^{n}.$$

But $\{g_i\}_{i=1}^{\infty}$ is not a frame sequence.

Theorem 3.7 *([14]) Let $\{f_i\}_{i=1}^{\infty}$ be a frame sequence with bounds A, B. Let $\{g_i\}_{i=1}^{\infty} \subseteq \mathcal{H}$ and suppose that there exists numbers $\lambda_1, \lambda_2, \in [0; 1]$ and $\mu \geq 0$ such that*

$$\left\|\sum_{i=1}^{n} c_i(f_i - g_i)\right\| \leq \lambda_1 \left\|\sum_{i=1}^{n} c_i f_i\right\| + \lambda_2 \left\|\sum_{i=1}^{n} c_i g_i\right\| + \mu \left(\sum_{i=1}^{n} |c_i|^2\right)^{\frac{1}{2}},$$

for all $c_1, \ldots, c_n (n \in N)$. Then $\{g_i\}_{i=1}^{\infty}$ is a Bessel sequence with upper bound $B(1 + \frac{\lambda_1 + \lambda_2 + \frac{\mu}{\sqrt{B}}}{1 - \lambda_2})^2$. If furthermore $\delta_N < 1$ and $\max(\lambda_2, \lambda_1 + \frac{\mu}{\sqrt{A}(1 - \delta_N^2)^{1/2}}) < 1$, then $\{g_i\}_{i=1}^{\infty}$ is a frame sequence with lower bound

$$A(1 - \delta_N^2)\left(1 - \frac{\lambda_1 + \lambda_2 + \frac{\mu}{\sqrt{A}(1 - \delta_N^2)^{1/2}}}{1 + \lambda_2}\right)^2.$$

Theorem 3.8 *([14]) Let $\{f_i\}_{i=1}^{\infty}$ be a frame sequence with bounds A, B. Let $\{g_i\}_{i=1}^{\infty} \subseteq \mathcal{H}$ and suppose there exist $\mu \geq 0$ such that*

$$\left\| \sum_{i=1}^{n} c_i(f_i - g_i) \right\| \leq \mu \left(\sum_{i=1}^{n} |c_i|^2 \right)^{\frac{1}{2}}$$

for all $c_1, \ldots, c_n (n \in N)$. Then $\{g_i\}_{i=1}^{\infty}$ is a Bessel sequence with upper bound $B(1 + \frac{\mu}{\sqrt{B}})^2$. If furthermore $\mu < \sqrt{A}(1 - \delta_R^2)^{1/2})$, then $\{g_i\}_{i=1}^{\infty}$ is a frame sequence with lower bound $A(1 - \delta_R^2)(1 - \frac{\mu}{\sqrt{A}(1 - \delta_R^2)^{1/2}})^2$.

Theorem 3.9 *([14]) Let $\{\lambda_n\}_{n=1}^{\infty}, \{\mu_n\}_{n=1}^{\infty} \subseteq R$. Suppose that $\{e^{i\lambda_n}\}_{n=1}^{\infty}$ is a frame for $L^2(-\pi, \pi)$ with bounds A, B. If there exists a constant $L < 1/4$ such that*

$$|\mu_n - \lambda_n| \leq L \text{ and } 1 - \cos \pi L + \sin \pi L < \sqrt{\frac{A}{B}},$$

then $\{e^{i\mu_n}\}_{n=1}^{\infty}$ is a frame for $L^2(-\pi, \pi)$ with bounds

$$A\left(1 - \sqrt{\frac{B}{A}}(1 - \cos \pi L + \sin \pi L)\right)^2, B(2 - \cos \pi L + \sin \pi L)^2.$$

4   Special Clases of Frames and The Projection Method

Theorem 4.1 *([7]) There exists a frame for $\mathcal{H}$ made up of norm one vectors, which has no subsequence which is a Riesz basis.*

This theorem is proved with the help of the following auxiliary propositions, of independent interest.

Lemma 4.1 *([7]) Let $\{e_i\}_{i=1}^{n}$ be an orthonormal basis for a finite dimensional space $\mathcal{H}_n$. Define*

$$f_j = \frac{1}{n} \sum_{i=1}^{n} e_i \text{ for } j = 1, \ldots n$$

$$f_{n+1} = \frac{1}{\sqrt{n}} \sum_{i=1}^{n} e_i.$$

Then

$$(1 - \frac{1}{\sqrt{2}})^2 \|f\|^2 \leq \sum_{j=1}^{n+1} |\langle f, f_j \rangle|^2 \leq 5\|f\|^2, \forall f \in \mathcal{H}_n.$$

Definition 4.1 *Given a sequence* $\{g_i\}_{i \in I} \subseteq \mathcal{H}$ *its* unconditional basis constant *is defined as the number*

$$\sup \{\| \sum_{i \in I} \sigma_i c_i g_i \| \setminus \| \sum_{i \in I} c_i g_i \| = 1 \ and \ \sigma_i = +1, -1, \forall i \}.$$

Lemma 4.2 *([7]) Define* $\{f_1, f_2, \ldots, f_{n+1}\}$ *as in lemma 4.1. Then any subset of* $\{f_1, f_2, \ldots, f_{n+1}\}$ *which spans* $\mathcal{H}_n$ *has unconditional basis constant greater than or equal to* $\sqrt{n-1} - 1$.

For each space $\mathcal{H}_n$ let $\{f_i^n\}_{i=1}^{n+1}$ be as in lemma 4.1, starting with the othonormal basis $\{e_{\frac{(n-1)n}{2}+1}, \ldots, e_{\frac{(n-1)n}{2}+n}\}$.

Lemma 4.3 *([7])* $\{f_i^n\}_{i=1,n=1}^{n+1,\infty}$ *is a frame for* $\mathcal{H}$, *with bounds* $(1 - \frac{1}{\sqrt{2}})^2$ *and* 5.

Lemma 4.4 *([7]) No subsequence of* $\{f_i^n\}_{i=1,n=1}^{n+1,\infty}$ *is a Riesz basis for* $\mathcal{H}$.

Definition 4.2 *A sequence* $\{f_i\}_{i=1}^{\infty}$, *in* $\mathcal{H}$ *is called* near-Riesz basis *if it is a frame plus finitely many elements*

Theorem 4.2 *([7]) Let* $\{f_i\}_{i=1}^{\infty}$ *be a frame for* $\mathcal{H}$ *with bounds* A, B. *Let* $\{g_i\}_{i=1}^{\infty} \subseteq \mathcal{H}$ *and assume that there exist* $\lambda, \mu \geq 0$ *such that*

$$\| \sum c_i(f_i - g_i) \| \leq \lambda \cdot \| \sum c_i f_i \| + \mu \sqrt{\sum |c_i|^2},$$

for all finite sequences $\{c_i\}$. *Then* $\{f_i\}_{i=1}^{\infty}$ *is a near-Riesz basis* $\Leftrightarrow$ $\{g_i\}_{i=1}^{\infty}$ *is a near-Riesz basis, in which case* $\{f_i\}_{i=1}^{\infty}$ *and* $\{g_i\}_{i=1}^{\infty}$ *have the same excess.*

Holub [32] shows that if a frame is norm-bounded below, then it is a near-Riesz basis if and only if it is *unconditional*, which means that if a series $\sum_{i \in I} c_i f_i$, converges, then it converges unconditionally.

Definition 4.3 *A frame* $\{f_i\}_{i=1}^{\infty}$ *is called a* Riesz frame *if every subfamily* $\{f_i\}_{i \in I}$, $I \subseteq N$ *is a frame for its closed span, with a lower bound* A *common for all those frames.*

Theorem 4.3 *([8]) Every Riesz frame contains a Riesz basis.*

Lemma 4.5 *([8]) Let $\{f_i\}_{i \in I}$ be a frame. Given $\epsilon > 0$ and a finte set $J \subset I$, there exists a finite set J' containing J containing I such that*

$$\sum_{i \in I - J'} |\langle f, f_i \rangle|^2 \le \epsilon \cdot \|f\|^2, \forall f \in span\{f_i\}_{i \in J}.$$

Theorem 4.4 *([8]) Let $\{f_i\}_{i \in I}$ be a frame with the property that every sub-family $\{f_i\}_{i \in J}$ is a frame for its closed span. Then there exists an $\epsilon > 0$ and a finite set $J \subset I$ with the following property: if $J'' \subset I - J$ and if $f \in span\{f_i\}_{i \in J''}$ satisfies $\sum_{i \in J''} |\langle f, f_i \rangle|^2 \ge \epsilon \cdot \|f\|^2$, then*

$$\sum_{i \in J} |\langle f, f_i \rangle|^2 \ge \epsilon \cdot \|f\|^2.$$

Theorem 4.5 *([8]) If every subset of $\{f_i\}_{i \in I}$ is a frame for its closed linear span, then $\{f_i\}_{i \in I}$ contains a Riesz basis.*

To prove that Theorem 4.5 really is an improvement of Theorem 4.3 one needs an example of a frame, where every subfamily is a frame for its closed span, but where there does not exist a common lower bound for all those frames. We present such an example now:

Example 3 ([8]) Let $\{e_i\}_{i=1}^{\infty}$ be an orthonormal basis for $\mathcal{H}$ and define

$$\{f_i\}_{i=1}^{\infty} := \{e_i, e_i + \frac{1}{2^i} e_1\}_{i=2}^{\infty}.$$

Theorem 4.6 *Let $\{f_i\}_{i=1}^{\infty}$ be a frame for the Hilbert space $\mathcal{H}$. Then $\{f_i\}_{i=1}^{\infty}$ is unconditional if and only if it is a near-Riesz basis.*

¿From the point of view of applications the problem with calculation of the frame coefficients is to invert S. A way to overcome this difficulty is by considering truncations: We look at finite subfamilies $\{f_i\}_{i=1}^{n}, n \in N$ and define the space $\mathcal{H}_n := span\{f_i\}_{i=1}^{n}$ and the corresponding frame operator

$$S_n : \mathcal{H}_n \rightarrow \mathcal{H}_n, S_n f = \sum_{i=1}^{n} \langle f, f_i \rangle f_i.$$

One can show that the orthogonal projection of $f \in \mathcal{H}$ onto H_n is given by

$$P_n f = \sum_{i=1}^{n} \langle f, S_n^{-1} f_i \rangle f_i = \sum_{i=1}^{n} \langle f, f_i \rangle S_n^{-1} f_i.$$

Since $P_n f \to f$ as $n \to \infty$, one can hope that the corresponding coefficients converges to the frame coefficients for f, i.e., that

$$\langle f, S_n^{-1} f_i \rangle \to \langle f, S^{-1} f_i \rangle \text{ as } n \to \infty, \forall_i \in N, \forall f \in \mathcal{H}. \qquad (5.8)$$

If (8) is satisfied we say that the *projection method* works. In this case the frame coefficients can be approximated as close as we want using finite dimensional methods, i.e., linear algebra, since S_n is an operator on the finite dimensional space $\mathcal{H}_n$. For a discussion of the projection method we refer to ([10]); in particular it is shown there that the projection method works if and only if

$$\forall j \in N \exists c_j : \|S_n^{-1} f_j\| \le c_j \text{ for } n \ge j.$$

If the projection method works the next natural question is how fast the convergence in (8) is. For example, one might wish that the set of coefficients $\{\langle f, S_n^{-1} f_i \rangle\}_{i=1}^n$ converges to the set of frame coefficients in the ℓ^2-sense, i.e., that

$$\sum_{i=1}^n |\langle f, S_n^{-1} f_i \rangle - \langle f, S^{-1} f_i \rangle|^2 + \sum_{i=n+1}^\infty |\langle f, S^{-1} f \rangle|^2 \to 0, \forall f \in \mathcal{H}. \quad (5.9)$$

We say that the *strong projection method* works if (8) is satisfied. Note again that this condition depends on the indexing of the elements. The second term $\sum_{i=n+1}^\infty |\langle f, S^{-1} f_i \rangle|^2 \to 0$ for every frame, $\forall f \in \mathcal{H}$, so we only need to show that

$$\lim_{n \to \infty} \sum_{i=n+1}^\infty |\langle S_n^{-1} P_n f, f_i \rangle|^2 = 0, \forall f \in \mathcal{H}.$$

A family $\{f_i\}_{i=1}^\infty$ is called a *conditional Riesz frame* if there are common bounds for all the frame sequences $\{f_i\}_{i=1}^n$, $n \in N$. In terms of the operators S_n this is equivalent to the condition $\sup_n \|S_n\|_{\mathcal{H}_n} < \infty$. Observe that this notion depends on the indexing of the frame elements. For example, if $\{e_i\}_{i=1}^\infty$ is an othonormal basis, then $\{e_i, \frac{1}{i} e_i\}_{i=1}^\infty$ is a conditional Riesz frame, but $\{\frac{1}{i} e_i, e_i\}_{i=1}^\infty$ is not.

To extend the definition above to the case of a frame indexed by an arbitrarily countable index set I, take a family $\{I_n\}_{n=1}^\infty$ of finite subsets of I such that

$$I_1 \subseteq I_2 \subseteq \dots I_n \nearrow I$$

and replace $\{f_i\}_{i=1}^n$ by $\{f_i\}_{i \in I_n}$. All the following results are true for general index sets with this modification.

Let T, $\{T_n\}_{n=1}^\infty$ be operators from $\mathcal{H}$ into the Hilbert space $(\mathcal{K}, \langle ., . \rangle_{\mathcal{K}})$. We say that $T_n \longrightarrow T$ in the *strong operator topology* if $T_n f \longrightarrow Tf$, $\forall f \in \mathcal{H}$, and that $T_n \longrightarrow T$ in the *weak operator topology* if $\langle f, T_n f \rangle_{\mathcal{K}} \longrightarrow \langle g, Tf \rangle_{\mathcal{K}}$, $\forall f \in \mathcal{H}$, $\forall g \in \mathcal{K}$.

Theorem 4.7 *([9]) Let $\{f_i\}_{i=1}^\infty$ be a frame. Then the following are equivalent:*

(a) *The strong proyection method works.*

(b) *$\{f_i\}_{i=1}^\infty$ is a conditional Riesz frame.*

(c) *$S_n^{-1} P_n \longrightarrow S^{-1}$ in the strong operator topology.*

(d) *$S_n^{-1} P_n \longrightarrow S^{-1}$ in the weak operator topology.*

(e) *$\lim_{n \to \infty} \sum_{i=n+1}^\infty |\langle S_n^{-1} P_n f, f_i \rangle|^2 = 0$, $\forall f \in \mathcal{H}$.*

Corollary 4.1 *([9]) If $\{f_i\}_{i=1}^\infty$ is a Riesz frame, then the strong projection method works for every subfamily $\{f_i\}_{i \in I}$, $I \subset N$, for every indexing of the frame elements.*

Theorem 4.8 *([9]) Let $U : \mathcal{H} \longrightarrow \mathcal{K}$ be an isomorphism. If the strong projection method works for the frame $\{f_i\}_{i=1}^\infty$, then it also works for the frame $\{U f_i\}_{i=1}^\infty$.*

Theorem 4.9 *([9]) Let $\{f_i\}_{i=1}^\infty$ be a frame. Then the following are equivalent:*

(a) *$\sum_{i=1}^n a_i S_n^{-1} f_i \to \sum_{i=1}^\infty a_i S^{-1} f_i$ for $n \to \infty$, $\forall \{a_i\}_{i=1}^\infty \in \ell^2(N)$.*

(b) *$S_n^{-1} \sum_{i=1}^n b_i f_i \to 0$ for $n \to \infty$ for all $\{b_i\}_{i=1}^\infty \in \ell^2(N)$ such that $\sum_{i=1}^\infty b_i f_i = 0$.*

(c) *$\{f_i\}_{i=1}^\infty$ is a conditional Riesz frame.*

5   Exponential Frames and Bases

In this section $r = (r_1, r_2, \cdots, r_d) \in Z^d$ is arbitrary but fixed, $k = (k_1, k_2, \cdots, k_d) \in Z^d$, $t = (t_1, t_2, \cdots, t_d)$ and $\lambda_k = (\lambda_{k_1}, \lambda_{k_2}, \cdots, \lambda_{k_d})$ are elements of $\mathcal{R}^d$, and $\| \cdot \|_r$ denotes the norm of $L^2(I_r)$, where $I_r := [\pi r, \pi(r+2)]$.

Theorem 5.1 ([26]) Assume that $|k_\ell - \lambda_{k_\ell}| \leq L, \ell = 1, \cdots, d$. If $L < 1/4$, then $\{e^{i\langle k,t\rangle} - e^{i\langle \lambda_k,t\rangle}, k \in Z^d\}$ is a Bessel sequence in $L^2(I_r)$ with bound $B_d(L)$.

Applying Theorem 3.2 we obtain:

Corollary 5.1 ([26]) Assume that $|k_\ell - \lambda_{k_\ell}| \leq L, \ell = 1, \cdots, d$, and that $L < 1/4$. If $B_d(L) < 1$, then $\{e^{i\langle \lambda_k,t\rangle}, k \in Z^d\}$ is a Riesz basis in $L^2(I)$ with frame bounds $[1 - B_d(L)^{1/2}]^2$ and $[1 + B_d(L)^{1/2}]^2$.

The requirement $B_d(L) < 1$ in the statement of the preceding corollary is somewhat cumbersome. The following proposition, which will be useful in the sequel, gives upper and lower bounds for $B_d(L)$, as well as a sufficient (but not necessary) condition for $B_d(L) < 1$ that is easier to verify:

Lemma 5.1 ([26]) We have:

(a) If $0 < B_1(L) < 1$, then $[1 + B_1(L)^{1/2}]^{2(d-1)} \leq B_d(L) \leq 9^{d-1}B_1$.

(b) Let $0 < \alpha \leq 1$. If

$$0 < L < \pi^{-1}\cos^{-1}(\frac{1 - \alpha\,9^{1-d}}{\sqrt{2}}) - \frac{1}{4},$$

then $B_d(L) < 1$.

Remarks:
Note that for $d = 1$, the condition in the preceding Lemma reduces to $0 < L < 1/4$.

It is easy to see that the inequality $A + B + C \leq B_1(L)$ mentioned in the proof of Theorem 5.1 can be established under the weaker assumption that $L < 1/2$. However, since $0 < L < 1/2$ and $B_1(L) < 1$ imply that $L < 1/4$, and, conversely, $0 < L < 1/4$ implies that $B_1(L) < 1$, nothing is gained by it. It is also clear that, for $d = 1$, the condition $B_d(L) < 1$ is redundant.

In [43, Lemma 3.1.6], it is shown that given $\varepsilon > 0$ there is a $\delta > 0$ such that if $|k - \lambda_k| \leq \delta$ for every $k \in Z^d$, then $\{e^{i\langle k,t\rangle} - e^{i\langle \lambda_k,t\rangle}, k \in Z^d\}$ is a Bessel sequence in $L^2(-\pi, \pi)$ with bound ε.

For $d = 1$, Corollary 5.1 is variously known as Kadec's 1/4-theorem ([44]), or the theorem of Paley-Wiener and Kadec ([24]). The latter also contains a multivariate version of the theorem for tensor product spaces. Note, however, that none of these results give frame bounds.

The condition $L < 1/4$ is known as the Kadec-Levinson condition ([3]). The study of the stability of $\{e^{int}, n \in Z\}$ was initiated by Paley and Wiener who showed that $\{e^{i\lambda_n t}, n \in Z\}$ is a Riesz basis in $L^2(-\pi, \pi)$ provided that $|\lambda_n - n| \le L < \pi^{-2}$. Eventually, Kadec showed that π^{-2} could be replaced by $1/4$.

It is well known that for $d = 1, L$ cannot equal $1/4$. Whether this remains true if $d > 1$ is still unknown: The existing one-dimensional proof ([44, p. 122]) is based on growth properties of entire functions of one complex variable, and in particular on the fact that, with the exception of the zero function, the zeros of such functions do not hace a finite limit point. This is not necessarily the case for entire functions of more than one complex variable (cf. e. g. [34]), making a straightforward generalization of the one-dimensional proof impossible.

There are a number of conditions guaranteeing that a sequence of exponentials is a frame in $L^2(-\pi, \pi)$. For example Jaffard [33] gives a complete characterization of such sequences in terms of properties of the sequence $\{\lambda_n\}$, thus generalizing earlier work of Duffin and Schaeffer [22]. However, none of these results gives any information about frame bounds either. Others do ([24, 25, 28, 29]), but are not applicable in this context.

6   Wavelet Frames, Bases, and Bessel Sequences.

In this section a will be a fixed positive real number, whereas b will be a positive real number which will occasionally be allowed to vary. Given a function $\phi : \mathcal{R}^d \to \mathcal{C}$, and $j, k \in \mathcal{Z}^d$, $\phi_{b,j,k}(x) := a^{dj/2}\phi(a^j x - bk)$, and $\Phi_b := \{\phi_{b,j,k}, j, k \in \mathcal{Z}^d\}$. Given a fixed sequence $\{\lambda_{j,k}, j, k \in \mathcal{Z}^d\} \subset \mathcal{R}^d$, $\phi_{b,j,k}^{\{p\}}(x) := a^{dj/2}\phi(a^j x - b\lambda_{j,k})$, and $\Phi_b^{\{p\}} := \{\phi_{b,j,k}^{\{p\}}, j, k \in \mathcal{Z}^d\}$. If there is no danger of ambiguity, the subscript "b" will be omitted. By abuse of notation, ϕ will be called the mother wavelet.

Theorem 6.1 *([16]) (Time-domain characterization).*
Let $\phi(t) \in L^2(-\infty, \infty)$ satisfy

$$|\phi(t)| \le C(1 + |t|)^{-1-\epsilon}, -\infty < t < \infty,$$

for some $\epsilon > 0$, where $C > 0$ is a constant independent of t, and that

$$\phi(t) \text{ is piecewise Lip } \alpha, \text{ for some } 0 < \alpha \le 1.$$

Then the following statements are equivalent.

a. $\{\phi_{b;j,k}\}$ *is a Bessel sequence;*

b. *the Fourier transform $\hat{\phi}(\omega)$ of $\phi(t)$ satisfies*

$$\int_{-\infty}^{\infty} \frac{|\hat{\phi}(\omega)|^2}{|\omega|} dw < \infty;$$

c. $\phi(t)$ *has zero mean, i.e.,*

$$\int_{-\infty}^{\infty} \phi(t)dt = 0.$$

Theorem 6.2 *(Frequency-domain characterization) ([16]) Let $\Theta(y)$ be any continuous function defined on $[0, \infty)$, that satisfies*

$$\Theta(y) \ge 0, \text{ for } 0 \le y < \infty,$$

$$\Theta(y) \text{ is nondecreasing near } 0,$$

$$\Theta(y) \text{ is nondecreasing near } + \infty$$

$$\int_0^a \frac{\Theta(y)}{y} dy < \infty, \text{ and}$$

$$\int_b^{\infty} \Theta(y) dy < \infty,$$

for some positive constants a and b. Then every $\phi(t) \in L^2(-\infty, \infty)$ that satisfies

$$|\hat{\phi}(\omega)| \le \Theta(|\omega|), -\infty < \omega < \infty,$$

generates a Bessel sequence $\{\phi_{b;j,k}(t)\}$ for any positive constant b.

Theorem 6.3 *(First oversampling theorem) ([16]) Let $\phi \in L^2(-\infty, \infty)$ generate a frame $\{\phi_{b;j,k}(t)\}$ for some $b > 0$, such that $\phi(t)$ satisfies either (21) - (22) or (24) (or both), where $\Theta(y)$ is constrained by (23). Then for any given positive integer N, the family $\{\phi_{b/N;j,k(t)}\}$ is also a frame of $L^2(-\infty, \infty)$.*

Theorem 6.4 *(Second oversampling Theorem) ([16]) Let $a \geq 2$ be any positive integer and $b > 0$. Also, let $\{\phi_{b;j,k}\}$ be any frame with frame bounds A and B, where $0 < A \leq B < \infty$. Then for every positive integer n which is relatively prime to a, the family*

$$\{\phi_{b/n;j,k}\}$$

is a frame with frame bounds nA and nB. Furthermore, if there is a function $\tilde{\phi} \in L^2$ that generates a dual frame of $\{\phi_{b;j,k}\}$ in the sense that for every $f \in L^2(-\infty, \infty)$

$$f = \sum_{j,k \in Z} \langle f, \phi_{b;j,k} \rangle \tilde{\phi}_{b;j,k} = \sum_{j,k \in Z} < f, \tilde{\phi}_{b;j,k} > \phi_{b;j,k},$$

then

$$\{n^{-1/2} \tilde{\phi}_{b/n;j,k}\}$$

is again a dual frame of $\{n^{-\frac{1}{2}}\phi_{b/n;j,k}\}$, provided, again, that n and a are relatively prime.

Theorem 6.5 *([16]) Let $\phi \in L^1 \cap L^2$ and set*

$$\Phi_\ell(\omega) := \sum_{j \in Z} |\hat{\phi}(2^j \omega)\hat{\phi}(2^j \omega + 2\ell\pi)|, \quad \text{and} \quad \beta(\ell) := \sup_{1 \leq |\omega| \leq 2} \Phi_\ell(\omega).$$

Then $\{\phi_{1;j,k}\}$ is a frame of L^2, if

$$\sum_{j \in Z} |\hat{\phi}(2^j \omega)|^2 \in L^\infty, \quad \text{and} \quad \operatorname{ess\,inf}_\omega \sum_{j \in Z} |\hat{\phi}(2^j \omega)|^2 > 2 \sum_{\ell=1}^{\infty} \sqrt{\beta(\ell)\beta(-\ell)}.$$

Consider $\phi \in L^2$ (but not necessarily in L^1), and for $a > 1$ and $b > 0$, set

$$\Phi_\ell(a, b; \omega) := \sum_{j \in Z} |\hat{\phi}(a^j w)\hat{\phi}(a^j \omega + 2\ell\pi b^{-1})|,$$

and

$$B_\ell(a, b) := \operatorname{ess\,sup}_\omega \sum_{j \in Z} \Phi_{-\ell}(a, b; w + 2\ell\pi b^{-1} a^{-j}) |\hat{\phi}(a^j w)\hat{\phi}(a^j \omega + 2\ell\pi b^{-1})|.$$

It is clear that if $\phi \in L^1 \cap L^2$ then

$$\beta(\ell)\beta(-\ell) \geq \min\{B_\ell(2, 1), B_{-\ell}(2, 1)\}, \qquad \ell \in Z.$$

Theorem 6.6 *([16]) Let $\phi \in L^2, a > 1$ and $b > 0$. Then $\{\phi_{b;j,k}\}$ is a frame of L^2, if*

$$\sum_{j \in \mathcal{Z}} |\hat{\phi}(a^j \omega)|^2 \in L^\infty$$

and

$$\operatorname*{ess\,inf}_{\omega} \sum_{j \in \mathcal{Z}} |\hat{\phi}(a^j \omega)|^2 > 2 \sum_{t=1}^{\infty} \sqrt{\min(B_\ell(a, b), B_{-\ell}(a, b))}.$$

We will say that $\phi \in L^2$ has the property P, if for any $x, y \in E :=$ $\{\omega : \hat{\phi}(\omega) \neq 0\}$, the inequality $x \neq y$ implies $x \not\equiv y (\mathrm{mod} 2\pi b^{-1})$. From Theorem 3.5, it follows that if $\phi \in L^2$ has the property P, then $\{\phi_{b;j,k}\}$ forms a frame with bounds A and B, if and only if

$$A \leq \frac{1}{b} \sum_{j \in \mathcal{Z}} |\hat{\phi}(a^j \omega)|^2 \leq B.$$

The next theorem studies the effect of perturbing the mother wavelet. This requires some care, since even a small perturbation may destroy the frame. Indeed, if a function ϕ generates a wavelet frame, then $x^{-1} \hat{\phi}(x)^2$ must be in $L(\mathcal{R}^d)$ (cf. [18, 19]). Thus, if ϕ generates a wavelet frame in $L^2(\mathcal{R}^d)$ and if, for instance, $\phi(x) :=$ $\phi(x) + \varepsilon x^{-1} \prod_{\ell=1}^{d} \sin x_\ell$, then ϕ will not generate a wavelet frame.

A double sequence $\{f_{j,k}, j, k \in \mathcal{Z}^d\} \subset L^2(\mathcal{R}^d)$ will be called *semiorthogonal* if for every pair of $j, m \in \mathcal{Z}^d$ with $j \neq m$, and any arbitrary choice of k and n in $\mathcal{Z}^d$, the functions $f_{j,k}$ and $f_{m,n}$ are orthogonal. (In other words, for every pair of $j, m \in \mathcal{Z}^d$ with $j \neq m$, and any arbitrary choice of k and n in $\mathcal{Z}^d$, the functions $exp(-a^{-j} 2\pi i \langle x, bk \rangle) \hat{f}(a^{-j} x)$ and $exp(-a^{-m} 2\pi i \langle x, bn \rangle) \hat{f}(a^{-m} x)$ are orthogonal.) This is consistent with the definition of semiorthogonality for wavelets introduced by Auscher and by Chui and Wang (cf. [18]).

Theorem 6.7 *([26]) Let Φ be a semiorthogonal sequence in $L^2(\mathcal{R}^d)$, and let ϕ be any function in the closure of the linear span of $\{\phi_{0,k}, k \in \mathcal{Z}^d\}$. If Φ is a wavelet frame (wavelet Riesz basis) in $L^2(\mathcal{R}^d)$ with bounds A and B and $\|\phi - \phi\| < A^{3/2}/B$, then $\Psi := \{\phi_{j,k}, j, k \in \mathcal{Z}^d\}$ is a semiorthogonal wavelet frame (wavelet Riesz basis) in $L^2(\mathcal{R}^d)$ with bounds*

$$([1 - (B/A^{3/2})]\|\phi - \phi\|)^2 A \qquad and \qquad ([1 + (B^{1/2}/A)]\|\phi - \phi\|)^2 B.$$

A function $\phi(x) \in L^2(\mathcal{R}^d)$ will be called *orthogonal with respect to* a, if for every pair of $j, m \in \mathcal{Z}^d$ with $j \neq m$ and any arbitrary choice of δ and λ in $\mathcal{R}^d$, the functions $\phi(a^j x + \delta)$ and $\phi(a^m x + \lambda)$ are orthogonal in $L^2(\mathcal{R}^d)$. This is the same as saying that for every pair of $j, m \in \mathcal{Z}^d$ with $j \neq m$, and any arbitrary choice of δ and λ in $\mathcal{R}^d$, the functions $exp(a^{-j} 2\pi i \langle x, \delta \rangle) \hat{\phi}(a^{-j} x)$ and $exp(a^{-m} 2\pi i \langle x, \lambda \rangle) \hat{\phi}(a^{-m} x)$ are orthogonal in $L^2(\mathcal{R}^d)$.

Theorem 6.8 *([26]) If* $\Phi := \{\phi_{j,k}, j, k \in \mathcal{Z}^d\}$ *is a semiorthogonal sequence in* $L^2(\mathcal{R}^d)$, *such that* $\hat{\phi}$ *is essentially bounded and* supp$\{\hat{\phi}\}$ *is contained in an interval* I *of the form* $[o, 1/b] + h$, *then* ϕ *is orthogonal with respect to* a.

Using Theorem 5.1 we can show that we may perturb a frame into a frame or a Riesz basis into a Riesz basis, without necessarily increasing the accuracy of the sampling sequence as $|j|$ increases:

Theorem 6.9 *([26]) Let* $\Phi := \{\phi_{j,k}, j, k \in \mathcal{Z}^d\}$ *be a wavelet frame (wavelet Riesz basis) in* $L^2(\mathcal{R}^d)$ *with bounds* A *and* B *such that* $\phi \in L^2(\mathcal{R}^d)$ *is orthogonal with respect to* a. *For* $r \in \mathcal{Z}^d$, *let*

$$I_r := [\pi r, \pi(r + 2)], \quad J_r := [r/(2b), (r + 2)/(2b)],$$

$$L := sup\{|k - \lambda_{j,k}|, j, k \in \mathcal{Z}^d\} < 1/4,$$

$$S_r := ess\,sup\{|\hat{\phi}(t)|, t \in J_r\} < \infty, \qquad M := \frac{B_d(L)}{(2\pi b)^d} \sum_{r \in Z} S_r^2,$$

and

$$M_1 := \frac{9^{d-1} B_1(L)}{(2\pi b)^d} \sum_{r \in Z} S_r^2.$$

If Φ *is a wavelet frame (wavelet Riesz basis) in* $L^2(\mathcal{R}^d)$ *with bounds* A *and* B, *and* $M < A$, *(in particular, if* $M_1 < A$,) *then* $\Phi^{\{p\}}$ *is a wavelet frame (wavelet Riesz basis) in* $L^2(\mathcal{R}^d)$ *with bounds* $[1 - (M/A)^{1/2}]^2 A$ *and* $[1 + (M/B)^{1/2}]^2 B$.

Example 4. Let $a \geq 2$ be an integer, and let $b > 0$ be given. Assume that $\Phi_b := \{\phi_{j,k}, j, k \in Z\}$ is an orthonormal basis in $L^2(\mathcal{R})$

such that $\hat{\phi}$ is essentially bounded and supp$\{\hat{\phi}\}$ is contained in an interval of the form $[\mathbf{0}, 1/b] + h$. Thus, Theorem 6.8 implies that $\phi(x)$ is orthogonal with respect to a. Let $n > 0$ be an integer relatively prime to a. By the multivariate version of the Second Oversampling Theorem [16, Theorem 8] (see also [17]), $\Phi_{b/n} := \{\phi_{b/n,j,k}, j, k \in Z^d\}$ is a wavelet frame with bounds $A = B = n^d$. Thus $\Phi_{b/n}$ and ϕ satisfy the hypotheses of Theorem 6.9.

7   References

[1] A. ALDROUBI Portraits of Frames, Proc. Amer. Math. Soc. 123 (1995), 1661-1668.

[2] J. J. BENEDETTO, Frame Signal Processing Applied to Bio-electric Data, in "Wavelets in Medicine and Biology", (A. Aldroubi and M. Unser, Eds.) pp. 493-512, CRC Press, Boca Raton, 1996.

[3] J. J. BENEDETTO, Irregular sampling and frames, in "Wavelets, a Tutorial in Theory and Applications," (C. K. Chui, Ed.) pp. 445-507, Academic Press, San Diego, 1992.

[4] J. J. BENEDETTO AND S. LI, Multiresolution analysis frames with applications, in "IEEE-ICASSP", Minneapolis, III (1993), pp. 304-307.

[5] J. J. BENEDETTO AND S. LI, Narrow band frame multireso-lution analysis with perfect reconstruction, in "IEEE-SP Int. Symp." Time-Frecuency and Time-Scale Analysis, Philadel-phia, 1994.

[6] J. J. BENEDETTO AND D. F. WALNUT, Gabor frames for L^2 and related spaces, in "Wavelets: Mathematics and Applications" (J. J. Benedetto and M. W. Frazier Eds.) pp. 97-162, CRC Press, Boca Raton, 1994.

[7] P. G. CASAZZA AND O. CHRISTENSEN, Frames containing a Riesz basis and preservation of this property under pertur-bations, preprint, 1995.

[8] P. G. CASAZZA AND O. CHRISTENSEN, Hilbert space frames containing a Riesz basis and Banach spaces which have no subspace isomorphic to c_0, J. Math. Anal. Appl. (to appear).

[9] P. G. CASAZZA AND O. CHRISTENSEN, Riesz frames and approximation of the frame coefficients, preprint, 1996.

[10] O. CHRISTENSEN, Frames and the projection method, *Appl. and Comp. Harm. Anal.* 1 (1993), 50-53.

[11] O. CHRISTENSEN, Frame perturbations, *Proc. Amer. Math. Soc.* (to appear).

[12] O. CHRISTENSEN, A Paley-Wiener theorem for frames, *Proc. Amer. Math. Soc.* (to appear).

[13] O. CHRISTENSEN, Moment problems and stability results for frames with applications to irregular sampling and Gabor frames, preprint, March 1994.

[14] O. CHRISTENSEN, Operators with closed range and perturbation of frames for a subspace, preprint, 1996.

[15] O. CHRISTENSEN AND C. E. HEIL, Perturbation of Banach frames and atomic decompositions, *Math. Nachr.* (to appear).

[16] C. K. CHUI AND X. SHI, Bessel sequences and affine frames, *Appl. and Comp. Harm. Anal.* 1 (1993), 29-49.

[17] C. K. CHUI AND X. SHI, $n\times$ oversampling preserves any tight affine frame for odd n, *Proc. Amer. Math. Soc.* 121 (1994), 511-517.

[18] C. K. CHUI, "An Introduction to Wavelets," Academic Press, San Diego, 1992.

[19] I. DAUBECHIES, "Ten Lectures on Wavelets," SIAM, Philadelphia, 1992.

[20] I. DAUBECHIES, The Wavelet transform, time-frequency localization and signal analysis, *IEEE Trans. Inform. Theory* 36 (1990), 961-1005.

[21] I. DAUBECHIES, A. GROSSMANN, AND Y. MEYER, Painless nonorthogonal expansions, *J. Math. Phys.* (27) (1986), 1271-1276.

[22] R. DUFFIN AND A. SCHAEFFER, A class of nonharmonic Fourier series, *Trans. Amer. Math. Soc.* 72 (1952), 341-366.

[23] G. G. EMCH AND G. C. HEGERFELDT, New classical properties of quantum coherent states, *J. Math. Phys.* 27 (1986), 2731-2737.

[24] H. G. FEICHTINGER AND K. GRÖCHENIG, Theory and practice of irregular sampling, in "Wavelets: Mathematics and Applications," (J. J. Benedetto and M. W. Frazier, Eds.) pp. 305-363, CRC Press, Boca Raton, 1994.

[25] H. G. FEICHTINGER AND K. GRÖCHENIG, Irregular sampling theorems and series expansions of band-limited Functions, *J. Math. Anal. Appl.* 167 (1992), 530-556.

[26] S. J. FAVIER AND R. A. ZALIK, On the stability of frames and Riesz bases, *Appl. and Comp. Harm. Anal.* 2 (1995), 160–173.

[27] K. H. GRÖCHENIG, Irregular Sampling of Wavelet and Short-Time Fourier Transforms, *Constr. Approx.* (9) (1993), 283-297.

[28] K. H. GRÖCHENIG, Acceleration of the frame algorithm, *IEEE Trans. Signal Proc.* 41 (1993), 3331-3340.

[29] K. H. GRÖCHENIG, A discrete theory of irregular sampling, *Lin. Alg. Appl.*, to appear.

[30] C. E. HEIL, "Wiener Amalgam Spaces in Generalized Harmonic Analysis and Wavelet Theory", Ph. D. dissertation, University of Maryland, College Park, Maryland, 1990.

[31] C. E. HEIL AND D. F. WALNUT, Continuous and discrete wavelet transforms, *SIAM Review* 31 (1989), 628-666.

[32] J. HOLUB, Pre-frame operators, Besselian frames and near Riesz bases, *Proc. Amer. Math. Soc.* 122 (1994), 147-160.

[33] S. JAFFARD, A density criterion for frames of complex exponentials, *Michigan Math. J.* 38 (1991), 339-348.

[34] S. G. KRANTZ, "Function Theory of Several Complex Variables", 2nd ed. Wadsworth & Brooks-Cole, Pacific Grove, California, 1992.

[35] E. KREYSZIG, "Introductory Functional Analysis with Applications," reprint, John Wiley and Sons, New York, 1989.

[36] J. R. KLAUDER AND B.S. SKAGERSTAM, "Coherent States - A pplications in Physics and Mathematical Physics," World Scientific, Singapore, 1985.

[37] R. E. A. C. Paley and N. Wiener, "Fourier Transforms in The Complex Domain", American Mathematical Society Colloquium Publications Vol. 19, American Mathematical Society, New York, 1934.

[38] J. R. RETHERFORD AND J. R. HOLUB, The stability of bases in Banach and Hilbert spaces, *J. Reine Angew. Math.* 246 (1971), 136-146.

[39] I. M. SINGER, "Bases in Banach Spaces I", Springer-Verlag, New York, 1970.

[40] H. S. SHAPIRO, "Topics in Approximation Theory", Lecture Notes in Mathematics No. 187, Springer - Verlag, New York, 1971.

[41] LI SHIDONG, General Frame Decompositions, Pseudo-duals and Applications, preprint.

[42] A. E. TAYLOR AND D. C. LAY, "Introduction to Functional Analysis", $2^{nd.}$ Ed., John Wiley and Sons, New York, 1980.

[43] D. WALNUT, "Weyl-Heisenberg Wavelet Expansions: Existence and Stability in Weighted Spaces", Ph. D. dissertation, University of Maryland, College Park, Maryland, 1989.

[44] R. M. YOUNG, "An Introduction to Nonharmonic Fourier Series," Academic Press, New York, 1980.

Chapter 6

Fourier Analysis of Petrov-Galerkin Methods Based on Biorthogonal Multiresolution Analyses

Sônia M. Gomes
Elsa Cortina

1 Introduction

When solving differential equations by means of a Galerkin approach, the approximating spaces are not only supposed to have good approximation properties, but also they must allow easy and fast computations. In addition, if the goal is the development of a multilevel method to detect and follow local singularities, or a multigrid scheme to solve the resulting linear systems, then hierarchical bases are necessary. As an example, we mention the finite element bases which have been widely used over the last three decades.

With the advent of wavelet analysis, a new approach has been successfully applied: the *wavelet Galerkin method*. Usually, a function Ψ is called an orthogonal *wavelet* if the scaled $(2^j, j \in \mathbf{Z})$ and integer translated versions $\Psi_{j,k}(x) = 2^{j/2}\Psi(2^j x - k)$ of Ψ form an orthonormal basis of $L^2(\mathbf{R})$. In a wavelet Galerkin method the approximating spaces are spanned by the multiscale wavelet bases $\{\Psi_{j,k}(x), k \in \mathbf{Z}\}$ for j up to a certain level $m - 1$. These subspaces V_m form what in wavelet analysis is called a *multiresolution analysis of* $L^2(\mathbf{R})$. They exhibit a natural hierarchy which, together with orthogonality, give nice and fast algorithms allowing zoom-in and zoom-out multilevel procedures (see Mallat [19] and also Meyer [20], Daubechies [10]). Beside of multilevel bases of wavelets, the spaces V_m also have other interesting orthonormal bases, namely the one-level nodal bases $\{\Phi_{m,k}(x), k \in \mathbf{Z}\}$, obtained from a basic function Φ called *scaling function*.

Good approximation properties can be achieved using wavelet series. Furthermore, the multiresolution structure of orthogonal wavelet bases provides a simple way to adapt computational refinements to the local regularity of the solution: high resolution computations are performed only in regions where singularities occur (cf. Bacry *et al.* [1], Glowinski *et al.* [11], Liandrat *et al.* [17], and Maday *et al.* [18]).

I. Daubechies has given in [9] a procedure to construct compactly supported wavelets and scaling functions with arbitrary regularity and zero moments (see also Daubechies [10]). However, the price for such nice properties is lack of symmetry and wide support. This drawback disappears in the context of *biorthogonal wavelets*, a concept introduced by Cohen, Daubechies and Feauveau in [3]. In this context two non-orthogonal basis functions $\Psi_{j,k}$ and $\Psi^*_{j,k}$, also called wavelets, are constructed from the scaled translates of two dual basic functions Ψ and Ψ^*. Related to these dual wavelets there are also dual scaling functions Φ and Φ^*. Instead of orthogonality, biorthogonality is required between the two families. In a biorthogonal setting one has more flexibility than in an orthogonal one. In fact, given a basic function, it is possible to construct infinitely many duals. This additional degree of freedom can be used to adapt the bases to the specific problem at hand. For instance, in Cohen *et al.* [3], starting from the B-splines, symmetry and compact support is obtained for a family of biorthogonal wavelets with increasing regularity and zero moments.

Unlike in the Galerkin schemes, where the same basic functions are used both as trial and test functions, in a Petrov-Galerkin method the test and trial functions may be different. In Petrov-Galerkin approximations by biorthogonal wavelets the idea is to use one of the families of basic functions as *trial* functions and one of its dual families as *test* functions. There is already in the literature some work in this direction. In [7] Dahlke and Weinreich constructed compactly supported biorthogonal wavelet bases adapted to some differential operators to obtain stiffness matrices with simple structure. Using these bases, Dahlke and Kunoth emphasise in [6] the construction of a biorthogonal two-grid method for the solution of the resulting linear systems. Preliminary results on the relevance of the biorthogonal Petrov-Galerkin approach when applied to evolution problems having local singularities are in Cunha and Gomes [5]. Our aim in this paper is to study the accuracy of such Petrov-Galerkin type schemes. For this purpose, instead of the multilevel basis of wavelets, it is better to expand the approximate solutions in terms of the one-level nodal basis of scaling functions $\Phi_{m,k}(x)$, the duals $\Phi^*_{m,k}(x)$ being the test functions.

Due to the convolution type of these approximation expansions, Fourier analysis is the appropriate approach in the study of their convergence. For Galerkin methods, where $\Phi = \Phi^*$, this procedure has already been applied in finite element and spline approximations (see Strang and Fix [26] and Thomeé [27]). It was also used for spline Petrov-Galerkin methods by Schoombie [24]. The same methodology can be used for wavelet based methods since the required hyphoteses are authomatically satisfied in the context of wavelet analysis. In fact, the basic function Φ is supposed to be r-regular, which means that Φ and all its derivatives up to order r should have fast decay at infinity. Φ also must satisfy the so called Strang-Fix condition which guarantees that smooth functions can be approximated with good accuracy from the approximating spaces. Regular scaling functions enjoy this property, as pointed out by Meyer [20]. For wavelet Galerkin schemes, an analysis of the convergence is done in Gomes and Cortina [12] (see also Cortina and Gomes [4]). We give here the extension of such results when different basic functions are used as trial and test functions. Some typical numerical results are also presented.

As in [24], we shall adopt as model problem the Korteweg de Vries (KdV) equation and its linearized version. The KdV equation

$$\frac{\partial u}{\partial t} + \mu u \frac{\partial u}{\partial x} + \epsilon \frac{\partial^3 u}{\partial x^3} = 0, \tag{6.1}$$

subject to the initial condition

$$u(x, 0) = u_0(x), \tag{6.2}$$

and several of its generalisations, have been used to model dispersive non-linear long wave phenomena. It has received considerable attention in the last decades not only because of its many applications but also because of its mathematical properties. Due to the counter-balance between the dispersive and the convective terms, interesting phenomena occur, such as the formation of permanent waves, called solitons which are fundamental quantities of the general solution to the KdV equation. For references concerning numerical solutions for the KdV equation, we refer to [24]and [21] and the bibliography mentioned there.

This paper is organized as follows. In Section 2 we give the notation used throughout the paper and we set the required hypotheses on the basic functions Φ and Φ^*. We briefly describe the Petrov-Galerkin method applied to the model problem in Section 3. Section 4 is devoted to study the convergence of the method when applied to the linearized KdV equation. For this case we analyse the accuracy of the method by using the Fourier transform technique. In Section 5 we discuss the particular case of biorthogonal spline scaling functions constructed in [3]. The corresponding numerical calculations are shown in Section 6.

2   Notation and some definitions

For functions $f, g \in L^2(\mathbf{R})$ the inner product and norm are denoted by

$$(f,g) = \int_{-\infty}^{\infty} f(x)\overline{g(x)}dx, \quad ||f||_{L^2} = (f,f)^{1/2},$$

and the Fourrier transform of f by

$$\widehat{f}(\xi) = \mathcal{F}f(\xi) = \int_{-\infty}^{\infty} e^{-ix\xi} f(x)dx.$$

For a function f in the Sobolev space $H^s(\mathbf{R})$ its norm is given by

$$||f||_{H^s} = \int_{-\infty}^{\infty} (1+|x|^2)^s |f(x)|^2 dx.$$

We shall also use the discrete ℓ^2 norm

$$||f||_{2,h} = (h \sum_j (|f(jh)|^2)^{1/2}.$$

To a given sequence $a \in \ell^2(\mathbf{Z})$ it is associated its discrete Fourier transform

$$\tilde{a}(\xi) = \sum_k a(k)e^{-ik\xi}.$$

Let us also introduce the following definitions

Definition 2.1 A function Φ is called r-regular if its derivatives satisfy

$$x^n \frac{d^l \Phi}{dx^l} \in L^2(\mathbf{R})$$

for all indices l such that $0 \leq l \leq r$ and all integers $n > 0$.

Definition 2.2 *(Strang-Fix condition)* A function Φ satisfies the Strang-Fix condition of order p if $\hat{\Phi}(0) \neq 0$ and $\hat{\Phi}(\xi)$ has zeros of order $p + 1$ at all points $\xi = 2k\pi, 0 \neq k \in \mathbf{Z}$, for some integer $p \geq 0$.

We shall refer as $\mathcal{H}_{r,p}$ the class of r-regular functions Φ satisfying the Strang-Fix condition of order p.

3 A Petrov-Galerkin method for the KdV equation

For $0 \leq \alpha \leq 3$ we consider the following weak form of (6.1)

$$\left(\frac{\partial u}{\partial t}, v\right) + \mu \left(u\frac{\partial u}{\partial x}, v\right) + \varepsilon(-1)^\beta \left(\frac{\partial^\alpha u}{\partial x^\alpha}, \frac{d^\beta v}{dx^\beta}\right) = 0 \qquad (6.3)$$

for all β-regular test functions v, where $\beta = 3 - \alpha$. In a Petrov-Galerkin method the main characteristic is that the test functions may be not the same as the trial functions. We describe in this section a Petrov-Galerkin method where the trial functions $\Phi_{h,k}(x)$, are the scale translates of a basic real function Φ. More precisely,

$$\Phi_{hk}(x) = h^{-1/2}\Phi(h^{-1}x - k), \; k \in \mathbf{Z},$$

where $h > 0$ is the mesh width. The approximating spaces $\mathcal{V}_h \subset \mathbf{L}^2(\mathbf{R})$ are then generated by $\{\Phi_{hk}(x), k \in \mathbf{Z}\}$, and we approximate the exact solution of (6.1) by

$$u_h(x, t) = \sum_k U_k(t)\Phi_{hk}(x). \qquad (6.4)$$

We assume that Φ is r-regular, $r \geq 1$. Similarly, the test functions also have the form $\Phi^*_{hk}(x)$, defined in terms of an r^*-regular dual basic real function Φ^*, with $r + r^* \geq 3$.

In the weak formulation (6.3) we chose $\alpha \leq r$ such that $\beta \leq r^*$. If we replace u by the trial solution (6.4) and v by each of the test functions $\Phi^*_{hl}(x)$, then the unknown coefficients $U_k(t)$ must be determined from the system of ordinary differential equations

$$\sum_k a(l - k)\frac{d}{dt}U_k \; + \; \mu h^{-3/2}\sum_s\sum_k b(s - k, l - k)U_kU_s -$$

$$+ \; \varepsilon h^{-3}\sum_k c(l - k)U_k = 0, \qquad (6.5)$$

where

$$a(k) = \int_{-\infty}^{+\infty} \Phi(x)\Phi^*(x-k)dx; \tag{6.6}$$

$$b(l,k) = \int_{-\infty}^{+\infty} \frac{d\Phi}{dx}(x)\Phi(x-l)\Phi^*(x-k)dx; \tag{6.7}$$

$$c(k) = (-1)^\beta \int_{-\infty}^{+\infty} \frac{d^\alpha \Phi(x)}{dx^\alpha}\frac{d^\beta \Phi^*}{dx^\beta}(x-k)dx. \tag{6.8}$$

The initial conditions $U_k(0)$, $k \in \mathbf{Z}$, are the coefficients of $u_h(x,0) = \mathbf{R}_h u_0 \in \mathcal{V}_h$, where $\mathbf{R}_h$ is some approximate initial scheme to be precised later.

The matricial version of (6.5) is

$$\frac{d}{dt}LU + \mu U^T MU + \varepsilon NU = 0, \tag{6.9}$$

where

$$U = (U_k),$$

and

$$\begin{aligned} L(l,k) &= a(l-k); \\ M(l,k,s) &= h^{-3/2}b(l-k,l-s); \\ N(l,k) &= h^{-3}c(l-k). \end{aligned}$$

The application of the trapezoidal rule to (6.5) with time step Δt yields the following nonlinear implicit scheme

$$L(U^{n+1} - U^n) + \Delta t\frac{G(U^{n+1}) + G(U^n)}{2} = 0, \tag{6.10}$$

where $U^n = U(n\Delta t)$, for $n \geq 0$ and $G(U) = \mu U^T MU + \varepsilon NU$. Equation (6.10) is then solved by a Newton iteration.

In the above description, we have only required some regularity properties on the basic functions Φ and Φ^*. However, for good approximating results additional hypotheses are necessary. This is the topic of the following section.

4   The linear case: Fourier analysis

We analyse the convergence of the Petrov-Galerkin method, described in the previous section, when applied to the linearised KdV equation

$$\frac{\partial u}{\partial t} + \mu \frac{\partial u}{\partial x} + \varepsilon \frac{\partial^3 u}{\partial x^3} = 0, \tag{6.11}$$

subjected to the same initial condition (6.2). The exact solution in this case can be represented as

$$u(x, t) = \mathbf{E}(t) u_0(x).$$

where $\mathbf{E}(t)$ is the bounded linear operator in $L^2(\mathbf{R})$ defined by

$$\mathbf{E}(t)v = \mathcal{F}^{-1}[\exp(\mathcal{W}(\xi)t)\hat{v}(\xi)],$$

Here, $\mathcal{W}(\xi) = i(\mu\xi - \varepsilon\xi^3)$.
   The equivalent to system (6.5) is

$$\sum_k a(l-k)\frac{d}{dt}U_k + h^{-1}[\mu d(l-k) + h^{-2}\varepsilon c(l-k)]U_k = 0, \tag{6.12}$$

where $d(k) = \int_{-\infty}^{+\infty} d\Phi(x)/dx \Phi^*(x-k)dx$. Similarly, the equivalent to system (6.10) is

$$\sum_k a(l-k)[U_k^{n+1} - U_k^n]+$$

$$+ h^{-1}\Delta t \sum_k [\mu d(l-k) + h^{-2}c(l-k)]\frac{U_k^{n+1} + U_k^n}{2} = 0. \tag{6.13}$$

Both (6.12) and (6.13) are in the form of discrete convolutions. Applying the discrete Fourier transform they are transformed to

$$\tilde{a}(\xi)\frac{d}{dt}\tilde{U}(\xi, t) + h^{-1}[\mu\tilde{d}(\xi) + h^{-2}\varepsilon\tilde{c}(\xi)]\tilde{U}(\xi, t) = 0,$$

and

$$\tilde{a}(\xi)[\tilde{U}^{n+1}(\xi) - \tilde{U}^n(\xi)] + h^{-1}[\mu\tilde{d}(\xi) + h^{-2}\varepsilon\tilde{c}(\xi)]\frac{\tilde{U}^{n+1}(\xi) + \tilde{U}^n(\xi)}{2} = 0.$$

The solutions of these equations are

$$\tilde{U}(\xi, t) = \tilde{U}(\xi, 0)\exp[\frac{-\mathcal{Q}_h(\xi)t}{h}]; \tag{6.14}$$

and

$$\tilde{U}^n(\xi) = \tilde{U}(\xi, 0)[\lambda_h(\xi)]^n, \tag{6.15}$$

respectively, where

$$\mathcal{Q}_h(\xi) = \frac{\mu \widetilde{d}(\xi) + h^{-2}\varepsilon\widetilde{c}(\xi)}{\widetilde{a}(\xi)};$$

$$\lambda_h(\xi) = \frac{1 - (\frac{\Delta t}{2h})\mathcal{Q}_h(\xi)}{1 + (\frac{\Delta t}{2h})\mathcal{Q}_h(\xi)}.$$

Therefore, the solution $u_h(x,t)$, as given in (6.4), has Fourier transform

$$
\begin{aligned}
\widehat{u}_h(\xi,t) &= \widetilde{U}(h\xi,t)h^{1/2}\widehat{\Phi}(h\xi) \\
&= \exp[\frac{-\mathcal{Q}_h(h\xi)t}{h}]\widetilde{U}(h\xi,0)h^{1/2}\widehat{\Phi}(h\xi) \\
&= \exp[\frac{-\mathcal{Q}_h(h\xi)t}{h}]\mathcal{F}(\mathbf{R}u_0)(\xi).
\end{aligned}
$$

Thus $u_h(x,t)$ can be written as

$$u_h(x,t) = \mathbf{F}_h(t)\left[\mathbf{R}_h u_0(x)\right],$$

where

$$\mathbf{F}_h(t)v = \mathcal{F}^{-1}\left[\exp[\frac{-\mathcal{Q}_h(h\xi)t}{h}]\widehat{v}\right].$$

Note that $\mathbf{F}_h(t)v$ can be expressed in the discrete convolution form

$$\mathbf{F}_h(t)v(x) = \sum_k f_k(t/h)v(x - kh),$$

where $f_k(t/h)$ are the Fourier coefficients of the exponential $\exp[-\mathcal{Q}_h(\xi)t/h]$. Analogously, the discrete solution

$$u_h^n(x) = \sum_k U_k^n \Phi_{hk}(x) \tag{6.16}$$

has Fourier transform

$$\widehat{u}_h^n(\xi) = [\lambda_h(h\xi)]^n \mathcal{F}(\mathcal{R}_h u_0)(\xi),$$

and consequently, can be written as

$$u_h^n(x) = \mathbf{G}_h^n\left[\mathbf{R}_h u_0(x)\right],$$

where

$$\mathbf{G}_h^n v = \mathcal{F}^{-1}[(\lambda_h(h\xi))^n \widehat{v}].$$

It is clear from the above considerations that this method is stable if

$$Re(\mathcal{Q}_h(\xi)) \geq 0 \quad \text{for all real } \xi. \tag{6.17}$$

In this case, the method is called conservative if $Re(\mathcal{Q}_h(\xi)) = 0$ for all ξ, or dissipative if $Re(\mathcal{Q}_h(\xi)) > 0$ on some interval.

4.1 Convergence results

We shall study here the pointwise convergence of the approximate solutions $u_h(x,t)$ and $u_h^n(x)$ at the mesh points $x = hk$. In order to avoid errors due to the approximation of the initial data, we shall assume that $\mathbf{R}_h u_0$ interpolates u_0 at the mesh points. In general, in order to define an interpolation operator in terms of the translates and dilates of a function Φ, the following interpolatory condition is required.

$$\sum_k \Phi(k)e^{-ik\xi} = \sum_k \hat{\Phi}(\xi + 2k\pi) \neq 0 \quad \text{for all real } \xi. \tag{6.18}$$

Spline interpolation has been extensively studied in the literature and we refer to Schoenberg [22] and [23]. It is well known that the B-splines satisfy the interpolatory condition (6.18)for each even index $N = 2l$ (for each odd index $N = 2l+1$, a similar contidion is satisfied at the points $\frac{k}{2}$). Numerical calculations suggest that it is also verified by Daubechies' scaling functions of compact support (see Gomes and Cortina [12]).

Since the interpolation operator $\mathbf{R}_h$ comutes with $\mathbf{F}_h(t)$, as well as with $\mathbf{G}_h^n$, then the approximate solution $u_h(x,t)$ and the finite difference solution $\mathbf{F}_h(t)u_0(x,t)$ (respectively $u_h^n(x)$ and $\mathbf{G}_h^n u_0(x)$) coincide on the mesh points.

Theorem 4.1 *Assume $\Phi \in \mathcal{H}_{p,r}$, satisfying the interpolatory condition, and $\Phi^* \in \mathcal{H}_{p^*,r^*}$, where $r \geq 1$, and $r + r^* \geq 3$. Assume also the stability condition (6.17). Then, for smooth initial data u_0 and for every $T > 0$ there exists a constant $C > 0$, independent of h, Δt and u_0, such that for $0 \leq t \leq T$ and $0 \leq n\Delta t \leq T$,*

$$
\begin{aligned}
\|u(.,t) - u_h(.,t)\|_{2,h} &= \|u(.,t) - \mathbf{F}_h(t)u_0(.,t)\|_{2,h} \\
&\leq Ch^{p+p^*-1}\|u_0\|_{H^{p+p^*+2}}, \tag{6.19}
\end{aligned}
$$

$$\|u(.,n\Delta t) - u_h^n\|_{2,h} = \|u(.,n\Delta t) - \mathbf{G}_h^n u_0\|_{2,h}$$
$$\leq C\left(h^{p+p^*-1} + \Delta t^2\right)\|u_0\|_{H^{p+p^*+2}} \quad (6.20)$$

The next lemma will be used in the proof of this theorem.

Lemma 4.1 *Suppose that* $\Phi \in \mathcal{H}_{r,p}$ *and* $\Phi^* \in \mathcal{H}_{r^*,p^*}$. *For* $0 \leq \alpha \leq r$, *and* $0 \leq \beta \leq r^*$ *define the* (2π) *- periodic function*

$$\mathcal{G}_{\alpha,\beta}(\xi) = \sum_k I_{\alpha,\beta}(k)e^{-ik\xi},$$

where

$$I_{\alpha,\beta}(k) = \int_{-\infty}^{\infty} \frac{d^\alpha\Phi}{dx^\alpha}(x)\frac{d^\beta\Phi^*}{dx^\beta}(x-k)dx.$$

Then $\mathcal{G}_{\alpha,\beta}(\xi)$ *defines a* C^∞-*function and*

$$\mathcal{G}_{\alpha,\beta}(\xi) = i^{\alpha-\beta}\xi^{\alpha+\beta}\hat{\Phi}(\xi)\overline{\hat{\Phi}^*(\xi)} + O(\xi^{p+p^*+2}), \quad (6.21)$$

as $\xi \to 0$.

Proof: By Parseval's relation we have

$$I_{\alpha,\beta}(k) = 2\pi\int_{-\infty}^{\infty} \hat{f}(\xi)\overline{\hat{g}(\xi)}d\xi$$
$$= 2\pi\mathcal{F}[\hat{f}\hat{g}](-k),$$

where $f(x) = d^\alpha\Phi/dx^\alpha(x)$ and $g(x) = d^\beta\Phi^*/dx^\beta(x)$. Poisson summation formula then gives

$$\mathcal{G}_{\alpha,\beta}(\xi) = \sum_k \hat{f}(\xi+2k\pi)\overline{\hat{g}(\xi+2k\pi)}. \quad (6.22)$$

Using the result in Lemma 1 of Lemarié [16], we conclude that $\mathcal{G}_{\alpha,\beta}(\xi)$ is a C^∞-function. Note that, $\mathcal{F}[d^\alpha\Phi/dx^\alpha](\xi) = i^\alpha\xi^\alpha\hat{\Phi}(\xi)$ and $\mathcal{F}[d^\beta\Phi^*/dx^\beta](\xi) = i^\beta\xi^\beta\hat{\Phi}^*(\xi)$. Therefore,

$$\mathcal{G}_{\alpha,\beta}(\xi) = i^{\alpha-\beta}\sum_k (\xi+2k\pi)^{\alpha+\beta}\hat{\Phi}(\xi+2k\pi)\overline{\hat{\Phi}^*(\xi+2k\pi)} \quad (6.23)$$
$$= i^{\alpha-\beta}\left\{\xi^{\alpha+\beta}\hat{\Phi}(\xi)\overline{\hat{\Phi}^*(\xi)} + \mathcal{R}_{\alpha,\beta}(\xi)\right\},$$

where

$$\mathcal{R}_{\alpha,\beta}(\xi) = \sum_{k\neq 0}(\xi+2k\pi)^{\alpha+\beta}\hat{\Phi}(\xi+2k\pi)\overline{\hat{\Phi}^*(\xi+2k\pi)}. \quad (6.24)$$

Since all terms of this sum have a zero of order $p+p^*+2$, and $\mathcal{R}_{\alpha,\beta}(\xi)$ is a C^∞-function as well, then

$$\mathcal{R}_{\alpha,\beta}(\xi) = O(\xi^{p+p^*+2}), \quad \text{as } \xi \to 0.$$

Proof of Theorem 4.1: Since $\tilde{a}(\xi) = \mathcal{G}_{0,0}(\xi)$, $\tilde{d}(\xi) = \mathcal{G}_{1,0}(\xi)$, and $\tilde{c}(\xi) = (-1)^\beta \mathcal{G}_{\alpha,\beta}(\xi)$, one obtains from (6.24) that

$$\mathcal{Q}_h(\xi) = \frac{i\sum_k \left[\mu(\xi + 2k\pi) - \epsilon h^{-2}(\xi + 2k\pi)^3\right] \widehat{\Phi}(\xi + 2k\pi)\overline{\widehat{\Phi^*}(\xi + 2k\pi)}}{\sum_k \widehat{\Phi}(\xi + 2k\pi)\overline{\widehat{\Phi^*}(\xi + 2k\pi)}}.$$

$$(6.25)$$

Then we conclude from Lemma 4.1 that $\mathcal{Q}_h(\xi)$ satisfies

$$\mathcal{Q}_h(\xi) = \frac{i\left[\mu\xi - \epsilon h^{-2}\xi^3\right] \widehat{\Phi}(\xi)\overline{\widehat{\Phi^*}(\xi)} + O(\xi^{p+p^*+2}) + h^{-2}O(\xi^{p+p^*+2})}{\widehat{\Phi}(\xi)\overline{\widehat{\Phi^*}(\xi)} + O(\xi^{p+p^*+2})}.$$

$$(6.26)$$

Consequently

$$\mathcal{Q}_h(\xi) - i[\mu\xi - \epsilon h^{-2}\xi^3] = O(\xi^{p+p^*+2}) + h^{-2}O(\xi^{p+p^*+2}). \quad (6.27)$$

By virtue of this assymptotic expansion, the proof of the theorem can be carried out exactly as in Schoombie [24], Theorem 2.1 (see also Gomes and Cortina [12]). $\square$

Note that for sufficient small Δt, the estimates (6.19) and (6.20) give superconvergence results. In fact, if $p^* \gg 2$ then the pointwise order of accuracy $O(h^{p+p^*-1})$ at the nodes turns to be much better than $O(h^{p+1})$, the best order of approximation from V_h in the global L^2-norm.

Remark 4.1 Suppose that $\widehat{\Phi}(\xi)\overline{\widehat{\Phi^*}(\xi)}$ is a real and even function, and that $\alpha + \beta$ is an odd integer. Then the function $\mathcal{R}_{\alpha,\beta}(\xi)$ defined in (6.24) is a real and odd function. If in addition $p+p^*$ is even, then the assymptotic result (6.21) in Lemma 4.1 can be improved giving

$$\mathcal{G}_{\alpha,\beta}(\xi) = i^{\alpha-\beta}\xi^{\alpha+\beta}\widehat{\Phi}(\xi)\overline{\widehat{\Phi^*}(\xi)} + O(\xi^{p+p^*+3}).$$

Therefore, under these hypotheses one obains

$$\mathcal{Q}_h(\xi) - i[\mu\xi - \epsilon h^{-2}\xi^3] = O(\xi^{p+p^*+3}) + h^{-2}O(\xi^{p+p^*+3}), \quad (6.28)$$

which again, according to Schoombie [24], implies pointwise convergence estimates (6.19) and (6.20) of order $O(h^{p+p^*})$ and $O(h^{p+p^*} + \Delta t^2)$, respectively.

4.2 Conditions for stability

Let us discuss some hypotheses that guarantee the stability condition (6.17). Assume that

$$\hat{\Phi}(\xi)\overline{\hat{\Phi}^*(\xi)} \quad \text{is real for all real } \xi. \tag{6.29}$$

Then equation (6.25) implies that $Re\mathcal{Q}_h(\xi) = 0$, which means that the method is stable and conservative. For instance, this happens if

- $\Phi = \Phi^*$ (Galerkin method);

- Φ and Φ^* are symmetric around the same point.

Suppose now that the test functions are obtained from $\eta_\lambda(x) = \Phi^*(x - \lambda)$, a shifted version of Φ^*, for some $\lambda > 0$. Since $\widehat{\eta_\lambda}(\xi) = e^{-i\lambda\xi}\widehat{\Phi^*}(\xi)$, in this case the fomula (6.25) becomes

$$\mathcal{Q}_{h,\lambda}(\xi) = \frac{i\sum_k \left[\mu(\xi + 2k\pi) - \epsilon h^{-2}(\xi + 2k\pi)^3\right]\hat{\Phi}(\xi + 2k\pi)\overline{\hat{\Phi}^*(\xi + 2k\pi)}w_\lambda}{\sum_k \hat{\Phi}(\xi + 2k\pi)\overline{\hat{\Phi}^*(\xi + 2k\pi)}w_\lambda},$$

where $w_\lambda = e^{i2k\lambda\pi}$.

We can easily see that, for $\nu = 1 - \lambda$,

$$\mathcal{Q}_{h,\nu}(\xi) = -\overline{\mathcal{Q}_{h,\lambda}(\xi)},$$

which implies

$$Re(\mathcal{Q}_{h,\nu}(\xi)) = -Re(\mathcal{Q}_{h,\lambda}(\xi)).$$

This property led us to the following conclusions for Φ and Φ^* satisfying (6.29):

- For $\lambda = 1/2$, $Re(\mathcal{Q}_{h,\lambda}(\xi)) = 0$ for all real ξ, and the method is stable and conservative.

- If for some $0 < \lambda < 1$ the method is stable and conservative then this is also true for the shift parameter $1 - \lambda$.

- If for some $0 < \lambda < 1$ the method is stable and dissipative, then it is unstable for the shift parameter $1 - \lambda$.

We shall see in the next section that condition (6.29) holds when Φ and Φ^* are one of the pairs of dual spline scaling functions introducced in Cohen et al. [3].

5 Biorthogonal framework

Let us recall some concepts from wavelet analysis. For the purpose of this paper, the following brief summary will suffice. For more details we refer to Mallat [19], Meyer [20] and Daubechies [10].

In wavelet analysis, the name scaling function is associated to a function Φ that satisfies a scale relation

$$\Phi(x) = 2 \sum_{l \in \mathbf{Z}} h(l) \Phi(2x - l), \tag{6.30}$$

where the coefficients $h(l)$ form a sequence in ℓ^2. For a scaling function it is also required that for any integer m the family

$$\Phi_{m,k}(x) = 2^{m/2} \Phi(2^m x - k),$$

$k \in \mathbf{Z}$, is a Riesz basis which generates a subspace $\mathcal{V}_m \subset L^2(\mathbf{R})$. These subspaces form what is usually called a multiresolution analysis of $L^2(\mathbf{R})$. In Meyer [20], Chap.II.10, is proved that an r-regular scaling function satisfies the Strang-Fix condition for p at least equal to r.

In the Fourier domain the scaling relation can be written as

$$\hat{\Phi}(\xi) = H(\xi/2)\hat{\Phi}(\xi/2), \tag{6.31}$$

where

$$H(\xi) = \sum_{l \in \mathbf{Z}} h(l) e^{-il\xi}$$

is the symbol associated with the sequence h. $H(\xi)$ is also called the scaling filter for Φ. It then follows that

$$\hat{\Phi}(\xi) = \prod_{l=1}^{\infty} H(2^{-l}\xi).$$

It is clear that the filter coefficients $h(l)$ cannot be any arbitrary sequence. Typically $H(0) = 1$ and $H(\pi) = 0$, the order $p + 1$ of the zero of H at π giving the order p of the Strang-Fix condition satisfied by Φ.

Let Φ^* be another scaling function and $\mathcal{V}_m^*$ the associated subspaces. Φ and Φ^* are called conjugate (or dual) scaling functions if, for any fixed m, the families $\{\Phi_{m,k}\}$ and $\{\Phi_{m,k}^*\}$ satisfy

$$(\Phi_{m,k}, \Phi_{m,l}^*) = \delta_{kl}.$$

As a result, the corresponding scaling filters $H(\xi)$ and $H^*(\xi)$ are such that

$$H(\xi)\overline{H^*(\xi)} + H(\xi+\pi)\overline{H^*(\xi+\pi)} = 1 \quad \text{for all real } \xi.$$

Let the spaces $W_m = V_m \cap V_m^{*\perp}$ and $W_m^* = V_m^* \cap V_m^{\perp}$ be the complements, not necessarily orthogonal, of V_m in V_{m+1} and V_m^* in V_{m+1}^*, repectively. Here the wavelets arise. There exist functions Ψ and Ψ^* such that the families

$$\Psi_{m,k}(x) = 2^{m/2}\Psi(2^m x - k),$$

$$\Psi_{j,k}^*(x) = 2^{m/2}\Psi^*(2^m x - k)$$

are Riesz bases of W_m and W_m^*, respectively. These functions are called dual wavelets.

5.1   Spline biorthogonal scaling functions

A family of biorthogonal scaling functions $\{\Phi, \Phi^*\}$, based on the B-splines, was constructed in Cohen *et al.* [3]. For even $N = 2l$, $\Phi = \Phi_N$ is the symmetric B-spline centered on 0 such that

$$H(\xi) = \left(\cos\frac{\xi}{2}\right)^N. \tag{6.32}$$

For each $N^* = 2l^*$ there exist a conjugate scaling function $\Phi^* = \Phi_{N,N^*}$. The corresponding scaling filter $H^*(\xi)$ is given by

$$H^*(\xi) = \left(\cos\frac{\xi}{2}\right)^{N^*} \sum_{k=0}^{l+l^*-1} \binom{l+l^*-1+k}{k} \left(\sin^2\frac{\xi}{2}\right)^k. \tag{6.33}$$

Similarly, for odd $N = 2l + 1$, $\Phi = \Phi_N$ is the symmetric B-spline centered on $\frac{1}{2}$ such that

$$H(\xi) = e^{-i\xi/2}\left(\cos\frac{\xi}{2}\right)^N. \tag{6.34}$$

For each odd index $N^* = 2l^* + 1$ there exists a conjugate scaling function $\Phi^* = \Phi_{N,N^*}$ such that

$$H^*(\xi) = e^{-i\xi/2}\left(\cos\frac{\xi}{2}\right)^{N^*} \sum_{k=0}^{l+l^*} \binom{l+l^*+k}{k} \left(\sin^2\frac{\xi}{2}\right)^k. \tag{6.35}$$

The basic functions Φ, Ψ as well as their duals Φ^* and Ψ^* have compact support in all the cases. Φ and Ψ are C^{N-2} piecewise polynomials of degree $N - 1$. Φ^* and Ψ^* have increasing regularity with increasing N^* (see Cohen *el al* [3] and also Daubechies [10]).

From equations (6.32), (6.33), as well as (6.34) and (6.35), it follows that the order of the zeros of $H(\xi)$ and $H^*(\xi)$ at $\xi = \pi$ are N and N^*, respectively. This means that Φ_N and Φ_{N,N^*} satisfy Strang-Fix conditions of orden $p = N - 1$ and $p^* = N^* - 1$, respectively.

Note also that for all the possibilities of N and N^*, Φ_N and Φ_{N,N^*} are symmetric around the same point, namely $x = 0$ if N is even and $x = \frac{1}{2}$ if N is odd. Therefore, for biorthogonal spline bases, the assertion (6.29) is true. This implies that a Petrov-Galerkin method based on these families applied to the linear KdV equation is stable and conservative.

Since $\hat{\Phi}_N(\xi)\hat{\Phi}_{N,N^*}(\xi)$ is a real and even function, and $N + N^*$ is allways even, then for odd $\alpha + \beta$ the estimate (6.21) can be improved given $O(\xi^{p+p^*+3}) = O(\xi^{N+N^*+1})$ (see Remark 4.1).

After the above considerations, the following result for the Petrov-Galerkin method, based on biorthogonal splines, applied to the linear KdV equation, follows as a direct application of Theorem 4.1 (and also Remark 4.1) in the previous section.

Corollary 5.1 *Let* $\Phi = \Phi_N$, $\Phi^* = \Phi_{N,N^*}$ *be such that* $N + r^* - 1 \geq 3$, *where* r^* *is the regularity index of* Φ_{N,N^*}. *Then for regular initial data* u_0 *and every fixed* $T > 0$, *and* $0 \leq t \leq T$ *and* $0 \leq n\Delta t \leq T$,

$$\|u(.,t) - u_h(.,t)\|_{2,h} = O\left(h^{N+N^*-2}\right),\tag{6.36}$$

$$\|u(.,n\Delta t) - u_h^n\|_{2,h} = O\left(h^{N+N^*-2} + \Delta t^2\right),\tag{6.37}$$

as $h \to 0$.

5.2 Algorithms for the calculation of $a(k), b(l, k)$ and $c(k)$

Note first that the biorthogonality relation implies that

$$a(k) = \int_{-\infty}^{\infty} \Phi(y)\Phi^*(y - k)dy = \delta_{k0}.\tag{6.38}$$

A procedure to calculate the integrals (6.7) and (6.8), which involve derivatives of Φ and Φ^*, is described in [2] and [8]. We compute these integral in a different way, by using an important fact about biorthogonal multiresolution analyses: their compatibliliy with derivatives and primitives. Differenciating $\Phi(x)$ and integrating $\Phi^*(x)$ we obtain the following formulae

$$\frac{d\Phi}{dx}(x) = \tilde{\Phi}(x) - \tilde{\Phi}(x-1),\tag{6.39}$$

and

$$\tilde{\Phi}^*(x) = \int_x^{x+1}\Phi^*(y)dy,\tag{6.40}$$

where $\{\tilde{\Phi}, \tilde{\Phi}^*\}$ are also conjugate dual scaling functions. It was proved in Gomes et al. [13] that, if $\{\Phi, \Phi^*\}$ are two conjugate spline scaling functions $\Phi = \Phi_N$ and $\Phi^* = \Phi_{N,N^*}$, then the associated pair $\{\tilde{\Phi}, \tilde{\Phi}^*\}$ satisfying (6.39) and (6.40) is given by

$$\tilde{\Phi}(x) = \begin{cases} \Phi_{N-1}(x) & \text{if } N \text{ is odd} \\ \Phi_{N-1}(x+1) & \text{if } N \text{ is even} \end{cases}$$

and

$$\tilde{\Phi}^*(x) = \begin{cases} \Phi_{N-1,N^*+1}(x) & \text{if } N \text{ is odd} \\ \Phi_{N-1,N^*+1}(x+1) & \text{if } N \text{ is even} \end{cases}$$

This property can be used to perform easy calculations of the integrals (6.7) and (6.8), that will be illustrated here for the particular case of the hat function

$$\Phi(x) = \Phi_2(x) = \begin{cases} x+1 & \text{if } -1 \le x < 0 \\ x-1 & \text{if } 0 \le x \le 1 \\ 0 & \text{elsewhere} \end{cases},$$

and its duals $\Phi^* = \Phi_{2,N^*}, N^* \ge 4$, which are supported on the intervals $[-N^*, N^*]$. From the above relations, it follows that

$$\Phi_{2,N^*}(x) = \int_x^{x+1}\Phi_{3,N^*-1}(y)dy,$$

which means that

$$\frac{d}{dx}\Phi_{2,N^*}(x) = \Phi_{3,N^*-1}(x+1) - \Phi_{3,N^*-1}(x).\tag{6.41}$$

It is also true that

$$\int_x^{x+1}\Phi_{2,N^*}(y)dy = \Phi_{1,N^*+1}(x+1).\tag{6.42}$$

According to our notation of Section 2, $\Phi_2 \in \mathcal{H}_{1,1}$. For sufficiently high values of N^*, the regularity parameter r^* of Φ_{2,N^*} increases, and eventually it will be greater than 2, satisfying the regularity requirements $(r + r^* \geq 3)$ for the application of the Petrov-Galerkin method. However, for small values of N^*, e.g. $N^* = 4$, $r^* < 2$. In fact, since $\Phi_{3,3} \in L^2(\mathbf{R})$ (cf. Cohen *et al.* [3]), and certainly does not have bounded derivatives, then we can only assure that $\Phi_{2,4} \in \mathcal{H}_{1,3}$. However, since $d^2\Phi_2/dx^2$ is a generalized function which can be expressed as

$$\frac{d^2}{dx^2}\Phi_2(x) = \delta(x+1) - 2\delta(x) + \delta(x-1), \qquad (6.43)$$

then the weak formulation (6.3) and the resulting system (6.5) still make sense with $\alpha = 2$, provided that the basic test function $\Phi^* = \Phi_{2,N^*}$ has continuous derivatives at the integers. In Lemma 4.1, the assymptotic behavior $O(h^{N^*+3})$ is still valid in this case if we take $\alpha = 2$ and $\beta = 1$ which, in turn, implies an accuracy of order $O(h^{N^*})$ of the approximate solutions for the linear case.

From equations (6.43), (6.41) and (6.42), we derive the following expressions for $c(k)$ and $b(l,k)$:

$$c(k) = \Phi_{3,N^*-1}(-1-k) - 3\Phi_{3,N^*-1}(-k) + 3\Phi_{3,N^*}(1-k) - \Phi_{3,N^*-1}(2-k),$$

and

$$b(l,k) = \begin{cases} 0 & \text{if } l \leq -2 \\ & \text{or } l \geq 2 \\[2mm] -I(-k-1) - k\Phi_{1,N^*+1}(-k) & \text{if } l = -1 \\[2mm] I(-k-1) + I(-k) + (k+1)\Phi_{1,N^*+1}(-k) + \\ +(k-1)\Phi_{1,N^*+1}(-k+1) & \text{if } l = 0 \\[2mm] -I(-k) - k\Phi_{1,N^*+1}(-k+1) & \text{if } l = 1, \end{cases}$$

where

$$I(k) = \int_k^{k+1} y\Phi_{2,N^*}(y)dy.$$

Due to the symmetry of $\Phi_{2,N^*}(x)$ around $x = 0$, which implies that $I(k) = -I(-k-1)$, and the symmetry of $\Phi_{1,N^*+1}(x)$ around $x = 1/2$, it follows that

$$b(-1,k) = -b(1,-k), \quad \text{and} \quad b(0,k) = -b(0,-k).$$

The ortogonality equation (6.38) applied to $\Phi = \Phi_2$ and $\Phi^* = \Phi_{2,N^*}$ gives the following relation

$$I(-k-1)-I(-k)+(k+1)\Phi_{1,N^*+1}(-k)+(1-k)\Phi_{1,N^*+1}(1-k) = \delta_{k0}.$$

This formula can be used to simplify the expression of $b(0,k)$ giving

$$b(0,k) = \delta_{0k} + 2I(-k) - 2(1-k)\Phi_{1,N^*+1}(k).$$

It also gives a straigtforward algorithm for the calculation of $I(k)$, as follows.

- $I(0) = \Phi_{1,N^*+1}(0) - 1/2.$

- For $k = 1, 2, \cdots, N^* - 1$

  $I(-k) = -I(k-1)$

  $I(k) = -I(-k) + (k+1)\Phi_{1,N^*+1}(-k) + (1-k)\Phi_{1,N^*+1}(k).$

- $I(-N^*) = -I(N^* - 1).$

To perform the above calculations it is necessary to have the values of Φ_{1,N^*+1} and Φ_{3,N^*-1} at the integers. In general, in order to compute the values of a scaling function of compact support at the integers, one can use the algorithm proposed by Strang [25] that begins with the dilation equation (6.30) and solves an eigenvalue problem for the matrix whose entries are $h(2l - k)$.

6   Numerical results

In order to evaluate the accuracy of the method we have chosen $\mu = 1, \varepsilon = 0.000484$ and

$$u_0 = 3c \; \cosh^{-2}(kx + d) \tag{6.44}$$

as the initial condition, where $c = 0.3$, $k = (c/4\varepsilon)^{1/2}$ and $d = -k$. In this case, the analytical solution of problem (6.1) is

$$u(x,t) = 3c \; \cosh^{-2}(kx - kct + d), \tag{6.45}$$

and it corresponds to a single soliton with amplitude 0.9. Over a time range $0 \leq t \leq 1$ the solution values are assumed to be zero outside the interval $(x_a, x_b) = (0, 2)$.

In the example presented here we have used the compactly supported spline biorthogonal scaling functions constructed in [3].

Numerical results obtained using $\Phi = \Phi_2$, $\Phi^* = \Phi_{2,4}$, $h = 2^j$, $j = -4, -5$ and -6, and an initial numerical scheme $R_h u_0$ that interpolates the initial data at the mesh points, are displayed in Table 6.1. The difference between the theoretical solution (6.45) and the numerical solution (6.16) is given in terms of the discrete ℓ_2 norm.

t	h	Error x 10^3
0.25	0.0625	28.35
0.50		30.50
0.75		32.19
1.00		33.01
0.25	0.03125	8.53
0.50		9.18
0.75		9.32
1.00		9.17
0.25	0.015625	1.08
0.50		1.35
0.75		1.57
1.00		1.86

TABLE 6.1.

7   Acknowledgements

The work of S . Gomes was partially supported by CNPq and FAPESP, Brasil. E. Cortina is Research Fellow of the CONICET–Argentina. Her work was also partially supported by the European Comission contract/Grant Nr. CTI* CT 91-0944.

8   References

[1] E. Bacry, S. Mallat and G. Papanicolaou. A Wavelet Based Space-Time Adaptive Numerical Mathod for Partial Differential Equations. *Num. Model. Math. Anal.* M^{2}AN 26:793–834, 1992.

[2] G. Beylkin. On the representation of operators in bases of compactly supported wavelets *SIAM J. Numer. Anal.* **29** (6): 1716-1740, 1992.

[3] A. Cohen, I. Daubechies and J.-C Feauveau. Biorthogonal bases of compactly supported wavelets. *Communications in Pure and Applied Mathematics*, **45**: 485-560, 1992.

[4] E. Cortina and S. Gomes. The wavelet-Galerkin method and the KdV equation. Instituto Argentino de Matemática, preprint 218. Buenos Aires, 1994.

[5] C. Cunha and S. Gomes. A high resolution method based on biorthogonal wavelets for the numerical simulation of conservation laws. *Proceedings of the XV Congresso Ibero Latino Americano sobre Métodos Computacionais para Engenharia, Belo Horizonte, December 1994*, 234–243.

[6] S. Dahlke and A. Kunoth. A biorthogonal wavelet approach for solving boundary value problems. Preprint. 1993.

[7] S. Dahlke and I. Weinreich. Wavelet-Galerkin methods: an adapted biorthogonal wavelet basis. *Constructive Approximations* **9**(2): 237-262, 1993.

[8] W. Dahmen and C. A. Michelli. Using the refinement equation for evaluating integrals of wavelets. SIAM Jr. Numer. Anal. 30 (2) 507-577, 1993.

[9] I. Daubechies. Orthonormal bases of compactly supported wavelets. *Communications in Pure and applied Mathematics*, **41**: 909-996, 1988.

[10] I. Daubechies. " Ten Lectures on wavelets". CBMS Lecture Notes, No. 61, SIAM, Philadelphia, 1992.

[11] R. Glowinski; W. Lawton; M. Ravachol and E. Tenembaum. Wavelet solution of Linear and Nonlinear elliptic, parabolic and hyperbolic problems in one dimension. AWARE, INC. AD890527.1, 1989.

[12] S. Gomes and E. Cortina. Convergence estimates for the wavelet-Galerkin method. *SIAM Jr. Numer. Anal.* 33 (1): 149 – 161, 1996.

[13] S. Gomes, E. Cortina and I. Moroz. Characterization of biorthogonal spline wavelets by means of derivatives and primitives. In: Approximation Theory VIII. C. K. Chui, L. L. Schumaker eds. World Scientific Publishing Co., Inc., p. 125–132, 1995.

[14] S. Jaffard and Ph. Laurençot. Wavelets and PDEs. In: Wavelets: A tutorial in theory and Applications. Edited by Ch. Chui. Academic Press, San Diego, 1992.

[15] A. Latto and E. Tenembaum. Compactly supported wavelets and the numerical solution of the Burgers' equation. *C.R. Acad. Sci. Paris* **311** Série I: 903-909, 1990.

[16] P.-G Lemarié, P. G. Functions a support compact dans les analyses multi-résolutions. *Revista Matemática Iberoamericana* **7** (2): 157-182, 1991.

[17] J. Liandrat; V. Perrier and Ph. Tchatmitchian. Numerical resolution of the regularized Burgers equation using the wavelet transform. In:*Wavelets and applications*, Y. Meyer ed. RMA, Masson, p 420–433, 1992.

[18] Y. Maday, V. Perrier and J. C. Ravel. Adaptivité dynamique sur bases d'ondelettes pour l'approximation d'équations aux dérivées partielle. *C. R. Acad. Sci. Paris* **312** Série I: 405-410, 1991.

[19] S. Mallat. Multiresolution approximations and wavelet orthonormal bases. *Trans. of the American Mathematical Society* **315**:69–87, 1989.

[20] Y. Meyer. "Ondelettes et Operateurs I". Hermann, Paris, 1990.

[21] J. M. Sanz-Serna and I. Christie. Petrov-Gelerkin methods for non-linear dispersive waves. *Journal of Computational Physics*, **39**: 94-102, 1979.

[22] I. J. Schoenberg. Cardinal interpolation and spline functions. *Journal of Approximation Theory*, **2**: 167-206, 1969.

[23] I. J. Schoenberg. Cardinal interpolation and spline functions II: interpolation of data of power growth. *Journal of Approximation Theory*, **6**: 404-420, 1972.

[24] S. W. Schoombie. Spline Petrov-Galerkin methods for the numerical solution of the Korteveg-de Vries equation. *IMA Journal of Numerical Analysis*, **2**: 95-109, 1982.

[25] G. Strang. Wavelets and dilation equations: a brief introduction. *SIAM Review*, **31** (4): 614-627, 1989.

[26] G. Strang; G. A. Fix. A Fourier analysis of the finite element method. In: Constructive Aspects of Functional Analysis. Edizioni Cremonese, Roma, 1973.

[27] V. Thomée. Convergence estimates for semi-discrete Galerkin methods for initial value-problems. *Lecture Notes in Mathematics*. **333**: 243-262, 1973.

Part II

Applications to Biomedical Sciences

Chapter 7

Fine Structure of ECG Signal using Wavelet Transform

Hervé Rix
Olivier Meste

1 Introduction

Since the two last decades, due to the development of electronic devices and computers, the observation of the fine structure of the electric cardiac signal through non-invasive measurements is more and more a realistic challenge. The technique allowing such measurements is called High Resolution Electrocardiography (HRECG) to make a difference with the classical Electrocardiography (ECG). In classical ECG, the paper records obtained from standardized amplificators, analogic filters and leads configurations enable a rapid observation of the macroscopic behaviour of heart activity: heart rate, rhythm, conduction intervals, morphological aspect of the ECG waves (P, Q, R, S, T) associated to the depolarization and repolarization of the auricles and ventricles. The dynamic of the input signal is a few millivolts.

The aim of HRECG is obtain an amplitude resolution of one microvolt with numerical data, in such a way that signal processing tools could be applied. The major drawback of this technique is the presence, at this resolution, of high level noise mainly due to muscle activity. In order to improve the Signal to Noise Ratio (SNR) the most popular method is signal averaging. The idea is to use the repetitive nature of the cardiac cycles to enhance the permanent waveforms (P, QRS, T) and cancel the random noise and other components which would not be synchronous with the heart rate. This technique gives good results when the basic hypotheses are verified that is when the signals to average are really repetitive and when the alignment process is very precise. Historically, the first objective was the observation of the His signal [1] which occurs between the P and Q waves and reflects the conduction from the auricles to the ventricles. Its observation by catheters is easy, and noise! free ... but rather invasive! Infortunately, even if observations of His signals have been published by several teams, the results were not sufficiently reproducible nor readable for physicians to lead to a routine method. The main reason is that the alignment process is synchronized on the QRS complex (where SNR is the highest) and that the time distance between the His signal and the fiducial point of the averaging process is not always constant.

The success of HRECG is mainly due to its ability to detect Ventricular Late Potentials [2]. These micropotentials (a few microvolts) appear, by averaging the ST segment, just after the QRS complex. In fact, the late potentials are a particular case of Delayed Inhomogeneous Depolarizations (DID) [3] which may occur elsewhere, for example under the QRS or after and under the P wave. The presence of late potentials (LP) and more particularly Ventricular Late Potentials (VLP) increases the probability of sudden death [4]. So any technique allowing their detection and their analysis is a precious help for the physicians. A typical way to analyze cardiac signals which may include such micropotentials is to take advantage of their non stationary character through a Time-Frequency analysis, and more particularly using Wavelet Transforms (WT).

As pointed out by Unser and Aldroubi [5], the prefered type of WT for signal analysis is the redundant one that is the Continuous Wavelet Transforms (CWT), in opposition to the non redundant type corresponding to the expansion on orthogonal or biorthogonal bases.

In this article we deal with two properties of the WT and their relevance for fine structure analysis of the human heart electric activity. The first property is the ability of WTs to localize events in the Time-Frequency plane, in the case of non-stationary signals. The second one is the invariance of WTs concerning the group of affine functions. This property is exploited to provide means of measuring shape variations of ECG waves, particularly in relation to the action of a drug.

2   Localization of Late Potentials by Time-Frequency Representations.

The detection and analysis of ventricular Late Potentials (VLP) sus-cited a lot of works, but only a few percentage is, at present, used in routine protocols by clinicians. The more recent attempt to stan-dardize VLP lead only to standardizing the acquisition protocol [6]. The most common algorithm implemented on available devices is due to Simson [4]. Based on signal averaging of the mean energy obtained from the three orthogonal derivations X, Y, Z (vector mag-nitude) and on a special filtering technique suited to the post QRS segment, the algorithm gives a positive answer (VLP detected) when two among three criterions are satisfied. This algorithm gives useful informations but its probability of bad detection remains too high. So a universally recognized standardization has not yet arrived [3]. The first features to extract for the physician are the amplitude and the localization both in time and frequency. It is well known that the Morlet WT gives the best compromise between time and frequency resolutions. Let us recall the reason of this property.
If we note $O_x(t, a)$ the WT of signal $x(t)$, where t is the time variable and a the scale variable:

$$O_x(t, a) = \frac{1}{\sqrt{a}} \int x(u) g^*(\frac{u - t}{a}) du \qquad (7.1)$$

where $g^*(t)$ is the conjugate complex of the mother wavelet $g(t)$ defined by:

$$g(t) = e^{i2\pi f_0 t} e^{-\frac{t^2}{2}} \qquad (7.2)$$

$O_x(t, a)$ can be written:

$$\begin{aligned} O_x(t, a) &= \frac{1}{\sqrt{a}} e^{i2\pi f_0 t} \int x(u) e^{-\frac{(u-t)^2}{2a^2}} e^{-i\frac{2\pi f_0}{a} u} du \\ &= e^{i2\pi f_0 t} \frac{1}{\sqrt{a}} \mathcal{F}[x w_a(t)] \left(\frac{f_0}{a} \right) \end{aligned} \qquad (7.3)$$

where $\mathcal{F}[x w_a(t)] \left(\frac{f_0}{a} \right)$ is the Fourier Transform of signal $x(u)$ multi-plied by the gaussian window centered in t and with standard devi-ation equal to a.
So the modulus of $O_x(t, a)$ is proportional to the modulus of this

Fourier Transform. The value of f_0 is chosen so as to verify the admissibility condition for g:

$$\int g(t)dt \approx 0 \quad or \quad \mathcal{F}(g)(0) \approx 0 \qquad (7.4)$$

The frequency shift f_0 produced by the modulation function $e^{i2\pi f_0 t}$, must be compared to the bandwidth of the gaussian spectrum of the mother wavelet. As defined in (7.2) the bandwidth is $\frac{1}{2\pi}$ in frequency. So, given a value of f_0, the condition (7.4) can be reached by scaling the gaussian of (7.2), that is introducing the new mother wavelet:

$$g_\sigma(t) = e^{i2\pi f_0 t} e^{-\frac{t^2}{2\sigma^2}} \qquad (7.5)$$

and the new WT:

$$
\begin{aligned}
O_{\sigma,x}(t,a) &= \frac{1}{\sqrt{a}} e^{i2\pi f_0 t} \int x(u) e^{-\frac{(u-t)^2}{2\sigma^2 a^2}} e^{-i\frac{2\pi f_0}{a} u} du \\
&= e^{i2\pi f_0 t} \frac{1}{\sqrt{a}} \mathcal{F}\left[x w_{\sigma,a}(t)\right] \left(\frac{f_0}{a}\right)
\end{aligned} \qquad (7.6)
$$

Large value of σ leads to large width temporal windows allowing a good resolution in the frequency domain. On the contrary small values of σ make it possible to track signal variations in the time domain to the prejudice of frequency resolution. It is shown in [7] how to take advantage of this property by multiplying the two modulus of $O_{\sigma_1,x}$ and $O_{\sigma_2,x}$ corresponding to the enhancement respectively in time (σ_1, smaller value) and in frequency (σ_2, larger value). Practically the sampling rate and the admissibility condition (7.5) imposes to σ_1 not to be too small: the bandwidth of gaussian spectrum is proportional to the inverse of σ. On the other hand, σ_2 cannot be too large in order to take into account the non stationary character of the involved signal. We clearly show, on a simulation example [7] the improvement of what we called the Modified Wavelet Transform (MWT) defined by:

$$O_{\sigma_1,\sigma_2,x}(t,a) = |O_{\sigma_1,x}(t,a)| \; |O_{\sigma_2,x}(t,a)| \qquad (7.7)$$

The application of this MWT to the comparison of healthy subjects and patients with tachycardia diagnoses gave Time-Frequency Representation (TFR) (some of them have been reproduced in the recent Biomedical handbook [8]). They put in light two major conclusions:

1. It's possible, by this technique to have reproducible representation of the beat to beat activity

2. The VLP cases differ from the healthy ones by the irregular and fragmented aspect of the contour plots, both under and just after the QRS complex. This conclusion is in good agreement with the "spectral turbulences" described in [9].

3   Signal Shape Differences

The first objective of Time-Frequency or Time-scale analysis was to provide 2D or 3D representations where some particular features could be seen. In a second step the problem is to give quantitative evaluations of features and of differences between representations. More generally, the question is how can Time-Frequency or Time-Scale Representations (obtained for example by a WT) be a tool for decision making ?

An application of this approach to ECG signals can be found in the work of Oficjalska [10] and Oficjalska and al. [11]. The basic practical problem was to provide an efficient tool to detect and measure (little) shape variations along time, of QRS, P and T waves. The first application we had in mind was the evaluation of the action of drug on cardiac activity through the measuring of changes in waves morphology.

In fact, studying ECG variations involves the measure of interval lengths, cardiac wave durations but also shape variations of these waves independently of shifts or scale changes. So it is necessary to define shape equality and to build a similarity criterion allowing to use a clustering algorithm to make shape classification. This will be developed in the following, first for 1D signals and, what is rather new, for 2D signals with a particular interest to TFRs obtained by WTs.

3.1   Shape classifications of 1-D signals

The classical definition of shape invariance refers to the affine group that is $s(t)$ and $v(t)$ are equal shape signals if we can write a relation of the form:

$$v(t) = ks(\frac{t-d}{a}) \;;\; a > 0 \qquad (7.8)$$

In the case of positive signals, the difference in shape is provided by the Distribution Function Method (DFM) [12]. Working on the normalized integrals $S(t)$ and $V(t)$, which are increasing distribution functions (since s and v are positive) (7.8) implies:

$$V(t) = S(\frac{t - d}{a}) \qquad (7.9)$$

So the affine function $t' = \frac{t-d}{a}$ appears when we plot t' in function of t, where t' is defined by:

$$V(t) = S(t') \qquad (7.10)$$

If the signals s and v may have different shapes, the function $\varphi(t)$ defined by:

$$V(t) = S(\varphi(t)) \qquad (7.11)$$

characterizes the shape variation and any criterion measuring the departure of $\varphi(t)$ from an affine function is a similarity criterion between s and v shapes. Several examples of such a criterion are given in [13] where applications of DFM to analytical chemistry are discussed. An application to signal averaging of HRECG has been presented in [14].

The extension to signals depending on two variables leads to what we called the Distribution Function in 2 dimension (DFM-2D). Such signals are for example images or, what is more connected with our purpose, TFRs.

3.2   Shape classification of 2-D signals

Let $f(x,y)$ and $g(x,y)$ be two positive signals integrables on $\mathbb{R}^2$. The equality of two shapes is defined by the existence of a relation of the form:

$$f(x,y) = kg\left(\frac{x - x_0}{\alpha}, \frac{y - y_0}{\beta}\right) \qquad (7.12)$$

In the general case, for example when f and g represent the intensities of two images, α and β are independent scale parameters. The shape identity notion can be restricted by linking α and β, e.g. imposing $\alpha = \beta$.

Measuring shape differences makes sue of the 2-D normalized integrals:

$$F(x,y) = \frac{\int_{-\infty}^{x} \int_{-\infty}^{y} f(u, v)dvdu}{\int_{-\infty}^{\infty} \int_{-\infty}^{\infty} f(u, v)dvdu} \qquad (7.13)$$

and a similar formula for $G(x,y)$. The sections at a fixed value of z give curves whose equations can be written as decreasing functions: $y = C_z(x)$ for F and $y = D_z(x)$ for G.

When signals $f(x,y)$ and $g(x,y)$ have the same shape, C_z and D_z are linked by:

$$C_z(x) = \beta D_z \left(\frac{x - x_0}{\alpha} \right) + y_0 \qquad (7.14)$$

Position parameters x_0 and y_0 are eliminated by making the two mean points to coincide, i.e. assuming $x_0 = y_0 = 0$. In the case where f and g are any two signals, measuring a shape difference needs the estimation of the scale parameter values α^* and β^* which give the best fitting between C_z^*, where

$$C_z^*(x) = \beta^* D_z \left(\frac{x}{\alpha^*} \right) \qquad (7.15)$$

and $C_z(x)$, according to the least mean square criterion. Then we have to compute the mean residual associated to each z. The "shape distance" $\Gamma(\alpha^*, \beta^*)$ is obtained by averaging these residuals for a series of z values going from 0 to 1. In the same way as for the DFM-1D, measuring a shape variation in the case of noisy signals needs a statistical determination of a threshold depending on the signal to noise ratio. Beyond this threshold a shape difference is significan t.

A particularly interesting case is that of 1-D signals which can be compared by DFM-2D using the modulus of any TFRs. Then α and β (scale parameters on time and frequency respectively) are linked by: $\beta = \frac{1}{\alpha}$. In this case the shape distance Γ is a function of only one parameter: α^*.

In [11] the presented results were obtained using the spectrogram in order to use the most popular tool for TFRs in biomedical applications. Other TFRs have been used in [10]. Here we would want to analyze the particularity of WT among TFRs.

If we want to detect and measure shape variations of TFRs which characterize shape variations of the initial 1-D signals, it is important to check that two equal shape 1-D signals lead to two equal shape 2-D TFRs. If we note $T_s(t, \nu)$ a particular TFR of signal s, and if v is another signal with the same shape, we must have the property:

$$T_v(t, \nu) = K T_s(\varphi_1(t), \varphi_2(\nu)) \qquad (7.16)$$

Where K is a constant coefficient, φ_1 and φ_2 two affine functions. This is the case when T is the Wigner-Ville transform. But this transform is very difficult to use in application like ours due to the presence of ghost frequencies produced by the cross-terms. In fact, these cross-terms are not a drawback for shape analysis but the physician needs to check and interpret any shape variation on an easy interpretable representation. To attenuate the influence of these interferences, the Cohen's class transforms make use of smoothing windows in the time domain (like the pseu do Wigner-Ville transform) or both in time and frequency domains (like the spectrogram) [15],[16]. For these transforms, (7.16) is not true. If we want to use such transforms, e.g. the spectrogram, it is necessary to adapt the width of the window to the width of the signal. This adaptation needs an estimation of the signal widths. In our application [11] this estimation was performed using the DFM-1D: the slope of the least mean square line fitted on function φ is an estimation of the width ratio.

This problem of adjusting the window to the signal is naturally resolved by the WT since the principle is to adapt the window (or filter) width to the frequency. In fact, assuming $v(t) = ks(\frac{t-t_0}{\alpha})$ with $k, \alpha > 0$, and noting $WT_s(t, \nu)$ the WT of signal s, we can write:

$$WT_v(t, \nu) = k^2 \alpha WT_s \left(\frac{t - t_0}{\alpha}, \alpha \nu \right) \qquad (7.17)$$

This relation has the form of (7.12) where $\beta = \frac{1}{\alpha}$.

The computation of the residual $\Gamma(\alpha^*)$ is a measure of the shape difference. In many applications we need to know if besides this shape difference exits or not a difference of the signal widths. We can, with the same algorithms, take into account "shape and width" differences, putting $\alpha = 1$.

For example, in measuring the action of drugs on ECG waves [10] it has been shown that:

(i) the action of Quinidine on the T-waves results in important width variations but only little shape variations: so any index of the broadening of the T-waves remains a good indicator of the drug action.

(ii) the action of Cibenzoline on the P-waves results in shape and width variations, where shape variations are preponderant.

The general conclusion of this work was that DFM-2D associated to TFR of 1D-signals, like WTs, give results correlated to the results obtained directly by DFM-1D, but with a higher sensitivity to small variations leading to better separated classes.

4   Conclusion

We have presented two applications of Continuous Wavelet Transform to the ECG signal. The first application was an attempt to make more evident the presence of Ventricular Late Potentials both after and under the QRS complex. The modified WT proposed in [7] gives images in the Time-frequency plane which can help the physician to detect the presence of VLPs, even when this detection is not possible from classical averaging of filtered temporal ECG derivations. At present this help comes from visual inspec tion of Time-Frequency images, contour plots for example, comparing a lot of healthy subjects and subjects with a Ventricular Tachycardia diagnosis. The comparison could also be done between records from the same patient before and after some critical event.

The further step is to quantify the differences between normal and pathological cases with the goal of building an automatic classification algorithm. The second application of Continuous WTs we have presented here could be a way to achieve such a classification. In fact, measuring shape differences between two images, and particularly two TFRs given by WTs, has been applied to the action of drug on heart activity. But the good results obtained on cases where visual inspection was quite inefficient allow to hope good results in our present problem. This hope is strengthened when we remark that a Late Potential is often observed as a narrow band component added to the ECG signal. The shape differences due to the addition of another component are typically the differences that our algorithms are able to detect.

In conclusion we feel that further works may be expected, combining the ability of continuous Wavelet Transforms to built images enhancing the localization of events in the Time-Frequency plane and methods which globally detect and measure shape variations of these images, independently of affine transformations of the coordinates.

5   References

[1] Varenne A. (1980) *High Amplification Electrocardiography, Vidal J. M. (eds.), Crenaf, Nice (France)*.(1980)

[2] Berbari E. J. *Critical Overview of late potentials recordings, J. Electrocardiology*, 125-127.(1987)

[3] Varenne A., Rix H. and Oficjalska B. *Late potentials: is it time to standardize, Int. Cong. on Electrocardiology, Warsaw, Poland.*(1991)

[4] Simson M. B., Falcone M. D. and Falcone R. A. *Late potentials, a non-invasive marker for ventricular tachycadia, 14th ICMBE and 6th ICMP, Espoo, finland,* 1464-1467.(1985)

[5] Unser M. and Aldroubi A. *A review of wavelets in biomedical applications, Proc. IEEE* **84**, no. 4, 626-638. (1996)

[6] Breithardt G., Cain M. F., El-sherif N., Flowers N., Hombach V., Janse M., Simson M. B. and Steinbeck G. *Standards for analysis of ventricular late potentials using high resolution or signal averaged electrocardiography, European Heart J.* **12**, 473-480. (1991)

[7] Meste O., Rix H., Caminal P., Thakor N. V. *Ventricular late potentials characterization in time-frequency domain by means of a wavelet transform, IEEE Trans. Biomed. Eng.* **41**, no. 7, 625-634. (1994)

[8] *The Biomedical Engineering Handbook* Bronzino J. D. *(ed.), CRC Press,* pp. 886-906. (1995)

[9] Henkin R. *Cardiac risk assesment from Holter ECG data, Del Mar monograph series* **2179**. (1990)

[10] Oficjalska B. *Morphologie de l'onde P du signal ECG. Analyse de forme des signaux bidimensionnels: mesure d'effets pharmacologiques sur les ondes P, QRS et T en représentation temps-fréquence. Thesis dissertation, Univ. of Nice.* (1994)

[11] Oficjalska B., Rix H., Chevalier E., Fayn J., Varenne A. *Measuring shape variations of ECG waves through time-frequency representations, Signal Processing VII, Theories ans Applications.* M. J. J. Holt, C. F. N. Cowan, P. M. Grant and W. A. Sandham *(eds.), Eurasip,* 70-73. (1994)

[12] Rix H., and Malengé J. P. *Detecting small variations in shape, IEEE Trans. on Syst. Man. and Cybern.* **10**, no. 2, 90-96. (1980)

[13] Rix H. *Le traitement des formes en chimie analytique, Anal. Chimica Acta* **191**, 467-472. (1986)

[14] Jesus S., Rix H. and Varenne A. *Signal averaging using shape classification, Proc. of EUSIPCO*, 1371-1374. (1986)

[15] Jeong J. and Williams W. J. *Kernel design for reduced interferences distributions, IEEE Trans. Signal Processing* **40**, no. 2, 402-412. (1992)

[16] Choi H. and Williams W. J. *Improved time-frequency representation of multicomponent signals using exponential kernels, IEEE Trans. Acoust., Speech, Signal Processing* **37**, no. 6, 862-871.(1989)

Chapter 8

Spectral Analysis of Cardiorespiratory Signals

Marcelo R. Risk
Jamil F. Sobh
Ricardo L. Armentano
Agustín J. Ramírez
J. Philip Saul

1   Introduction

The study of the rhythmic and nonrhythmic oscillations of the arterial blood pressure (ABP) was first described by Hales [19] two centuries ago. Twenty seven years later, Albrecht von Haller described fluctuations of the cardiac rhythm. In 1847, Carl Ludwig [31], by mean of continuous recordings of physiological events in horses and dogs, was able to graph the rhythmic fluctuations of the ABP. The motivation behind these experiments was to clarify the spontaneous behavior and to overcome the lack of interpretation for these oscillations. It is interesting to remind the first description relating an evident correlation with respiratory fluctuations, mainly because the ease of its visualization, both in laboratory animals and humans beings.

The interpretation of these oscillations began with Carl Ludwig [31], who proposed that the oscillations of ABP coincided with respiration. These oscillations could be originated from compression and stretching of the vessels caused by changes in the intrathoracic pressure during expiration and inspiration. Thus suggesting that the collateral effects overcame the normal conditions of circulation. However, the systematic analysis of mechanical influences of respiration on circulation began in the middle of XIX century with Donders [15] in 1853, and Einbrodt [16] in collaboration with Ludwig in 1860. The studies made then demonstrated the misunderstanding of the of the respiratory influences on ABP. They demonstrated that a 5 mmHg change in the magnitude of intrathoracic pressure is small compare to the magnitude of respiratory oscillations modulating ABP, which normally reaches values two to four times greater than intrathoracic pressure. After that no information of this subject was reported until the middle of this century in which it was demonstrated that the mechanical effects of the low pressure system are more likely related to the production of ABP fluctuations, than the mechanical effects on the intrathoracic vessels. However, it remained still in doubt whether this respiratory action on the ABP was a direct interaction originating in the central nervous system (between the nuclei responsible ABP and/or respiratory control), or if it is an entraining effect induced by interaction between the activity of an oscillator (respiration) with the activity developed by a signal generator (nucleus responsible of vascular tone) [50] [51]. This last proposal suggested that both centers, in normal oscillatory activity, compatibilize the discharge frequencies to hold the equilibrium of the controlling system. This last concept was verified by several authors [25] [26] who demonstrated that the oscillatory components of ABP and HR signals can be split in different frequency bands:

- A high frequency band coinciding with respiration.

- A middle frequency band that seems to be related to the vasomotor tone and/or baroreceptor activity.

- A low frequency band related to thermoregulation.

The interpretation of these findings was facilitated by the description of Erich von Holst [21] in 1939. He described the coordination rules between the motor rhythms generated at central level, demonstrating the existence of the entrainment phenomenon [21]. In 1958, it was demonstrated [18] that the rules of relative coordination between the respiratory rhythm and the wave of ABP can be observed in humans.

The different values which are obtained when ABP or HR are measured beat to beat or several times in the same subject, allows to incorporate the concept of HR and ABP variability. This concept can be quantified through the evaluation of the standard deviation of the found mean value or its variance [35] [36]. Today it is known that this variability contains regular oscillations of different frequencies [34] [41] [13] [17], related to different phenomena as mentioned above. However, although it is suggested that the 0.25 Hz oscillatory components (Traube-Hering waves) are related to the respiratory activity, the mechanisms of generation together with the physiological

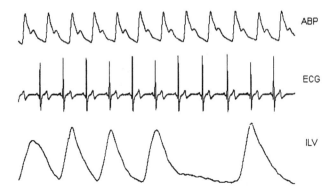

FIGURE 1. Arterial Blood Pressure (ABP), electrocardiogram (ECG) and instantaneous lung volume (ILV), from a human subject in basal condition.

means of these oscillations are still unknown [14] [20]. The oscillations with a frequency of 0.1 Hz, denominated Mayer's waves, were related with the vasomotor activity induced by the baroreflex activity [12] [54] [53]. Less consensus exists in relation to the nature of the 0.05 Hz oscillations, still when several authors suggest a relation with termoregulatory system activity [8] [22] [37]. The analysis in the time domain of the 24 hour HR and ABP variability in humans or in long term studies in laboratory animals, has proved to be an important tool in:

1. Studying the mechanisms responsible of neural control of the regulation of ABP and HR [40] [28] [29] [43].

2. Improving the diagnostic of arterial hypertension [28] [49] [32].

3. Verification of the therapeutic effectiveness in this pathology [33].

The aim of this work is to present the commonly use algorithms in the study of cardiorespiratory signals from the parameter extraction to time and frequency domains analysis in short and long term data.

2   Short term variability

2.1   Spectral Analysis Methods

In this work we used records from the database which was developed in collaboration between the Harvard Medical School, the Massachusetts Institute of Technology, and the Favaloro Foundation Medical School (HMS-MIT-FFMS) [48]. The records contains three channels, ECG, BP, and RESP signals sampled at 360 Hz after antialaising at 180 Hz. Intra-arterial blood pressure was measured invasively, lung volume changes were measured with an inductance plethysmograph (Respitrace), and ECG recordings used was lead II (Figure 1). Parameter Extraction

The RP and BP signals were digitally filtered and decimated at 3 Hz to yield ILV and mean arterial blood pressure (MAP) signals. R-waves were identified from the ECG leads by a peak detection program [44] and smoothed instantaneous HR beat and time series were constructed at 3 Hz. Systolic and diastolic blood pressures (SBP and DBP) were identified from each beat of the non decimated BP signal, and the values were splined and decimated at 3 Hz [2]. A pulse pressure (PP) signal was formed by subtracting DBP from SBP.

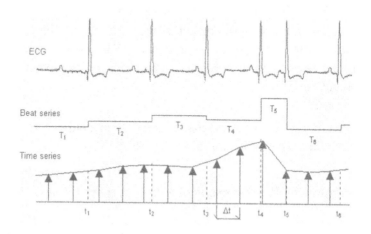

FIGURE 2. ECG signal; beat series of HR derived from ECG; $T(t)$ in dotted lines and with arrows the instantaneous HR sampled at a constant interval t.

The beat series of HR is named tachogram [9], the SP beat series, systogram, and the DP beat series diastogram. It is possible to represent it as a function of the interval number:

$$T(i) = A(t_i - t_{i-1}) \qquad (8.1)$$

where $i = \cdots, -2, -1, 0, 1, 2, \cdots$ and A is a constant.

The beat series can be used like data for to calculate the power spectra density, but since the points of its samples are unevenly spaced in time, it is more accurate to use time series, who are evenly spaced in time, and therefore represents the instantaneous variations [52].

The first step in calculating of time series is to take into account the moment, in real time, when the events are produced. After that, to obtain the time series, the beat series is sampled at a fixed frequency of 3 Hz (Figure 2). The values of the variable in study must be identified interpolating the intermediate real values. The necessary condition of the interpolation algorithm it must be symmetric, to avoid phase errors in the time series [30]. The representation of the heart rate versus time can be expressed with the following equation:

$$T(t) = B \sum_{i=-\infty}^{\infty} T(t_i - t_{i-1})\delta(t - t_{i-1}) \qquad (8.2)$$

where B is a constant and δ is the impulse function. The time series derived from ABP can be appreciated in Figure 3.

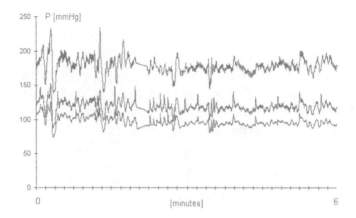

FIGURE 3. record of 6 minutes showing the time series of SP, MP and DP.

2.2   Spectral Analysis

The power spectra was calculated using the following methods:

- Autoregressive and moving average modeling.

- Fast Fourier Transform.

- Blackman-Tukey method.

2.3   Autoregressive and moving average modeling

The autoregressive and moving average modeling are based on the theory of linear estimation. The scope is to find the parameters of a linear model, which fit the data signal. When the model has poles only, it is denominated autoregressive (AR), and when the model has poles and zeros, it is denominated autoregressive and moving average (ARMA). In Figure 4 we can appreciate the block diagram of fitting algorithm. The fitting algorithm consist in to excite the choose model with white noise, in others words to excite with a broadband signal. The output of the model, that is the estimated signal, and the observed signal, both are used as inputs to an adder, the output of this adder brings the difference between the two inputs, call the error signal. This error signal is used to modify the parameters of the model, calculating by this way the new parameters with an algorithm which finds the minimum error signal. The model may be represented with the following recurrence equation:

$$y(n) = \sum_{i=0}^{p} b_i x(n - i) + \sum_{k=1}^{m} a_k y(n - k) \qquad (8.3)$$

$b_0 = 1$, where $x(n)$ is the input signal to the model (white noise), and $y(n)$ is the output of the model. The order of the model (p and m) are chosen by taking into account the data signal [1]. The recurrence equation may be expressed in the z plane:

$$Y(z) = \sum_{i=0}^{p} b_i x X(z) z^{-i} + \sum_{k=1}^{m} a_k Y(z) z^{-k} \qquad (8.4)$$

The coefficients which are multiplying the input samples (actual and delayed inputs) are called the zeros of the model (b_i), and the coefficients which are multiplying the output samples and the successive feedback samples are called poles of the model (a_k). One of the most used algorithm to find the minimum error signal to modify the parameters of the model is the Anderson test [7], that analyzes the autocorrelation of the error signal. The main characteristics of the ARMA model is the possibility to obtain a good spectral estimation having one cycle of the signal only, and the possibility to construct the spectrum at any resolution. In the Figure 5 it is possible to appreciate the result of the spectral estimation using the autoregressive modeling.

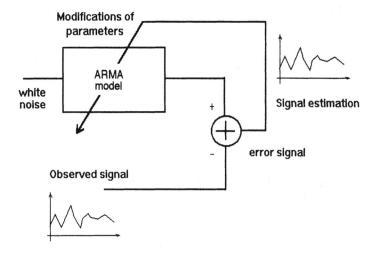

FIGURE 4. Block diagram of ARMA modeling.

2.4 Fast Fourier Transform method

The power spectral density estimation using Fast Fourier Transform (FFT) is the optimized implementation of the discrete Fourier transform (DFT), using the redundant operation and reducing the numbers of computations [10] [27]. The DFT, in its direct and inverse form are defined:

$$X(f) = F_f[x(n)] = \sum_{n=0}^{M-1} x(n)e^{j2\pi fn/M} \tag{8.5}$$

$$x(n) = F_n^{-1}[X(f)] = \frac{1}{M} \sum_{f=0}^{M-1} X(f)e^{-j2\pi fn/M} \tag{8.6}$$

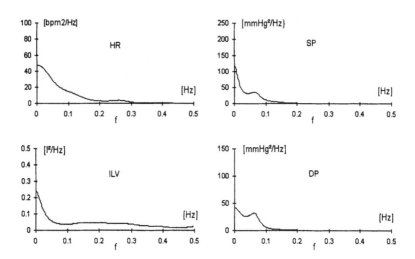

FIGURE 5. Spectral estimation with autoregressive modeling of HR, ILV, SP and DP.

where M is the number of samples. The power spectral estimation may be implemented calculating the spectrum of the autocorrelation function of the signal, which is defined in the domain time as:

$$R_{xx}(k) = \frac{1}{N - |k|} \sum_{n=0}^{N-|k|-1} x(n)x(n + |k|) \qquad (8.7)$$

where $x(n)$ are the samples without mean value (harmonic zero) of the signal and N is the total number of samples. The autocorrelation function may be calculated in the frequency domain:

$$R_{xx} = \frac{1}{N - |k|} [X^*(f)X(f)] \qquad (8.8)$$

where F is the DFT operator, $X(f)$ and $X^*(f)$ are the Fourier transform of $x(n)$ and its conjugate transform respectively. To obtain identical results of the equation (7) using equation (8), the number of points of the DFT must be twice the number of samples ($M = 2N$). Figure 6 shows the result of spectral estimation using the FFT method.

2.5   Blackman-Tukey method

The Blackman-Tukey method was originally developed to analyze communication systems [6] [3]. It was based on multiplying the time

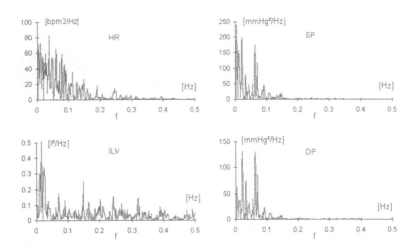

FIGURE 6. Spectral estimation using FFT of HR, ILV, SP and DP.

series by a function different from the rectangular window, whose effect is to trunk the signal. The window $w(k)$ used in this method is a gaussian function whose spectrum has attenuated lateral lobs, and is better for minimizing the trunk effect. The advantage of this method is to preserve the Parseval's identity even after multiplied by a gaussian function. The Blackman-Tukey method first step is to calculate the autospectra using the autocorrelation function $R_{xx}(k)$, with this equation:

$$S_{xx}(f) = \Delta t F_f [R_{xx}(k)w(k)] \tag{8.9}$$

where $w(k)$ must be chosen like the best compromise between frequency resolution and variance estimator. The gaussian function used is:

$$w(k) = e^{-(k\Delta t)^2/2\sigma_t^2} \tag{8.10}$$

where σ_t is an adjustable parameter which represent the half of the gaussian window in the time domain. Figure 7 shows the result of the Blackman-Tukey method.

2.6 Transfer Function

The transfer function of a system can be characterized with any spectral estimation method, assuming that the system is linear. A

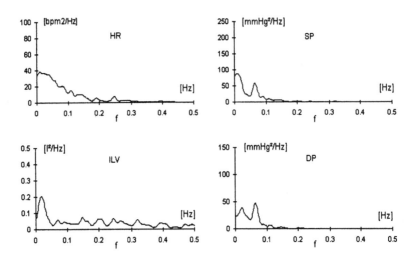

FIGURE 7. Spectral estimation using Blackman-Tukey of HR, ILV, SP and DP.

system can be considered linear if the following conditions apply:

- Additive: if the response to $x_1(t)$ is $y_1(t)$ and to $x_2(t)$ is $y_2(t)$, then $x_1(t) + x_2(t) = y_1(t) + y_2(t)$.

- Homogeneous: if the response to $x(t)$ is $y(t)$, then the response to $cx(t)$ is $cy(t)$, where c is an arbitrary constant.

Thus, it is assumed the invariance in time of the system, and the system must be causal too. The method used is denominated complex transfer function:

$$H(f) = \frac{S_{xy}(f)}{S_{xx}(f)} \tag{8.11}$$

where $S_{xx}(f)$ is the autospectrum (spectrum of the autocorrelation function) and $S_{xy}(f)$ is the crosspectra of the output and the input of the system. The complex transfer function can be decomposed in modulus $|H(f)|$ and phase $\Theta(f)$, from its real $H_R(f)$ and imaginary $H_I(f)$ parts:

$$|H(f)| = \left[H_R^2(f) + H_I^2(f)\right]^{1/2} \tag{8.12}$$

$$\Theta(f) = arctg\left[\frac{H_I(f)}{H_R(f)}\right] \tag{8.13}$$

The magnitude of the transfer function $|H(f)|$ shows the degree of influence of each harmonic in the input on the output of the system. The phase $\Theta(f)$ reflects the delay, like a portion of one period, between each harmonic component in the input on the output of the system. It is possible to calculate the quality of the transfer function quantifying the linear dependence between each harmonic component of input and output; such quantification can be calculated with the coherence function, and its expression is:

$$\gamma^2(f) = \frac{|S_{xy}(f)|^2}{S_{xx}(f)S_{yy}(f)} \tag{8.14}$$

When the coherence function is one, the linearity is maximum; if not (values between zero and one) it mainly reflects the lack of linearity or noise. Figure 8 shows the transfer function taking as input the mean pressure and as output the HR, characterizing the baroreflex system.

3   Long term variability

In the study of the long term variability we used time and frequency domain analysis. These methods give results which allows to predict mortality risk after cardiac infarction and sudden death [23] [11] [39] [4]. Other works support the idea that short term variability can be used to predict mortality risk after cardiac infarction [5] and congestive heart failure [46]. The study of the long term variability is divided in two groups:

1. The group of time domain methods.

2. The group of frequency domain methods.

Both groups of methods use beat and time series derived from ECG recordings, arterial blood pressure and respiration. In such cases where there is not respiration recordings, like in Holter tapes of 24 hours, the respiratory signal may be derived from the relationship of areas under the QRS complex of ECG [38], using the following expression:

$$r = arctg\left(\frac{A_X}{A_Y}\right) \tag{8.15}$$

where r is the derived respiration, A_x and A_Y are the areas under QRS complex from two different leads. Figure 8 shows the results of the R wave detection algorithm and derived respiration.

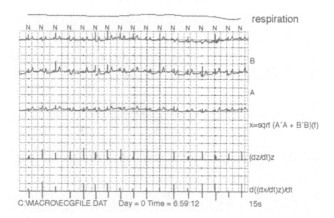

FIGURE 8. Record segment of 15 seconds from 24 hours of ECG, showing the peak location algorithm and derived respiration.

The algorithms for R wave detection used in 24 hours ECG recordings must have high noise rejection, and beat classification to discard premature beats [44]. Figure 9 shows the time series from an ECG record of 24 hours, composed by more than one hundred thousand beats.

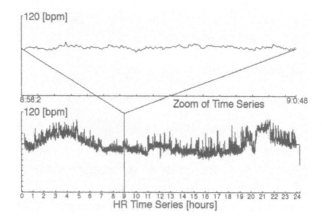

FIGURE 9. Time series of ECG (24 hours) and zoomed at 9 hours.

3.1 Time domain methods

Of all the time domain methods used, calculated standard deviation of all normal RR intervals (SDNN index) was the best predictor of all cause mortality after myocardial infarction [23] [11] [39] [4] [5]. The prediction capacity of the SDNN index is independent of others risk factors such as pulmonary problems originated in coronary unit, ventricular ejection fraction and ventricular arrhythmia, among others.

The table 1 shows the most commons indexes used in the clinical practice.

On the other way, when analyzing spectral analysis data, it could be seen that the very low frequency fluctuations of HR, affect strongly the parameters mentioned in table 1. These fluctuations are due to mechanisms that regulate the circadian rhythm of cardiovascular functions. Other very useful representation of HR variability in the time domain is the Poincare map, whose representation is also called dispersion map. The Poincare map is constructed drawing the RR intervals versus the next RR interval, resulting an X-Y graph that gives the possibility to appreciate the dynamics characteristics of each beat respect the precedent one [24], as it is shown in the Figure 10.

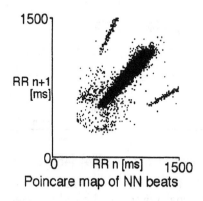

FIGURE 10. Poincare map of RR n versus RR n+1 intervals.

3.2 Frequency domain methods

The spectrum derived from the 24 hour time series of HR in normal subjects shows the 1/f behavior when it is drawn as the logarithm

Table 1

Variable	Domain Units	Definition
Night-day	time–ms	Difference between the average of all the normal RR intervals at night (14:00 to 5:00) and the average of all the normal RR intervals during the day (7:30 to 21:00).
SDNN	time–ms	Standard deviation of all normal RR intervals in the entire 24 hour ECG recording.
SDANN index	time–ms	Standard deviation of the average normal RR intervals for all 5 minute segments of a 24 hour ECG recording (each average is weighted by the fraction of the 5 minutes that has normal RR intervals).
SDNN index	time–ms	Mean of the standard deviations of all normal RR intervals for all 5 minutes segments of a 24 hour ECG recording.
r-MSSD	time–ms	Root-mean-square successive difference (the square root of the mean of the squared differences between two adjacent normal RR intervals over the entire 24 hour ECG recording).
pNN50	time–percent	Percentage of differences between adjacent normal RR intervals that are greater than 50 ms computed over the entire 24 hour ECG recording).
Total power	freq.–ms^2	The energy in the heart period power spectrum up to 0.4 Hz.
Ultra low frequency power	freq.–ms^2	The energy in the heart period power spectrum up to 0.0033 Hz.
Very low frequency power	freq.–ms^2	The energy in the heart period power spectrum between 0.0033 and 0.04 Hz.
Low frequency power	freq.–ms^2	The energy in the heart period power spectrum between 0.04 and 0.15 Hz.
High frequency power	freq.–ms^2	The energy in the heart period power spectrum between 0.15 and 0.4 Hz.
LF/HF ratio	freq.–none	The ratio of low to high frequency power.

of the power spectral density [24] [45], like is shown in Figure 11. In the left upper panel it can be appreciated the single data of the power spectral density and, in the right upper panel, the same spectra smoothed where it is easily appreciate the 1/f slope. Finally, in the lower panel it is shown the time series used to calculate the above mentioned spectra. In normal subjects the spectra slope tends to -1, this reference is interesting due to the fact that in different pathologies like congestive cardiac failure or cardiac transplant, the slope is greater [47], but the total power remains at the same values, so the total power was not different between the subjects being the slope the only parameter altered. Additionally, it was also demonstrated that the 75 in the frequency domain is found in the frequency band from 0.0033 to 0.01 Hz, which is below of the traditionally bands in the studies of short term variability [45]. During the studies of 24 hour is very useful to calculate, together with the spectral power, the spectra of segments of 5 minutes. This enables us to visualize spectral changes due to different activities during the day of study. The Figure 12 shows a sequential representation of 5 minutes spectra blocks along the 24 hour, and the mean and standard deviation of all blocks area bands under the spectra. For each segment of 5 minutes, the same algorithms can be used in short term variability studies, to calculate the transfer function between HR and RESP, to compare different situations in the same subject. The Figure 13 shows the result analyzing one segment of 5 minutes.

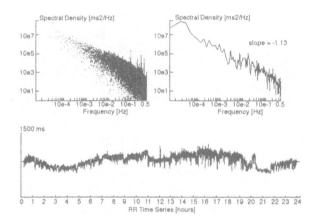

FIGURE 11. Total logarithmic spectral (A) and average (B) of the 24 hours RR time series (C).

4   Discussion

The analysis of the variability of cardiorespiratory signals, in the time and the frequency domains, provides information about the mechanisms of regulation of these signals. This information can be used to compare conditions of health and sickness [42]. However, the spectral estimation methods are still subject of controversy, and also the conditions in which the measurement was made. This is so

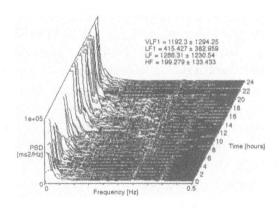

FIGURE 12. Power spectral estimation of 24 hours RR intervals in segments of 5 minutes.

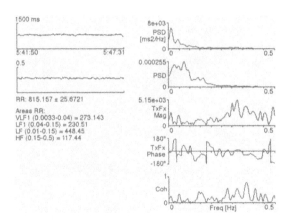

FIGURE 13. Power spectral estimation of RR intervals and respiration of a 5 minutes segment, and the transfer function of both signals.

since the reproducibility of such studies are not ever the optimal. For example, the autoregressive method assume a stochastic nature of the signals, splitting by this way, the different frequency bands in a spectra with superimposed broad-band wide noise. On the other hand, the FFT and Blackman-Tukey assume a deterministic nature of the signals and, in the case of Blackman-Tukey method allows to change the spectral resolution by means and adjustable parameter. The transfer function is able to show the influence that the signals have between them, allowing by this way, to quantify the coherence between each output oscillation with its respective input.

The measurements in time and frequency domains of the HR can be correlated in two different ways: 1) To develop the spectral short term spectral analysis of the HR in isolated segments (from 2 to 15 minutes) and the total 24 hour recording; in this way it is possible to compare spectral parameters such as high and low frequency powers and the ratio between them. 2) To calculate a single spectra from the 24 hour recording (Figure 11) and to calculate the power of determined bands [4]. Between the findings were noted the high correlation between the SDNN with the square root of the total area under the spectra; this finding can be explained by the Parseval's theorem because the area under the spectra is equal to the variance of the time domain data and the variance is equal to the square root of the standard deviation. On the other hand, Bigger found the 82 percent of the power is found mainly in the frequencies below 0.0033 Hz.

Up to now, the studies made in a patients with congestive cardiac failure suggest that the short term analysis is not a good predictor of mortality [46]. However, recent studies have shown that the values of very low, low and high frequency areas and the ratio LF/HF calculate in segments from 2 to 15 minutes, seems to be good predictors of post infarction mortality [4]. In addition it was found that the result of LF and HF is correlated with the results found with the whole record of 24 hours.

Taking into account that the parameters which quantify the long term variability are mortality predictors, and that this parameters have a moderated correlation with short term variability analysis which quantify the physiological control of the cardiovascular system, it is suggested that the prediction of mortality based in this parameters is not due to the autonomic control of the HR.

In conclusion, the analysis of the cardiorespiratory signals in time and frequency is able to give a better knowledge of the mechanisms responsible of the neural control and the regulation of the cardiovascular system, it could allow us to improve the diagnostic and treatment of arterial hypertension and the associated therapeutic.

5   References

[1] Akaike H. Statistical predictor identification. Am. Inst. Statist. Math., (22):203–217, 1970.

[2] Berger RD, Akselrod S, Gordon D, Cohen RJ. An Efficient Algorithm of Heart Rate Variability. IEEE Transactions in Biomedical Engineering, (9):900–904, 1986.

[3] Berger RD, Saul JP, Cohen RJ. Transfer function analysis of autonomic regulation I. Canine atrial rate response. Am. J. Physiol. 256: H142-H152, 1989.

[4] Bigger JT, Fleiss JL, Steinman RC, Rolnitzky LM, Kleiger RE, Rottman JN. Frequency Domain Measures of Heart Period Variability and Mortality After Myocardial Infarction. Circulation, 85, 164-171. 1992.

[5] Bigger JT, Fleiss JL, Rolnitzky LM, Steinman RC. The Ability of Several Short-Term Measures of RR Variability to Predict Mortality After Myocardial Infarction. Circulation, 88, 927-934. 1993.

[6] Blackman RB, Tukey JW. The measurement of power spectra from the point of view of Communications Engineering. New York, Dover. 1959.

[7] Box GEP, Jenkins GM. Time series analysis: forecasting and control. Holden-Day. San Francisco. 1976.

[8] Burton AC: The range and variability of the blood flow in the human finger and the vasomotor regulation of body temperature Am J Physiol 127: 437-453, 1939.

[9] Cerutti S, Baselli G, Bianchi A, Signorini MG. Spectral Techniques for Analysis for Blood Pressure and Heart Rate Signals. Blood Pressure and Heart Rate Variability. M. Di Rienzo et al, Eds. IOS Press, 1992.

[10] Cooley JW, Tukey JW. An algorithm for the machine computation of complex Fourier series. Mathematics of Computation. 19, pp. 297-301, abril 1965.

[11] Cripps TR, Malik M, Farrell TG, Camm AJ. Prognostic Value of Reduced Heart Rate Variability After Myocardial Infarction: Clinical Evaluation of a New Analysis Method, Br. Heart J, 65, 14-19. 1991.

[12] DeBoer RW, Karemaker JM and Strackee J: Hemodynamic fluctuations and baroreflex sensitivity in humans: A beat to beat model. Am J Physiol 253: H680-H689, 1987.

[13] DiRienzo M, Castiglioni P, Mancia G, Parati G and Pedotti A: 24h sequential spectral analysis of arterial blood pressure and pulse interval in free-mooving subjects. IEEE Transactions on Biomed Engin 36: 1066-1075, 1989.

[14] Doenhorst AC, Howard P and Leathart GL: Respiratory variations in blood pressure. Circulation 6: 553-558, 1952.

[15] Donders FG: Zf. Nat Med 3: 287-319, 1853

[16] Einbrodt A: Uber den Einfluss der Atembewegungen auf Herzschlag und Blutdruck. Akad Wiss Wien Math Naturwiss Kl 2 Abt 40: 361-418, 1860.

[17] Furlan R, Guzzetti S, Crivellaro W, Dassi S, Tinelli M, Baselli G, Cerutti S, Lombardi F, Pagani M and Malliani A: Continuous 24h assesment of the neural regulation of systemic arterial pressure and RR variabilities in ambulant subjects. Circulation 81: 537-547, 1990.

[18] Golenhofen K and Hildebrandt G: Die Beziehungen des Blutdruchrhythmus zu Atmung und peripherer Durchblutung. Pf—gers Arch 267: 27-45, 1958.

[19] Hales S: Statical Essays, vol 2: Haemastatics. London Ed. Innys and Manby, 1733.

[20] Hirsch JA and Bishop B: Respiratory sinus arrhythmia in humans: How breathing pattern modulates heart rate. Am J Physiol 241: H620-H629, 1981.

[21] Holst E: Die relative Koordinatio als Phenomen und als Methode Zentralnervoser Funktionsanalyse. Erg Physiol 42: 228-306, 1939.

[22] Hyndman BW, Kitney RI and McA. Sayers B: Spontaneous rythms in physiological control systems Nature 233: 339-341, 1971.

[23] Kleiger RE, Miller JP, Bigger JT, Moss AR. Multicenter post-Infarction Research Group: Decreased Heart Rate Variability and its Association with Increased Mortality after Acute Myocardial Infarction. Am. J. Cardiol, 59, 256-262. 1987.

[24] Kobayashi M, Musha T. 1/f Fluctuations of Heartbeat Period. IEEE Trans. On Biomed. Eng, BME-29, 456-457. 1982.

[25] Koepchen HP and Thurau K: Untersuchungen —ber zusammenhange zwischen Blutdruckwellen und Ateminnervation Pfl—gers Arch 267: 10-26, 1958.

[26] Koepchen HP: Die Blutdruckrhythmik. Dr D Steinkopff Verlag, Darmstadt, 1962.

[27] Kraniauskas P. A Plain Man's Guide to the FFT. IEEE Signal Processing Magazine, vol. 11, no. 2: 24-35, abril 1994.

[28] Littler WA, Honour J, Pugsley DJ and Sleight P: Continuous recording of direct arterial pressure in unrestricted patients. Its role in the diagnosis and management of high blood pressure. Circulation 51: 1101-1106, 1975.

[29] Littler WA, West MJ, Honour AJ and Sleight P: The variability of arterial pressure. Am Heart J 95: 180-186, 1978.

[30] Luczak H, Laurig W. An Analysis of Heart Rate Variability. Ergonomics 16, 85-97, 1973.

[31] Ludwig C: Beitrage zur Kenntnis des Einflusses der Respirationsbewegungen auf den Blutumlauf im Aortensystem. Arch Anat Physiol, 242-302, 1847.

[32] Mancia G, Ferrari A, Gregorini L, Parati G, Pomidossi G, Bertinieri G, Grassi G, DiRienzo M, Pedotti A and Zanchetti A: Blood pressure and heart rate variabilities in normotensive and hypertensive human beings. Circ Res 53: 96-104, 1983.

[33] Mancia G, Ferrari A, Pomidossi G, Parati G, Bertinieri G, Grassi G, Gregorini L, DiRienzo M and Zanchetti A: Twenty four hour blood pressure profile and blood pressure variability in untreated hypertension and during antihypertensive treatment by once a day nadolol. Am Heart J 108: 1078-1083, 1984.

[34] Mancia G, Bertinieri G, Cavallazzi A, DiRienzo M, Parati G, Pomidossi G, Ramirez AJ and Zanchetti A: Mechanisms of blood pressure variability in man. Clin Exp Hypert: Theory and Practice 17(2-3): 167-178, 1985.

[35] Mancia G, ParatiG, Pomidossi, G Casadei R, DiRienzo M and Zanchetti A: Arterial baroreflexes and blood pressure and heart rate variabilities in humans. Hypertension 8: 147-153, 1986.

[36] Mancia G and Zanchetti A: Blood pressure variability In Handbook of Hypertension vol 7: Pathophysiology of hypertension, cardiovascular aspects Ed by Zanchetti A and Tarazi C: Amsterdam, Elsevier pp: 125-152, 1986.

[37] McA.Sayers B: Analysis of heart rate variability Ergonomics 16: 17-32, 1973.

[38] Moody GB, Mark RG, Zoccola A, Mantero S. Derivation of respiratory signals from multi-lead ECG's. Computers in Cardiology 1986: pp 511-516.

[39] Odemuyiwa O, Malik M, Farrell T, Bashir Y, Poloniecki J, Camm J. Camparison of the Predictive Characteristics of Heart Rate Variability Index and Left Ventricular Ejection Fraction for All-Cause Mortality, Arrhythmic Events and Sudden Death after Acute Myocardial Infarction. Am. J. Cardiol, 68, 434-439. 1991.

[40] Parati G, DiRienzo M, Bertinieri G, Pomidossi G, Casadei R, Groppelli A, Pedotti A, Zanchetti A and Mancia G: Evaluation of the baroreceptor-heart rate reflex by 24h intra-arterial blood pressure monitoring in humans Hypertension 12: 214-222, 1988.

[41] Parati G, Castiglioni P, DiRienzo M, Omboni S, Pedotti A and Mancia G: Sequential spectral analysis of 24h blood pressure and pulse interval. Hypertension 16: 414-421, 1990.

[42] Parati G, Saul JP, Di Rienzo M, Mancia G. Spectral Analysis of Blood Pressure and Heart Rate Variability in Evaluating Cardiovascular Regulation, A Critical Appraisal. Hypertension. 1995;25:1276-1286.

[43] Ramirez AJ, Bertinieri G, Belli L, Cavallazzi A, DiRienzo M, Pedotti A and Mancia G: Reflex control of blood pressure and

heart rate by arterial baroreceptors and by cardiopulmonary receptors in the unanesthetized cat. J Hypertens 3: 327-335, 1985.

[44] Risk MR, Sobh JF, Barbieri R, Saul JP. A Simple Algorithm for QRS Peak Location: Use On Long Term Recordings from the HMS-MIT-FFMS Database. IEEE Engineering in Medicine and Biology 17th Annual Conference. 1995.

[45] Saul JP, Albrecht P, Berger RD, Cohen RJ. Analysis of Long-Term Heart Rate Variability: Methods, 1/f Scaling and Implications. In: Proc. Computers in Cardiology. IEEE Computer Society Press, Los Alamitos, CA, pp. 419-422. 1987.

[46] Saul JP, Arai Y, Berger RD, Lilly LS, Colucci WS, Cohen RJ. Assessment of Autonomic Regulation in Chronic Congestive Heart Failure by Heart Rate Spectral Analysis. Am. J. Cardiol, 61: 1292-1299. 1988.

[47] Saul JP. Heart Rate Variability During Congestive Heart Failure: Observations and Implications. In: Di Rienzo M, Mancia G, Parati G, Pedotti A, Zanchetti A (Eds), Blood Pressure and Heart Rate Variability, Computer Analysis, Methodology and Clinical Applications. IOS Press, pp 266-275. 1993.

[48] Sobh JF, Risk MR, Barbieri R, Saul JP. Database for ECG, Arterial Blood Pressure, and Respiration Signal Analysis: Feature extraction, Spectral Estimation and Parameter quantification. IEEE Engineering in Medicine and Biology 17th Annual Conference. 1995.

[49] Sokolow M, Perloff D and Cowan R: Contribution of ambulatory blood pressure to the assesment of patients with mild to moderate elevation on office blood pressure. Cardiovasc Rev Rep 1: 295-303, 1980.

[50] Valentinuzzi ME, Geddes LA. The central component of the respiratory heart-rate response. Cardiovascular Research Center Bulletin. Vol. XII, No. 4. Pp 87-103, 1974.

[51] Valentinuzzi ME, Baker LE, Powell T. The heart rate response to the valsalva manoeuvre. Medical and Biological Engineering. 12(6): 817-822, November 1974.

[52] Voorde BT. Modelling the Baroreflex, a system analysis approach. PhD Tesis. CopyPrint 2000, Enschede, 1992.

[53] Wesseling KH, Settels JJ, Walstra HG, Van Esc HJ and Donders JJH: Baromodulation as the cause of short term blood pressure variability In Proceedings International Conferences Applied Physics Medical Biology. Eds. Alberi G, Bajzer Z and Baxa P. World Scientific Publishing Co pp 247-276, 1983.

[54] Wesseling KH and Settels JJ: Baromodulation explains short-term blood pressure variability in Psychophysiology of cardiovascular control. Ed: JF Orlebeke G, Mulder G and VanDoornen LJP. New York: Plenum pp 69-97, 1985.

Chapter 9

Characterization of Epileptic EEG Time Series (I): Gabor Transform and Nonlinear Dynamics Methods

Susana Blanco
Silvia Kochen
Rodrigo Quian Quiroga
Luis Riquelme
Osvaldo A. Rosso
Pablo Salgado

1 Introduction

It has been well over a century since it was discovered that the mammalian brain generates a small but measurable electrical signal. The electroencephalogram (EEG) of small animals was measured by Caton in 1875, and in man by Berger in 1925. It had been thought by the mathematician N. Wiener, among others, that *generalized harmonic analysis* would provide the mathematical tools necessary to penetrate the mysterious relations between the EEG time series and the functioning of the brain. The progress along this path has been slow however, and the understanding and interpretation of EEG's remain quite elusive.

The traditional EEG tracing is now interpreted in much the same way it was done fifty years ago. More channels are used now and much more is known about clinical implication of the waves, but the basic EEG display and quantification of them are quite similar to its predecessors of a half century ago. There is no taxonomy of EEG patterns which delineates the correspondence between those patterns and brain activity. The clinical interpretation of EEG records is made by a complex process of visual pattern recognition and association on the part of clinician and significantly more often in the last years (with the introduction of the personal computers) through the use of Fourier transform.

Quantitative EEG analysis as a field, includes a wide variety of techniques. These are: frequency analysis (spectral analysis), topographic mapping, compressed spectral arrays, significance probability mapping and other complex analytical techniques [1 − 3]. A new recently approach to the problem of the quantification of the EEG series, has been presented by the nonlinear dynamics [4 − 7].

The morphology and topography of sharp transients have been correlated with seizure type and therapeutic response to different medications and surgery. An essential component of the traditional visual interpretation of the clinical EEG is the characterization of infrequent, morphologically variable transient events, especially those associated with the epilepsies ("spikes", "spikes and waves", etc.) [1 − 3]. Accordingly, a great deal of energy has been spent over the years in efforts to automatically search long recordings for these phenomena and epileptiform transient detection, but with different results [8 − 10]. Anyway the most diffused quantitative method in the clinical practice is the spectral analysis together with a vissual assessment [1 − 3].

The methods mentioned above are applied to the activity analysis in a single channel independently of the activity in the other channels. The most common methods of studying interactions between two channels are the cross-correlation and the cross-spectral analysis [11, 12]. The average amount of mutual information and nonlinear correlation are recently developed methods [13 − 16]. These methods all try to determine whether two channels have a common activity and, often, whether one channel contains activity induced by the activity in the other channel. Clearly proving causality is extremely difficult, but it can sometimes be inferred by measuring time differences.

From another point of view, an electroencephalogram may be considered as a time series measured on a dynamical system that represents the brain activity. This subject has caught the attention of several researchers on this field, finding the important feature that the variability of the EEG signals is not noise and presents an attractor [5 − 7, 17 − 23]. The treatment of EEG series under the approach of nonlinear dynamic systems opened new possibilities to the brain dynamic knowledge. However, the aims are not limited to this, but also to obtain new forms to quantify differences in the EEG series that have some kind of clinical application [24 − 27].

In this work we review a method based in a Gabor's old idea [28] for the simultaneous treatment in the time-frequency space of a signal recently introduced by us [29 – 32].

This method let us analyze the time evolution of the different rhythm of an EEG signal, and visualize the frequency engagement during epileptic activity as well as paroxismal activities. The correlation between the obtained frequency evolution series, for the different channels and bands can be used to obtain some knowledge about the interaction and consequently causality between channels and bands.

The different time behaviors identified by this method can be verified by the corresponding phase portraits obtained from the associated EEG epochs and characterize evaluating their geometrical and dynamical parameters using nonlinear metrics tools [26, 29]. This information can be used as a first step in the formulation of dynamical models of the epileptic seizure and its propagation.

2 Experimental Setup and Clinical Data

Our methods were applied to the EEG recording from interseizure and seizure brain activity of refractory epileptic patients prone to surgical treatment. This information is simultaneously correlated with clinical symptomatology. For each patient the strategy for the use of implanted electrodes is planned in relation with the spatial and temporal organization of the epileptic discharges.

The EEG time series were obtained from a male patient of 21 years old, during nine hours, with 12 multilead depth electrodes of 1 mm thickness implanted stereotactically. Each electrode carried 5 to 15 cylindrical contacts of nickel-chromium alloy with a lengh of 2 mm and intercontact distance of 1.5 mm. These electrodes were placed in the pileptogenic zone and in the propagation brain areas. In Fig. 1 we display a schematic diagram of the electrodes position and notation.

Each signal was amplified and filtered using a $1 - 40\ Hz$ band-pass filter. A 4 pole Butterworth filter was used as low-pass filter, serving as an anti-aliasing scheme. After 10 bits A/D conversion the EEG data was written continuously onto a disk of a data acquisition computer system with a sampling rate of 256 Hz per channel. Selected artifact free EEG data sets of ictal and interictal activity were stored for subsequent off-line analysis.

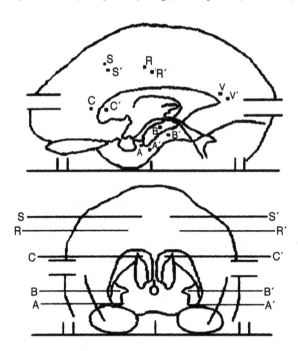

FIGURE 1. Schematic location of depth electrodes. Abbreviations: **A** (**A'**) Right
(Left) Amygdalin Nucleus (10 contacts); **B** (**B'**) Right (Left) Hypocampus
(10 contacts); **V** (**V'**) Right (Left) Temporal - Occipital (5 contacts); **S**
(**S'**) Right (Left) Supplementary Motor Area (10 contacts); **C** (**C'**)
Right (Left) Gyrus Cingular (15 contacts); **R** (**R'**) Right (Left) Rolandic
Cisure (5 contacts).

According to the visual assessment of the EEG seizure recording,
this patient presented an epileptogenic focus in the left hemisphere
correponding to Hypocampus with immediate propagation to Gyrus
Cyngular and Motor Supplementary Area and to the right contralateral homologous area.

In this work we present two intracraneal registers or stereo EEG
signals (Fig. 2 and Fig. 3) that were selected because they present
a different morphology in their trace. In Fig.2 we display the *EEG
Signal I* for 16 *sec* corresponding to a depth electrode in the left
Hypocampus (B'_2). In this sample we can see an isolate paroxysm
at 2 *sec* and a paroxismal activity which starts arround the 4*th* second and finished arround the 13*th* second. The *EEG Signal II* for 64
sec corresponding to the same electrode in the epileptogenic region
is shown in Fig.3 and display an epileptic seizure. From visual inspection it is clear that around the 10*th* second the epileptic seizure
starts, and finishes arround the 54*th* second.

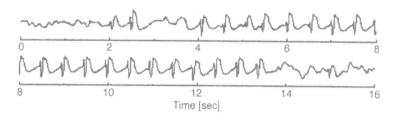

FIGURE 2. Recording of the EEG signal corresponding to a contact in the epileptogenic region, left Hypocampus ($\mathbf{B}'_2$). Paroxysmal activity (*EEG Signal I*).

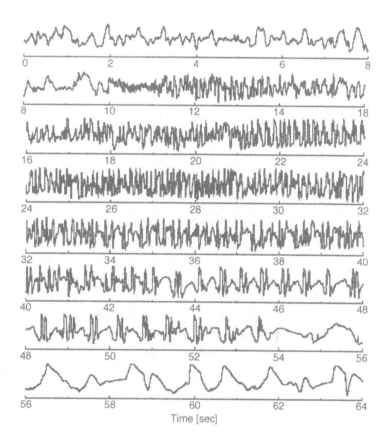

FIGURE 3. Recording of the EEG signal corresponding to a contact in the epileptogenic region, left Hypocampus ($\mathbf{B}'_2$). Epileptic seizure (*EEG Signal II*).

The use of depth electrodes provides records where the noise and artifact contamination effects (usually present in the EEG series obtained with scalp electrodes) are minimized. Anyway the applicability of the proposed methods is not restricted to the use of this kind of EEG records.

3   Armonic Analysis of EEG Data

Analysis of EEG signals always involves the queries of quantification, i.e., the ability to state objective data in numerical and/or graphic form. Without such measures, EEG appraisal remains subjective and can hardly lead to logical systematization [1 − 3].

The EEG is a complex signal whose statistical properties depends on both space and time. Regarding the temporal characteristics, the EEG signals are nonstationary and from a dynamical point of view they are chaotics. Nevertheless, they can be analytically subdivided into representative epochs where stationarity hypothesis is accomplished [33].

In the last years the use of Fourier transform, with the introduction of the personal computers, has been generalized in the analysis of EEG signals. Spectral decomposition of the EEG by computing the Fourier Transform has been used since the very early days of electroencephalography. The rhythmic nature of many EEG activities leads itself naturally to this analysis. Fourier Transform allows separation of various rhythms and an estimation of their frequencies independently of each other, a difficult task to perform visually if several rhythmic activities occur simultaneously. Spectral analysis can also quantify the amount of activity in a frequency band.

The Spectral Analysis of EEG signals has proved to be quite useful in comparing short samples of data from patients against age-matched normative values, as well as in sleep stage analysis, and quantification of drug, metabolic effects and various disease states [8]. But important information about peak timing is lost. The inclusion of the time evolution in the quantification of the EEG series is being an open problem.

Morphology and topography of sharp transients have been correlated with seizure type and therapeutic response to different medications and surgery. When it is working in the frequency domain it is useful to divide EEG activities into three different categories [11]:

(i) spontaneous non-paroxysmal or background, *(ii)* spontaneous paroxysmal activity, and *(iii)* activity evoked by external sensory stimulation. Consequently, it is quite obvious that in the frequency domain representation, rhythmic components are relatively enhanced at corresponding frequencies, whereas transients (for example, epileptic spikes, isolated paroxysm, etc.), are smeared over the spectrum and, therefore, are no longer recognizable. From this, it follows that the principal field of spectral analysis is the background activity, that means the first category mentioned above, whereas in the other two categories there exist only special cases to which standard spectral analysis can be successfully applied [8 − 12].

The results of EEG spectral analysis are often grouped into the traditional frequency bands, i.e., delta (less than 3.5 Hz), theta (3.5 − 7.5 Hz), alpha (7.5 − 12.5 Hz), beta (12.5 − 30 Hz) and gamma activity (above 30 Hz). There is much physiological and statistical evidence for independence of several of these bands, but their boundaries can vary a little according to the particular experiment being considered, and they can be adjusted as required [1 − 3].

The most popular way of performing frequency analysis has been to apply the Fast Fourier Transform (FFT) algorithm directly to a short (usually 1 − 4 sec) segment of digitized data [8 − 12]. In the methods mentioned before the time evolution is not taken into account or as in the case of the compress spectra, they only provide a visual tool with a difficult interpretation of the contained information.

4   Time-Frequency Analysis Based on Gabor Transform

Recently we introduced a method for the description of the EEG signals in a combined time-frequency space based on Gabor Transform [29 − 32]. For this purpose, we followed the Gabor's ideas introduced in 1946 [28], as a basic element of the proposed method. The Gabor Transform is equivalent to wavelets algorithms, with a fixed window. Moreover in the analyzed frequency range both methods are equivalent using an appropriate window [30]. The Gabor Transform is easily comprehensible by its analogies with the Fourier transform. This topic becomes relevant in the comparison with bibliography and in the application of the method by the physician team.

Then, we performed the Gabor Transform of the EEG signal denoted by $S(t)$ as follow:

$$\mathcal{G}_D(\omega_o, t_o) = \int_{-\infty}^{\infty} S(t)\, g_D^*(t - t_o)\, e^{i \cdot \omega_o \cdot t}\, dt \ , \qquad (9.1)$$

where $\mathcal{G}_D(\omega, t)$ symbolizes the Gabor Transform. ω represents the frequencies, t the time and $*$ denotes complex conjugation respectively. The parenthesis notation indicates that Gabor Transform varies with frequency and time.

The subscript D is used to denote the length of the epoch under consideration. $g_D(t - t_o)$ represents the temporal window being considered i.e., having a width D and localized temporarily in t_o.

From a practical point of view, the Gabor Transform can be thought as a Fourier Transform with a temporal window over short epochs. This window slides along the entire signal thereby providing information on the frequency changes with time. We used for $g_D(t)$ a slide Gaussian window with width D, as has been suggested by Gabor. In this way, the introduction of false frequencies is avoided when performing the Gabor Transform. This Gaussian window is recommended for the time frequency analysis in order to achieve maximal concentration in time and frequency.

According to this algorithm, one dimensional signals are represented in a combined time-frequency space. This function is situated on a lattice in this combined space, with clearances t_o and ω_o in the time and frequency axes respectively. If we represent the associated intensity to each point in the time-frequency lattice (as for example, the graph that corresponds to the *EEG Signal I* display in Fig. 2) we obtain a tridimensional pattern which is shown in Fig. 4 or its corresponding density level diagram (Fig. 5). In these two graphs, we normalized to the maximum intensity. In Fig. 5, eighty levels have been considered.

In this figure the high rhythmic activity corresponding to 2 Hz which is associated with the spike-wave paroxysmal activity of *EEG Signal I*.

These graphs, in particular the density level diagram with color codes, can be used as a visual tool for a qualitative description of the time evolution of the different spectral frequencies contained in the EEG recording. Again, it is difficult to extract the non-obvious information contained in both kinds of graphs. In this sense the most

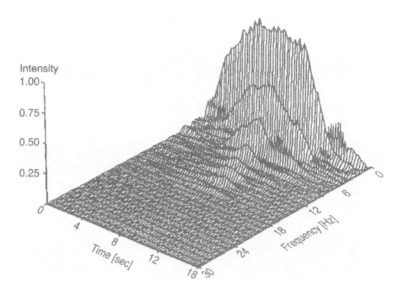

FIGURE 4. Associated intensity in the time frequency lattice for the *EEG Signal I* (Fig. 2). The dimension of the unitary shell was $\Delta\omega = 0.25 \ Hz$ and $\Delta t = 0.25 \ sec$.

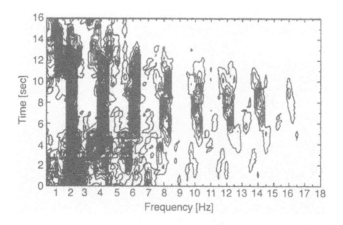

FIGURE 5. Density level diagram corresponding to the tridimensional pattern shows in Fig. 4. Eighty levels are displayed.

important advantage of our method is its systematic instrumentation and the rigorous mathematical background in comparison with the compress spectra and other similar methods [11]. Moreover its results are interpreted in easy manner.

Our method is based on the following definitions. We defined the evolution of the *spectral frequency content*, $\mathcal{B}^{(i)}$, for the frequency band i defined in the frequency interval ($\omega^{(i)}{}_{min}$, $\omega^{(i)}{}_{max}$) as:

$$\mathcal{B}^{(i)}(\omega, t) = \mathcal{G}_D^*(\omega, t) \cdot \mathcal{G}_D(\omega, t) \quad \forall \;\; \omega^{(i)}{}_{min} \leq \omega \leq \omega^{(i)}{}_{max} ; \tag{9.2}$$

then, the power spectral intensity for the i-band as a time function will be

$$I^{(i)}(t) = \int_{\omega^{(i)}{}_{min}}^{\omega^{(i)}{}_{max}} \mathcal{B}^{(i)}(\omega, t) \; d\omega , \tag{9.3}$$

and consequently the total spectral power intensity is

$$I_T(t) = 2 \int_0^\infty \mathcal{B}(\omega, t) \; d\omega , \tag{9.4}$$

where the spectral intensity content is defined in the frequency interval $(-\infty, \infty)$. Then the power spectral intensity per band relative to the total intensity will be

$$\rho^{(i)} = (\; I^{(i)}/I_T \;) \times 100 . \tag{9.5}$$

For the subsequent analysis of the EEG signal we define for the different bands the *mean weight frequency* value $\tilde{\omega}$ at time t as

$$\tilde{\omega}^{(i)}(t) = \left[\int_{\omega^{(i)}{}_{min}}^{\omega^{(i)}{}_{max}} \mathcal{B}^{(i)}(\omega, t) \; \omega \; d\omega \right] / \; I^{(i)}(t) \tag{9.6}$$

and the *main peak frequency* in the i-band at time t, $\omega_M^{(i)}$ as the frequency value for which $\mathcal{B}^{(i)}$ takes its maximum value in the frequency interval ($\omega^{(i)}{}_{min}$, $\omega^{(i)}{}_{max}$). That is,

$$\mathcal{B}^{(i)}(\omega_M, t) > \mathcal{B}^{(i)}(\omega, t) \quad \forall \;\; \omega \neq \omega_M \in (\omega^{(i)}{}_{min} , \omega^{(i)}{}_{max}) . \tag{9.7}$$

The Fourier spectrum will be represented by only one sharp peak at one frequency, when we are in presence of a mono-frequency signal. For this case, if we evaluate the mean weight frequency and the main peak frequency, they both will be the same. Therefore, when $\tilde{\omega}(t)$ is approximately equal to $\omega_M(t)$ *during a appreciable time interval* in some band, we shall say that we are in presence of a *quasi-monofrequency engagement,* in that band. We stress that in our formalism a signal will be quasi-monofrequency in this band if this engagement is observed during a *reasonable period,* related with the total time seizure duration (typically about 5% of time seizure duration).

Now, we introduce a new parameter, $\Delta^{(i)}$, and call it *monofrequency deviation.* This parameter, as a function of time, gives us an idea about the periods in which the engagements are relevant:

$$\Delta^{(i)}(t) = \mid \tilde{\omega}^{(i)}(t) - \omega_M^{(i)}(t) \mid . \qquad (9.8)$$

Moreover, in order to compare these new time series, for different bands and channels we normalized each one to its maximum value

$$(\Delta_N^{(i)}(t) = \Delta^{(i)}(t)/\Delta_{max}^{(i)}) . \qquad (9.9)$$

The importance of having introduced these new time series $\rho(t)$, $\tilde{\omega}^{(i)}(t)$, $\omega_M^{(i)}(t)$ and $\Delta_N^{(i)}(t)$ is that they allow us to characterize the paroxysmal activity and epileptic seizure as well as its evolution with time by means of quantifiable magnitudes which are independent of the signal's morphology. Also, throughout this formalism, valuable dynamical information about the epileptic seizure can be extracted (see section V).

In order to avoid the nonstationarity problems usually present in the EEG signal we worked with a gaussian window width $D = 4\ sec$ (1024 data). In this way, the stationary hypothesis was accomplished. The sample rate of the EEG signals was 256 data per second and the slide Gaussian window was displaced 64 data. Then, the resolution in the time-frequency space was: $\Delta\omega = 0.125\ Hz$, and $\Delta t = 0.25\ sec$.

For the analysis of both signals we considered the traditional frequency bands: δ ($0.5 - 3.5\ Hz$); θ ($3.5 - 7.5\ Hz$); α ($7.5 - 12.5\ Hz$); and the β-band was divided in two, β_1 ($12.5 - 18\ Hz$) and β_2 ($18 - 30\ Hz$).

4.1 Analysis of EEG Signal I

In Fig. 2, we display the EEG signal for 16 *sec* corresponding to a depth electrode in the hypocampus. In this sample, we can see an isolate paroxysm at 2 second and a paroxysmal activity (spike-wave) which starts around the *4th* second and finishes around the *13.5th* second. In Fig. 6 the spectral intensity per band relative to the total intensity, $\rho^{(i)}$ for *EEG Signal I* is shown.

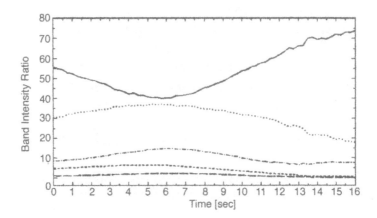

FIGURE 6. Power spectral intensity per band relative to the total intensity, as a function of time for the *EEG Signal I* showed in Fig. 2. Delta band (solid line); Theta band(dots line); Alpha band (dot dash line); Beta-1 (short dash line); Beta-2 (long dash line).

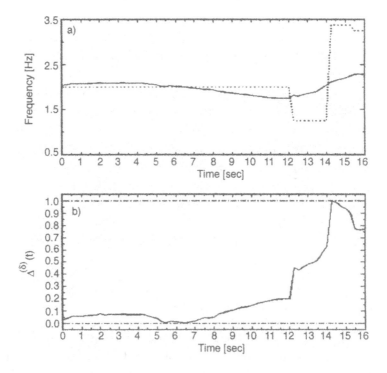

FIGURE 7. a) Time evolution of the mean frequency $\tilde{\omega}$ (solid line) and main peak frequency ω_M (dots line) for Delta band (*EEG Signal I*). b) Normalized monofrequency deviation Δ_N for Delta band (*EEG Signal I*).

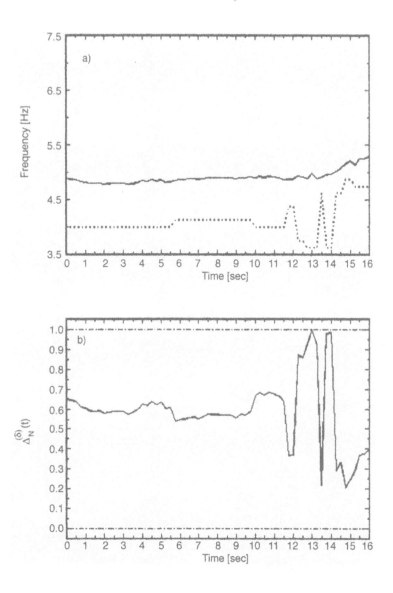

FIGURE 8. Same of Fig. 7, for Beta-1 band (*EEG Signal I*).

In Figs. 7 – 11.a we show the time evolution of mean weight frequency value, $\tilde{\omega}$, and main peak frequency, $\omega_M^{(i)}$, for the different bands considered in this work.

In Figs. 7 – 11.b we display the time evolution of the normalized monofrequency deviation $\Delta_N^{(i)}(t)$ for the different bands.

Looking at the time evolution of this quantities a good agreement

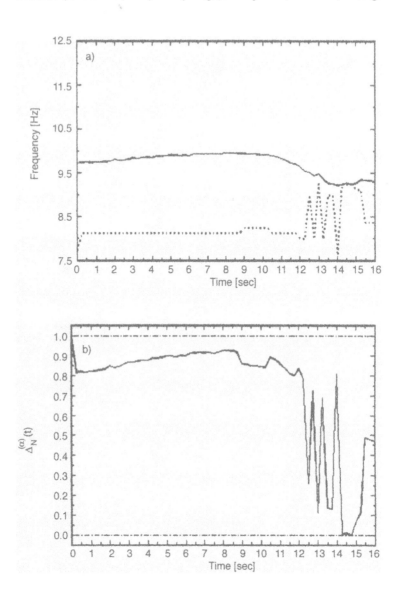

FIGURE 9. Same of Fig. 7, for Beta-2 band (*EEG Signal I*).

between the changes in $\rho^{(i)}$ and in the signal morphology can be established. In these figures we can see that when $\tilde{\omega} \sim \omega_M$ or equivalently when $\Delta_N^{(i)}(t) \sim 0$ a quasi mono-frequency band behavior has evolved. This can be understood as a quasi-frequency engagement.

From Fig. 6 this paroxistical activity can be characterized by the predominance of low frequencies (δ and θ bands). In particular from

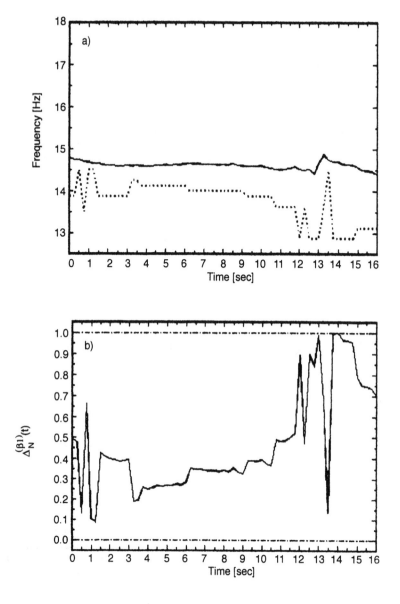

FIGURE 10. Same of Fig. 7, for Theta band (*EEG Signal I*).

Fig. 7, we can see that this signal presents a strong engagement in the δ-band between $0 - 12$ *sec* and β_1 weak engagements in the same interval (Fig. 10) which correspond to the wave and the spike activity respectively.

To conclude, this paroxysmal activity, *EEG Signal I*, can be characterized by a Delta dominant monofrequency activity.

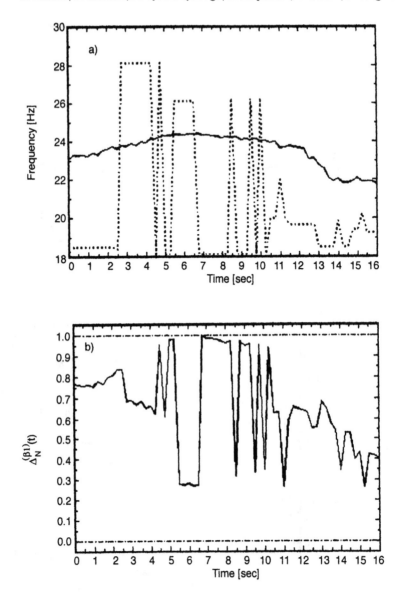

FIGURE 11. Same of Fig. 7, for Alpha band (*EEG Signal I*).

4.2   Analysis of EEG Signal II

In Fig. 3 the *EEG Signal II* for 64 *sec* corresponding to one depth
electrode in the left hypocampus region is shown. From a visual inspection, it is clear that around the 10*th* second the epileptic seizure
starts, and finishes around the 54*th* second. In order to discuss this
kind of time-frequency analysis, we divided the signal shown in Fig.

3 in different time intervals. These intervals were suggested by the modifications of the signal pattern. By visual inspection the intervals are: *0 to 10 sec* - slow background EEG activity; *10 to 12 sec* - fast activity and decrease of amplitude; *12 to 23 sec* - low and clonic activity with increasing amplitude; *23 to 29 sec* - spike discharges; *29 to 42 sec* - spikes and waves discharges with frequencies around $5 - 6\ Hz$; *42 to 54 sec* - polispike-wave complexes; *54 to 64 sec* - very slow activity.

In Fig. 12 we show the spectral intensity per band relative to the total intensity, $\rho^{(i)}$, as a function of time for the *EEG Signal II* shown in Fig 3.

In Fig. 13 to 17 we display the time evolution of $\Delta_N^{(i)}(t)$ for the different bands. Again, In these figures we can see that when $\Delta_N^{(i)}(t) \sim 0$ a quasi mono-frequency band behavior has evolved. This can be understood as a *quasi-monofrequency engagement* if its duration is greater than 2 *sec* and its value is $\Delta_N \leq 0.2$.

We want to stress that these new time series provide quantifiable an objective information about the frequency content and their relative intensities present in each interval of the EEG signal. In this way hidden information can be put in evidence.

0 to 10 sec : Up to 5 *sec* the predominance of the slow frequencies is observed, where the δ rhythm is the principal component. From 5 *sec* on the fast frequencies increase and the α and θ rhythm are the most important near 10 *sec* (Fig. 12). From the Fig. 3 we can observe that around to 10 *sec*, it is clear that the epileptic seizure starts. Between $8 - 10$ *sec* engagements in the θ ($\Delta_N^{(\theta)} < 0.1$), and β_1 ($0.2 < \Delta_N^{(\beta_1)} < 0.3$) bands are observed (Figs. 14 and 16). Note that this engagement in the β_1 band, are quite stable.

10 to 12 sec : The seizure start. The θ activity is dominant and α activity is the second in importance, with increecment of δ activity and a contribution about 20% and 10% in the β-bands (see Fig. 12). In Fig. 13 can be observed a very strong engagement in δ-band ($\Delta_N^{(\delta)} \approx 0.0$). Quite stable engagement in β_1-band ($\Delta_N^{(\beta_1)} \sim 0.25$) and tendency to engagement in β_2-band can be observed too (see Figs. 13, 16 and 17).

12 to 23 sec : It is clear the predominance of the α rhythm, with components within the θ rhythm. From the middle of this interval an increase in the intensity of the β_2 rhythm is observed (see Fig.

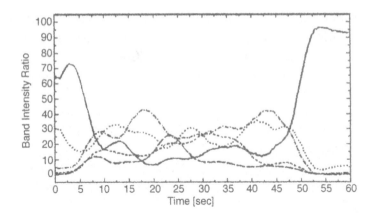

FIGURE 12. Power spectral intensity per band relative to the total intensity, as a function of time for the *EEG Signal II* showed in Fig. 3. Delta band (solid line); Theta band (dots line); Alpha band (dot dash line); Beta-1 (short dash line); Beta-2 (long dash line).

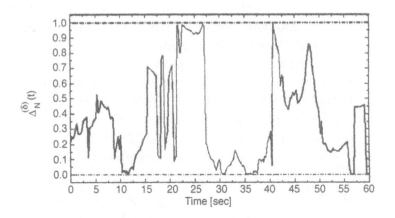

FIGURE 13. Time evolution for the normalized monofrequency deviation Δ_N for Delta band (*EEG Signal II*).

12). Clear engagements in δ ($\Delta_N^{(\delta)} < 0.2$), θ ($\Delta_N^{(\theta)} \sim 0.2$) and α ($\Delta_N^{(\alpha)} \sim 0.2$) and very strong engagements in β_1 ($\Delta_N^{(\beta_1)} \approx 0.05$) are observed (see Figs. 13 – 16).

A trend to engagement in β_2 band is observed at the beginning of the interval (see Fig. 17).

23 to 29 sec : The epileptic seizure is totally developed. The frequencies in the α and β_1 bands are predominant and coexist with low frequencies in the θ band (see Fig. 12). A strong δ engagement

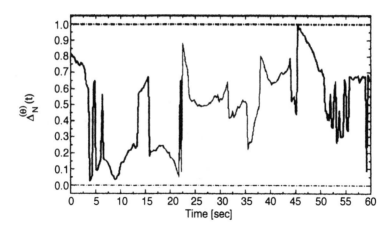

FIGURE 14. Same of Fig. 13, for Theta band (*EEG Signal II*).

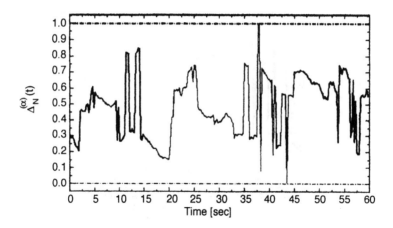

FIGURE 15. Same of Fig. 13, for Alpha band (*EEG Signal II*).

($\Delta_N^{(\delta)} < 0.1$) and short engagements in β- bands can be seen (see Figs. 13 and 16 – 17).

29 to 42 sec : We can see a strong decreasement in the relative intensity of the β rhythms and an increase for the θ and α rhythm. The strong δ engagement holds in the interval.

42 to 54 sec : The epileptic seizure is finishing (54 *sec*). The δ rhythm is completely dominant around the middle of the interval. Engagements in the δ band are observed at the end of the interval ($\Delta_N^{(\delta)} \sim 0.15$), and short engagements in the β-bands (see Fig. 13 and 16 – 17).

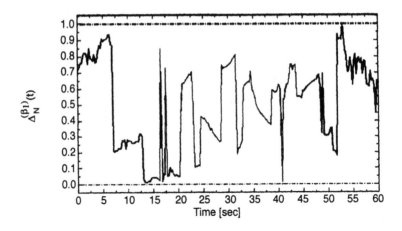

FIGURE 16. Same of Fig. 13, for Beta-1 band (*EEG Signal II*).

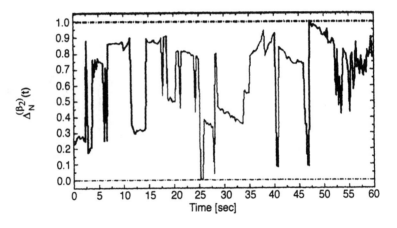

FIGURE 17. Same of Fig. 13, for Beta-2 band (*EEG Signal II*).

54 to 60 sec :

The seizure is over. The dominance of δ-band is total. Strong engagements are observed in δ-band and short in α-band (see Figs. 13 and 15)

The previous analysis suggests that this epileptic seizure, *EEG Signal II*, can be characterized during the first seconds by a strong engagement in the high frequency band (β_1) and another engagement in low frequency ones (δ). The low frequency engagement holds during all the seizure. A similar treatment can be done for all the available signals of the epileptogenic and propagation zones.

4.3 Information Transfer Analysis

As we said in the introduction, the most common methods of studying interactions between two channels are the cross-correlation and the cross-spectral analysis [11, 12]. The average amount of mutual information and nonlinear correlation are recently developed methods [13 − 16]. These methods all try to determine whether two channels have a common activity and, often, whether one channel contains activity induced by the activity in the other channel. Clearly proving causality is extremely difficult, but it can sometimes be inferred by measuring time differences.

The frequency engagements, pre and during seizure, can be used as a method for detecting the information transference between channels (brain zones), involved in the epileptogenic zone and the other zones of the brain. When we speak about *information transfer* we do not pretend to explain the underlying neurophysiology envolved in the seizure propagation but to how the dynamical behavior of the seizure in a band and in a determined channel of the epileptogenic zone, affect the dynamical behavior in other bands in channels that are not so close by.

The $\Delta^{(i)}$ parameter, as a time function, gives us an idea about the periods in which the engagements are relevant. It is interesting to note that we were not observed $\Delta^{(i)} \leq 0.2$ extended in time for the EEG background activity except for isolated paroxysms.

If we evaluate the linear correlation among the time series $\Delta^{(i)}$ arised from two different channels and corresponding band, as a function of a time shift τ, we can investigate when an engagement in one channel and band induces similar behavior in another channel and band. This means that the correlation will be large at some value of time lag τ if the first time series $(\Delta_N^{(i)})$ is a close copy of the second $(\Delta_N^{(j)})$ but lags in time by τ, i.e., if the first time series is shifted to the right of the second. The time τ is a measure of the delay in which the dynamical behavior in a channel and a band is copied by other channel or band. In consequence only the first seconds of delay are relevant.

From now on we will analyze the Delta-Delta, Delta-Theta and Delta-Alpha correlations because the engagements in the δ-band were identified as the relevant ones in the seizure. In Figs. 18 − 20 we display the lag correlation for τ between 0 − 3 *sec* for the

$\Delta_N^{(\delta)} \to \Delta_N^{(\delta)}, \Delta_N^{(\delta)} \to \Delta_N^{(\theta)}$ and $\Delta_N^{(\delta)} \to \Delta_N^{(\alpha)}$ respectively. In these Figs. the origin of the time shift corresponds to the $10th$ second, that is when the seizure starts.

All Figs. correspond to the correlation between a contact in the epileptogenic focus, left Hypocampus ($\mathbf{B'_2}$), and three signals provided by contacts located in: left Hypocampus ($\mathbf{B'_1}$), the same epileptogenic region; right Gyrus Cingular region ($\mathbf{C_2}$), the epileptogenic focus contralateral homologous anatomical region; and other in the left Amygdala ($\mathbf{A'_2}$), out of the immediate propagation region.

We observe in Fig. 18 very low correlation Delta - Delta for $\mathbf{B'_2} \to \mathbf{B'_1}$.

Also we can see low and almost constant correlation for $\mathbf{B'_2} \to \mathbf{C_2}$ and $\mathbf{B'_2} \to \mathbf{A'_2}$. Fig. 19 displays the Delta - Theta correlations. We observe very low and almost constant correlation between $\mathbf{B'_2} \to \mathbf{A_2}$. High correlations with decreasing values from the last second are observed for the other two contacts. A similar behavior is observed in Fig. 20, for Delta-Alpha correlations, that is high correlations with the contacts in the epileptogenic focus and contralateral anatomical homologous region, and low correlation with the contact that is out of the immediate propagation region. Although the mathematical linear correlation was done among all frequency bands of all available signals (contacts), those which present a meaningful behavior were the bands with the strongest engagements.

From these results, we can observe that an engagement in a band can induce engagements in another channel and in another bands, shedding light on a special characteristic of the seizure. Moreover, higher correlations are observed between signals in the epileptogenic region and the immediate propagation one and a less correlation with decreasing values correspond to signals in the out of immediately seizure propagation region (Figs. 18 - 20).

The behavior observed with these tools allows us to make the dynamic assumption that the epileptogenic zone (focus) acts as a global pacemaker with some characteristic frequency, determined by the frequency engagements.

Even though here we only show a few correlations, all of them has been evaluated. The agreement between the correlation behavior described above and the localization of the epileptic focus and the propagation areas was very good. Similar analysis was made for three

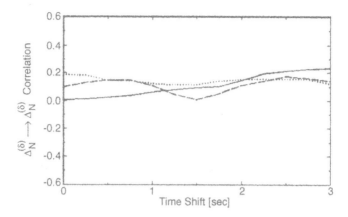

FIGURE 18. Delta-Delta lag correlation between monofrequency deviation time series (*EEG Signal II*) between a contact in the epileptogenic focus and contacts in: the same epileptogenic region $\mathbf{B'}_2 \rightarrow \mathbf{B'}_1$ (solid line); the epileptogenic focus contralateral anatomical region $\mathbf{B'}_2 \rightarrow \mathbf{C}_2$ (dots line); out of the immediate propagation region $\mathbf{B'}_2 \rightarrow \mathbf{A'}_2$ (dot dash line). The origin of the time shift correspond to the 10*th* second (seizure start) of Fig. 3.

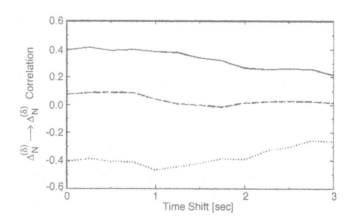

FIGURE 19. Same as Fig. 18 for Delta-Theta lag correlation (*EEG Signal II*).

different patients prone to surgery treatment with highly coincidence between both points of view, that is the physician localization and engagements band correlation [29 − 32]. Then, this technique is a possible tool to distinguish contacts in an epileptogenic region from another one.

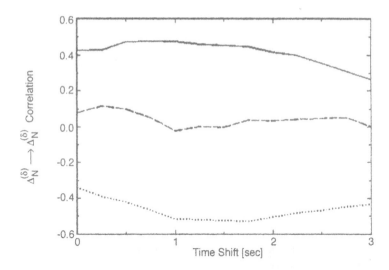

FIGURE 20. Same as Fig. 18 for Delta-Alpha lag correlation (*EEG Signal II*).

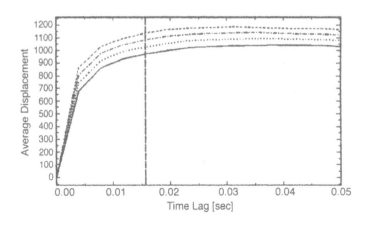

FIGURE 21. Average displacement $< S_D >$ versus time lag τ^* for the *EEG Signal II* segment $36 - 44$ sec. The curves correspond to embedding dimensions of $D_e = 11$ (solid line), $D_e = 12$ (dots line), $D_e = 13$ (dot dash line) and $D_e = 14$ (short dash line). The value of optimal time lag $\tau^* = \tau \cdot t_s = 0.0156$ sec is distinguished using a dashed vertical line.

5 Nonlinear Dynamics Analysis

The application of the nonlinear dynamic techniques to the analysis of EEG time series can be justified taken into account the characteristic behavior of the neuronal activity. The neurons are the constitutive elements of the brain, there are about 10^{11} and their interconnections are about 10^{14}. The neurons are highly nonlinear elements, then we can spect that the EEG activity, that can be though as the result of the temporal and spatial average of the postsynaptic potentials, present similar nonlinear characteristic or sometimes chaotic behavior [6, 7].

As we mentioned in the introduction, the treatment of EEG series under the approach of nonlinear dynamic systems has opened new possibilities for to the knowledge of the brain dynamic [5 − 7, 17 − 23]. However, the aims are not limited to this, but also to obtain new ways to quantify differences in the EEG series which have some kind of clinical application [24 − 27]. The insight gained by the concept of deterministic chaos for the EEG is that a seemingly disordered process may be governed by a relatively few simple laws, which could be determined [6, 7].

The metric approach, usually employed in nonlinear dynamical analysis, is based on distances and assumes the stationarity of the data sets. Distances between points in appropriate embedding dimension of the data are used to compute a set of metric properties. These quantities are difficult to compute, require large data sets and degrade rapidly with additive noise [34].

5.1 Stationarity

The terms non-stationarity or time-varying mean that the characteristic of a time series, such as mean, variance and spectral characteristic, change with time. Statistical tests of stationarity have revealed a variety of results depending on conditions, with estimates of the amount of time during which the EEG is stationary ranging from several seconds to several minutes [35 − 37]. However, as a practical matter, whether or not the same data segment is considered stationary, depends on the problem being studied, the type of analysis being performed, and the measured (features) used to characterize the data. This problem was not well studied and it has brought about a great variety of results exposed by different authors [6, 7].

Due to the great number of data of the EEG series that are necessary for the nonlinear metric treatment (satisfying all the mathematics hypothesis) a criterion of stationarity is almost impossible to satisfy in practice. Consequently, if the time series are nonstation-

ary, the metric algorithms *must not be used;* in other words, the
calculated magnitudes via these procedure will give wrong results.

If the probability distribution $f(x)$ associated with the measurements is normal or Gaussian, it can be completely characterized by
its mean (m) and its variance (σ^2). So, if we have a multivariate
normal distribution, the existence of fixed mean and variance should
be enough to ensure the stationarity of a Gaussian process.

A less restrictive requirement, called "weak stationarity" of order
n, is that the moments up to some order n are fairly stable with time.
Then the second order stationarity ($n = 2$), plus an assumption of
normality, is enough to produce complete stationarity [38]. Then, in
order to assure the stationarity of the EEG zones to be used for the
dynamical analysis, we employ the following procedure, based in the
stationarity criteria defined above [33].

a) We divide the total EEG time series in bins with a fixed length.
The election of the number of data in the bin must be large enough
to give reliable statistics in the mean and variance calculations and
will depend of the analyzed activity and the sample frequency.

b) The mean and variance are evaluated for each bin, and we look
for zones where these values did not change significatively for at least
some consecutive bins that provide enough data. From a practical
point of view, we impose that the fluctuation (δP) of the parameter
$P \equiv m$ or σ^2 satisfy $| \delta P | \leq \sigma_P^2$, where σ_P^2 is the variance of the
parameter P taken over all the bins.

c) We constructed the corresponding histogram for this zone, and
verified the normality of the obtained distribution.

After the stationary EEG time series portions has been chosen
(the portions that could give reliable results), we can evaluate
the metric parameters in a robust way, satisfying all mathematical
hypothesis.

5.2   Dynamical Systems

A dynamical system may be though as any set of equations, giving
the time evolution of the system from a knowledge of its previous
history. The dynamical behavior of most deterministic systems is
generally described by a finite set of differential equations. These
equations can be constructed when precise knowledge is available
about the elements of the system and about the type of interactions
that take place among these elements. When the equations can be

solved (either by analytical or numerical methods) the solution describes the systems behavior as function of the time under the conditions defined by the values of the parameters [39 − 41].

For a continuous-time dynamics we could write a set of m-ordinary differential equations for the m-independent variables $\mathbf{X}(t) = (x_1(t), x_2(t), \cdots, x_m(t))$, as

$$\frac{dx_1(t)}{dt} = G_\mu^1(x_1(t), x_2(t), \cdots, x_m(t))$$

$$\frac{dx_2(t)}{dt} = G_\mu^2(x_1(t), x_2(t), \cdots, x_m(t))$$

$$\ddots$$

$$\frac{dx_m(t)}{dt} = G_\mu^m(x_1(t), x_2(t), \cdots, x_m(t)) \qquad (9.10a)$$

or

$$\frac{d\mathbf{X}(t)}{dt} = \mathbf{G}_\mu(\mathbf{X}(t)) \qquad (9.10b)$$

where $\mathbf{X}(t) \in \mathrm{Re}^m$ is the state of the system and $\mathbf{G} : \mathrm{Re}^m \rightarrow \mathrm{Re}^m$ is called the vector field and it is always taken to be differentiable as often as needed. The index μ indicates that the system depends on a parameter μ (often it will be several parameters).

The state of the system at any point in time can be completely characterized by the set of observables $x_1, x_2, \cdots, x_m$. Furthermore the evolution of this system is determined by $x_1(t), x_2(t), \cdots, x_m(t)$, that is to say by a trace in a m-dimensional space spanned by $x_1, x_2, \cdots, x_m$, where m is the number of degree of freedom of the system. Such a system is called deterministic because its evolution is always defined by it current state. The laws governing this evolution are given by eq. (9.10) and express the rate of change of the state in terms of the observables [39 − 40].

In realistic situations observations are only sampled every t_s (sampling time) intervals of time, in consequence the time is considered as a discrete variable. In this case the evolution is given by a map from vectors in Re^m to other vectors in Re^m. Each one of these vectors is labeled by a discrete time

$$\mathbf{X}(n) = \mathbf{X}(t_0 + n\,t_s) \in \mathrm{Re}^m \qquad \text{with} \qquad n = 0, 1, 2, \ldots \quad (9.11)$$

and the evolution of the vector is given by the map

$$\mathbf{X}(n+1) \ = \ \mathbf{F}_\mu(\mathbf{X}(n)). \qquad (9.12)$$

The continuos-time and the discrete-time views of the dynamics can be connected if we remember that the time derivative can be approximated by

$$\frac{d\mathbf{X}(t)}{dt} \approx \frac{\mathbf{X}(t_0 + (n+1)\, t_s) \ - \ \mathbf{X}(t_0 + n\, t_s)}{t_s} \qquad (9.13)$$

which could be written as

$$\mathbf{F}_\mu(\ \mathbf{X}(n)\ ) \ \approx \ \mathbf{X}(n) \ + \ t_s\, \mathbf{G}_\mu(\ \mathbf{X}(n)\ ). \qquad (9.14)$$

5.3 Attractors

One gains a good understanding of the global behavior of the system by plotting the evolution of $\mathbf{X}(t)$ in m-dimensional space. The set of all possible states of the system is called the *state space* or *phase space*. In the continuous case the temporal evolution (or dynamical evolution), then leads to a curve in this space called *trajectory* or also orbit. In the discrete case, a sequence of points is obtained, usually called an *orbit*. One property of these systems is that the trajectories never cross [39 − 40].

The set of trajectories that represents the behavior of the system in dynamics equilibrium, after transient has died out, is called an *attractor* of the system. Examples are fixed points (equilibrium), limit cycles (periodic behavior), k-torus (quasi periodic with k frequencies), strange chaotic and non-chaotic attractors. An important property of the strange attractor is its *dimension*, which usually is *fractal* not integer. A further property is that the strange attractors posses self-similar structure. The *basin* of attraction for an attractor is the set of initial conditions (points in the state space) moving upon evolution toward a given attractor.

Some different concepts had been defined to describe properties of attractors, and their basins of attraction in the state space. Some of them are the generalized *Rényi dimensions, Lyapunov exponents, entropies*, between others [39 − 40].

5.4 Attractor Reconstruction

In mathematical models of physical dynamicals systems, the dynamic evolution is visualized in a state space whose dimension is given by the number of independent variables. In experiments the state space is usually not known beforehand and often only one variable of the system can be measured. Thus only a projection of a trajectory of the system with a usually high dimensional state space onto a single coordinate axis is given. It can be shown that one variable already contains most of the information about the total system and not just a minor part [40, 41]. In fact, the single variable considered develops in time not due to it own isolated dynamical law but (usually) is coupled to all the other independent variables of the system. Its dynamics, therefore, reflect the influence of all the other variables which in term react to the influence of the variable consider. This mutual interaction lets a single variable contains the dynamic of all the other ones.

From the discussion of dynamical system we see that if we were able to stablish values for time derivatives of measured variable, we could imagine finding the connection among the various derivatives and the state variables, namely, the differential equation, which produced the observations [40]. If we examine the formula for different derivatives, by example: $d\mathbf{X}(n) \, / \, dt$, $d^2\mathbf{X}(n) \, / \, dt^2$, $\ldots$, we see that at each step we adding the information already contained in the measurement $\mathbf{X}(n)$ measurements at other time step t_s (see eq. (9.14)). Packard et all. [42] introduced the idea of using time-delay coordinates to reconstruct the state space of an observed dynamical system.

The main idea is that we really do not need the derivatives to form a coordinate system in which to capture the structure of orbits on phase space, but that we could directly use the lagged variables $x(n + \tau) \; = \; x(t_0 + (n + \tau) \, t_s)$, where τ is some integer to be determined.

From the set of observations, multivariate vectors in D_e dimensional space

$$\mathbf{Y}^\tau(n) = (\, x(n), \; x(n+\tau), \; \cdots , \; x(n + (D_e - 1)\tau) \,) \qquad (9.15)$$

are used to trace out the orbit of the system. From now on, the superscript τ will be omitted except when it is necessary. Time evolution of the vectors $\mathbf{Y}$ is given by $\mathbf{Y}(n) \to \mathbf{Y}(n+1)$. In practice, the obvious questions of what time delay τ and what embedding dimension D_e to use in this reconstruction have had a variety of answers [40].

The purpose of this time delay embedding is to unfold the projection back to a multivariate state space that is representative of the original system. The general topological result of Mañé and Takens [43, 44] states that when the attractor has dimension D_A, all self-crossings of the orbit will be eliminated when one chooses $D_e > 2D_A$. These self-crossing of the orbit are a result of the projection, and the embedding process seeks to undo that. One may view the method of delays as a bridge between the temporal fluctuations of a single observable variable and the spatial characteristics of a dynamical system that it represents. The Mañé and Takens result is only a sufficient condition and provides no assistance when selecting a reconstruction delay for experimentally obtained data.

The identification of the minimum embedding dimension, $D_e^{(min)}$, is the subsequent step in our nonlinear dynamical analysis. From the mathematical point of view the attractor will be unfolded if we use the minimum embedding dimension $D_e^{(min)}$ or any $D_e > D_e^{(min)}$. In practice, working in any dimension larger than the minimum required by the data leads to excessive computation when we evaluate any metric parameters that we are interested in. It also enhances the problem of contamination by round-off or instrumental error since this "noise" will populate and dominate the additional $D_e - D_e^{(min)}$ dimensions of the embedding space where no dynamics is operating [40].

To choose the appropriate time delay τ and the minimum embedding dimension $D_e^{(min)}$ we follow the methods introduced by Rosenstein et al. [45] and Kennel et al. [40, 46] respectively. These methods were chosen because, according our criterium, they are robust, based on geometrical properties of the reconstructed attractor and quite independent of the signal to be analyzed even when it is noisy. In the following subsections we describe the implementation of these methods.

5.5   Choosing the Optimal Time Delay

The method of delays, the most widespread approach for the attractor reconstruction, depends upon the delay parameter τ. Here the choice of the time lag τ is almost arbitrary. However, in practice there are limitations. A value of τ that is too small, originates in little information gain between successive delay coordinates (that is, the vectors will have components which are almost identical) - *redundance* - , resulting in a reconstructed attractor which will be very close to the "diagonal", or identity line of the embedding space. On the other hand, if τ is very large, successive delay coordinates may become causally unrelated - *irrelevance* -, and trajectories on the attractor appear to wander all around phase space such that the structure is hard to detect. These effects increase their influence when we work with experimental time series and/or noisy finite data sets.

Recently Rosenstein *et al.* proposed a geometry-based method for choosing best delay times based on the optimal tradeoff between redundance and irrelevance [45]. Their method is based on the quantification of the expansion from the main diagonal by measuring the average displacement, $\langle S_{D_e} \rangle$, of the embedding vectors in dimension D_e from their original locations on the identity line.

They evaluated $\langle S_{D_e} \rangle$ as a function of τ by:

$$\langle\, S_{D_e}(\tau)\,\rangle \;=\; \frac{1}{M} \sum_{k=1}^{M} \| \mathbf{Y}^{\tau}(k) - \mathbf{Y}^{0}(k) \| \,, \tag{9.16}$$

where the superscripts denote the time delay between successive embedding components, M is the number of vectors for the corresponding dimension D_e and $\| \; \|$ denotes the Euclidean norm in this embedding space. $\langle S_{D_e} \rangle$ quantifies the decrease in redundance error and the increasing of the irrelevance error with increasing τ. Following the prescription given by Rosenstein *et al.*, we chose the best time lag as the point where the slope of the curve given by eq. (9.16) decreases less than 40% for the first time.

5.6   Choosing the Minimum Embedding Dimension

One of the important features of an attractor is that it is often a compact object in the phase space. Hence, points of an orbit on the attractor acquire neighbors in this phase space. The utility of this neighbors, among other things, is that they allow the information on how phase space neighborhoods evolve. This information can be used to generate equations for the prediction of the time evolution of new points on, or near the attractor. They also allow accurate computations of the Lyapunov exponents of the system [40, 47].

In an embedding dimension D_e that is too small to unfold the attractor, not all points that lie close to one another will be real neighbors because if we increase the embedding dimension they will be far one to another. Kennel *et al.* developed a method from geometrical considerations alone, using this criterium, in which a value for the minimum embedding dimension $D_e^{(min)}$ can be found [40, 46].

In dimension D_e each vector $\mathbf{Y}(k)$ has a nearest neighbor $\mathbf{Y}^{NN}(k)$, in the sense of Euclidean distance. Let $\mathbf{R}_{D_e}(k)$ the distance between the vectors $\mathbf{Y}(k)$ and $\mathbf{Y}^{NN}(k)$,

$$\mathbf{R}_{D_e}^2(k) = \| \mathbf{Y}(k) - \mathbf{Y}^{NN}(k) \|^2$$

$$= \sum_{j=0}^{D_e-1} [x(k+j\tau) - x^{NN}(k+j\tau)]^2 . \qquad (9.17)$$

In dimension $D_e + 1$ this distance betweem nearest neighbors will be

$$\mathbf{R}_{D_e+1}^2(k) = \mathbf{R}_{D_e}^2(k) + [x(k+D_e\tau) - x^{NN}(k+D_e\tau)]^2 . \qquad (9.18)$$

If $\mathbf{R}_{D_e+1}(k)$ is large, we can presume it is because the closeness of the two points being compared is due to the projection from some higher-dimensional attractor down to dimension D_e. By going from dimension D_e to dimension $D_e + 1$, we have "unprojected" these two points away from each other. Some threshold size R_T is required to decide when neighbors are false.

Then if

$$\frac{| x(k+D_e\tau) - x^{NN}(k+D_e\tau) |}{\mathbf{R}_{D_e}(k)} > R_T \qquad (9.19)$$

the nearest neighbors at time point k (or $t_0 + k\tau$) are considered false. In practice, for values of $10 \leq R_T \leq 50$ the number of false neighbors identified by this criterion is usually constant. With such a broad range of independence of R_T one has confidence that this is a workable criterion.

But as Kennel *et al.* [46] have observed, if one applies it to data from a pure stochastic system (noisy or very-high-dimensional system), it indicates that this set of observations can be embedded in a small dimension. If one increases the number of points analyzed, the apparent embedding dimension rises. The problem is that when one tries to populate uniformly (as "noise" will try to do) an object in D_e dimensions with a fixed number N of points, the points must move further and further apart as D_e increases because most of the volume of the object is at large distances. In fact, the mean distance between points will be proportional to N^{1/D_e}. If there are an infinite quantity of data, there would be no problem, but with finite quantity of data eventually all points have "near neighbors" that do not move apart very much as dimension is increased. As it happens, the fact that points are nearest neighbors does not mean they are close on a distance scale set by the approximate size R_A of the attractor.

If the nearest neighbor to $\mathbf{Y}(k)$ is not close, so that $\mathbf{R}_{D_e}(k) \approx R_A$, then the distance $\mathbf{R}_{D_e+1}(k)$ will be about 2 $\mathbf{R}_{D_e}(k)$. This means that apparent neighbors will be stretched to the extremities of the attractor when they are unfolded from each other.

Taken into account this fact, Kennel *et al.*, gave a second criterion for falseness of nearest neighbors that if

$$\frac{\mathbf{R}_{D_e+1}(k)}{R_A} \geq 2 \ , \tag{9.20}$$

then $\mathbf{Y}(k)$ and its nearest neighbor are false nearest neighbors. As a measure of R_A one can use the standard deviation of the data set, that is $R_A = \sigma(x_1, \cdots, x_N)$.

Following the prescription of Kennel *et al.* [46] nearest neighbors failing either of these two criteria are designated as false. Then data from a pure stochastic system are now identified as high dimensional. Scalar data from low-dimensional chaotic systems are identified as low dimensional.

5.7 Correlation Dimension

As we mentioned above, the embedding dimension is a very rough description of the "amount of space" an attractor occupies. Several kinds of dimensions have been introduced (Rényi dimensions) in order to give more precise description of the "complexity" of the system [40, 41]. Anyway, the correlation dimension, D_2, has become the most widely used measure in the literature.

When a dynamical system does not approach either equilibrium or a periodic state, the correlation dimension can be used as a measure by means of which it is possible to distinguish whether the system's apparently "random" behavior can be ascribed to the existence of a strange attractor or whether it may be due to external noise. If the attractor is chaotic then since nearby trajectories separate exponentially in time, we expect that most pairs of points $\mathbf{Y}(j)$, $\mathbf{Y}(k)$ with $j \neq k$ will be dynamically uncorrelated. Even though these points may appear to be essentially random, they all lie on the same attractor and therefore are correlated in phase space [40].

The algorithms for the evaluation of the correlation dimension assume the stationarity of the time series. In particular, the algorithm of Grassberger and Procaccia [48, 49] needs an estimation of the embedding dimension. The Grassberger-Procaccia algorithm estimates dimensions by examining the scale properties of the correlation integral $C(r; D_e)$ of the reconstructed attractor. For a given embedding dimension, D_e, the correlation integral is approached by

$$C(r; D_e) = \frac{2}{M(M-1)} \sum_{j \neq k}^{M} \theta(\,r\,-\,\|\,\mathbf{Y}(j)\,-\,\mathbf{Y}(k)\,\|\,)\,, \quad (9.21)$$

where $\theta(\ )$ is the Heavyside step function and M is the number of reconstructed vectors.

Therefore, $C(r; D_e)$ is interpreted as the fraction of pairs of points that are separated by a distance less than or equal to r. Grassberger and Procaccia showed that the correlation dimension D_2 can be given by

$$D_2 = \lim_{r \to 0} \lim_{D_e \to \infty} D_2(r; D_e)\,, \quad (9.22)$$

where $D_2(r; D_e)$ is the slope of log-log plot of $C(r; D_e)$ versus r,

$$D_2(r; D_e) = d[ln(\,C(r; D_e)\,)] \,/\, d[ln(\,r\,)] \quad (9.23)$$

In calculations involving experimental data, neither of the limits in eq. (9.22) can be taken. Small values of r are blurred by noise and by limitations on experimental accuracy, while large values of D_e are precluded by practical limitations on data set sizes and computing times. In practice, one infer the existence of an attractor when we observe a "plateau" area in the plot of $D_2(r; D_e)$ versus $ln(r)$. The flatness of the plateau is interpreted as evidence for a self-similar scaling region of some dimension and the convergence of the slopes with increasing embedding dimension indicates the reconstruction of a topological unique attractor. If the dimension is fractal and larger than two, it is taken as evidence of chaotic motion on a strange attractor.

5.8 Lyapunov Exponent

The presence of chaos in a dynamical system may be confirmed by measuring the largest Lyapunov exponent. It measures the average local rate of divergence of neighboring trajectories in the phase-space embedding and estimates the amount of chaos in the system [39−41]. If a system is known to be deterministic, a positive Lyapunov number can be taken as definition of chaotic system.

Recently Rosenstein et al. [47] have developed a new method for the evaluation of the largest Lyapunov exponent. Their method follows directly from the definition of the largest Lyapunov exponent and can be used with not very large and noisy data set.

In a dynamical system, two randomly chosen initial conditions will diverge exponentially at a rate given by the largest Lyapunov exponent [39 − 41]. This means that one can expect that a random vector of initial conditions will converge to the most unstable manifold since exponential growth in this direction quickly dominates growth (or contraction) along the other Lyapunov directions. Thus, the largest Lyapunov exponent is defined by

$$d(t) = C \, e^{\lambda_1 t} , \qquad (9.24)$$

where $d(t)$ is the average divergence at time t and C is a constant that normalizes the initial separation.

After the attractor is reconstructed, "nearest neighbor" of each point on the trajectory are localized imposing the additional constraint that nearest neighbors have a temporal separation greater than the mean period of the time series ($| \, k \, - k^* \, | \, > \, \hat{T}$). The mean period, $\hat{T}$, can be estimated, as the reciprocal of the mean frequency of the power spectrum. This allows us to consider each pair of neighbors as nearby initial conditions for different trajectories.

The largest Lyapunov exponent is then estimated as the mean rate of separation of the nearest neighbors.

From the definition of λ_1, eq. (9.24) , one assumes that the k-th pair of nearest neighbors diverge approximately at a rate given by the largest Lyapunov exponent,

$$d_k(i) \; = \; \| \; \mathbf{Y}(k+i\ t_s) \; - \; \mathbf{Y}(k^*+i\ t_s) \; \| \; \approx \; C_k \; e^{\lambda_1 \; (i \; t_s)} \quad (9.25)$$

and taking the logarithm of both side,

$$ln(\ d_k(i)\ ) \; \approx \; ln(\ C_k\ ) \; + \; \lambda_1 \; (i \; t_s) \; . \quad\quad (9.26)$$

Eq.(26) represents a set of approximately parallel lines (for $k = 1, 2, \cdots, M$), each with a slope roughly proportional to λ_1. Then, the largest Lyapunov exponent can be easily calculated as the slope (using a least-squares fit) of the "average" line defined by

$$< \; ln(\ d(i)\ )\ > \; = \; < \; ln(\ C\ )\ > \; + \; \lambda_1 \; (\ i \; t_s\ ) \; , \quad\quad (9.27)$$

where $\langle\ \rangle$ denotes the average over all values of k. This process of averaging is the key to calculate accurate values of λ_1 using small, noisy data sets, developped by Rosenstein $et\ al.$

One of the main problems found in the measurement of dimensions and Lyapunov exponents using embedded experimental time series is the amount of available data. Eckman and Ruelle [50] stablish the relation between the data-set size and the fundamental limitations for estimating dimensions and Lyapunov exponents in dynamical systems. In particular, they have shown that the Grassberger and Procaccia algorithm will not produce reliable dimensions larger than

$$D_2^{max} \; = \; 2 \; ln(N) \; / \; ln(1/\rho) \quad\quad (9.28)$$

with $\rho = r/(2R_{D_e}) \ll 1$. Here r is the radius of the ball containing the nearest neighbors, for which the platoe occurs, R_{D_e} is the radius of the reconstructed attractor, and N is the data-set size. For the Lyapunov exponent they said that the number of points needed to estimate it is about the square of the needed to estimate the dimension. In particular the minimum number will be:

$$N_{min} \; \gg \; (\ \frac{1}{\rho}\ )^{D_2} \quad\quad (9.29)$$

This last criteria is very difficult to satisfy in practice or when one work with experimental time series. Then in the cases for which the requirement for the recording time series will be beyond our scope, that is when only shorter data sets are available, it will be necessary to verify the convergence of values of the exponents as the number of points increase [22].

5.9   Analysis of Signal II

In order to study changes in the dynamic behavior in the *EEG Signal II* we divided the signal in intervals of 8 *sec* as follows: (*i*) preseizure (0 − 8 *sec*); (*ii*) start of the seizure (10 − 18 *sec*); (*iii*) full development of the seizure (21 − 29 *sec*) and (29 − 37 *sec*); (*iv*) end of the seizure (36 − 44 *sec*) and (44 − 52 *sec*).

For these intervals, we tested the stationarity following the criteria described above. The length of the bins used was 512 data and we used 4 bins consecutive. In the second interval (10 − 18 *sec*) these criteria of stationarity were not satisfied due to the fast changes in the *EEG Signal II* morphology. The time delay and the minimum embedding dimension were estimated following the method introduced by Rosenstein *et al.* [45] and Kennel *et al.* [46], and presented in subsections V.5 and V.6.

In Fig. 21 we present the average displacement $\langle S_{D_e}\rangle$ as a function of time lag for the EEG interval (36 − 44 *sec*) for different embedding dimensions $(D_e = 11, 12, 13, 14)$. The vertical line represents the optimal time lag ($\tau^* = \tau \cdot t_s = 0.0156\ sec$).

In Figs. 22.a and 22.b we show the percentage of false nearest neighbors (FNN) as a function of the embedding dimension for the same EEG interval and for the corresponding optimal time lag $\tau^* = 0.0156\ sec$. Each point represents the percent of FNN that fail in some of the two criteria stablish in subsection V.6. From Fig. 22.b we can see that for embedding dimensions grater than 11 the FNN percentage is less than 1%. In particular for this portion of EEG we chose $D_e^{(min)} = 11$ as the optimal embedding dimension and its correspondent percentage of FNN is about 0.5% (see Table 1).

In Table 1 we give the corresponding stationary portions of the *EEG Series II* considered, its length N, optimal time lags τ^*, minimum embedding dimensions $D_e^{(min)}$ and the FNN percentage associated. The corresponding approximate size of the attractor R_A and

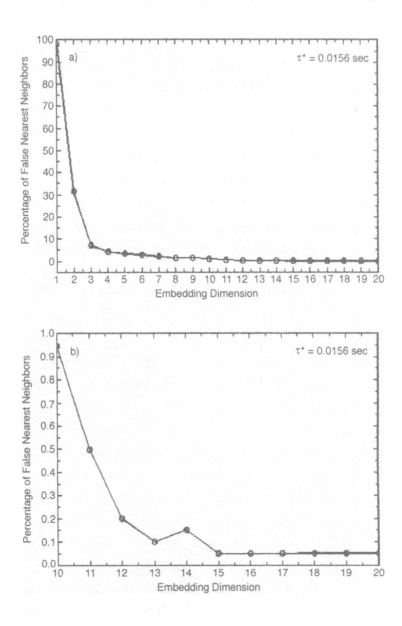

FIGURE 22. Percentage of false nearest neighbors as a function of embedding dimension for the *EEG Signal II* segment 36 − 44 *sec* evaluated with optimal time lag $\tau^* = 0.0156$ *sec*.

the threshold radio used R_T in the evaluation of the FNN are also consigned in Table 1.

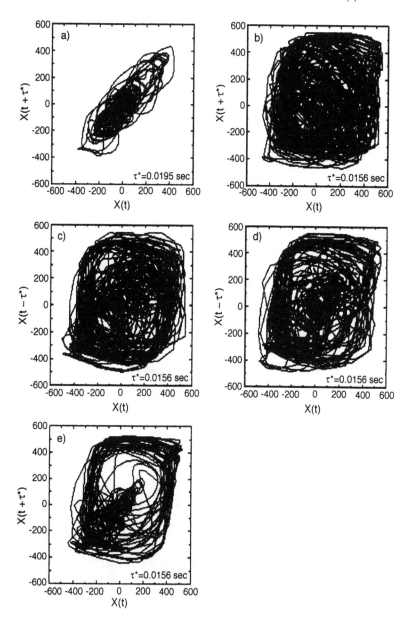

FIGURE 23. Bidimensional projection of the phase portrait corresponding to the *EEG Signal II* segments: a) $0 - 7$ *sec* (pre seizure); b) $21 - 29$ *sec* and c) $29 - 37$ *sec* (full develop of the seizure); d) $36 - 44$ *sec* and e) $44 - 52$ *sec* (end of seizure). τ^* is the time lag employed.

In Fig. 23 we display the two dimensional projection of the phase portrait corresponding to the *EEG Signal II* (Fig. 2) for the selected intervals (see Table 1). We must stress that although the seizure

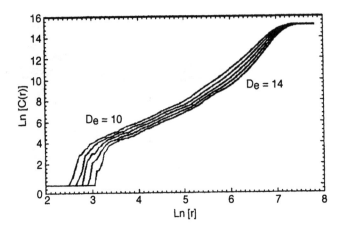

FIGURE 24. Plot of the logarithm of Correlation integral $ln[\,C(r)\,]$ versus $ln(r)$ for the *EEG Signal II* segment $36 \;-\; 44\;sec$ with time lag $\tau^* = 0.0156\;sec$ and embeddings dimensions between $D_e = 10$ to 14. The slope of this curves for $3.5 \leq ln(r) \leq 5.5$ gives the correlation dimension D_2.

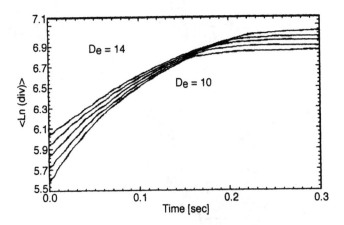

FIGURE 25. Plot of the average displacement $<\,ln(\,div)\,>$ as a function of time for the *EEG Signal II* segment $36 \;-\; 44\;sec$ with time lag $\tau^* = 0.0156\;sec$ and embedding dimensions between $D_e = 10$ to 14. The slope of this curves for $0.1 \leq t \leq 0.17\;sec$ gives the maximum Lyapunov exponent λ_1.

seems with a predominance of high frequencies, the underlying Delta frequency establishes the attractor limit cycle (see subsection IV).

Starting from the attractors reconstruction of each EEG portions, we characterized them by their corresponding correlation dimension D_2 and maximum Lyapunov exponent λ_1. In Fig. 24 we display the $ln[\,C(\,r\,)\,]$ versus $ln(r)$ for different embedding dimensions $(11 \leq D_e \leq 14\,)$ for the *EEG Signal II* interval $36 \;-\; 44\;sec$. Note

TABLE 9.1. Optimal value of time lag τ^* (in sec), minimum embedding dimension $D_e^{(min)}$, approximate attractor size R_A, threshold ratio R_T, percentage of false nearest neighbors $\%FNN$, correlation dimension D_2, and maximum Lyapunov exponent λ_1 for the indicated *EEG Signal II* of 8 sec segments (Data-set size $N = 2048$) respectively.

EEG Interval	τ^*	$D_e^{(min)}$	R_A	R_T	$\%FNN$	D_2	λ_1
0 − 8 sec	0.0195	8	158.654	30	0.000	4.30	4.6
21 − 29 sec	0.0156	13	256.136	30	0.851	2.60	4.0
29 − 37 sec	0.0156	10	247.417	30	0.697	2.50	4.0
36 − 44 sec	0.0156	11	234.087	30	0.499	2.05	3.5
44 − 52 sec	0.0156	13	214.861	30	0.701	2.15	3.0

that between $3.5 \leq ln(r) \leq 5.5$ all curves have almost the same slope.

In Fig. 25 we display the average displacement $\langle ln(div) \rangle$ as a function of time respectively, for the same embedding dimensions, again the curves present similar slopes for times $0.1 \leq t \leq 0.17$.

The radio of the attractor reconstructed in dimension D_e can be estimated as the average euclidean distance of the points in the attractor to the origin. For this portion of the *EEG signal II* $(36 − 44\ sec)$ we have $R_{D_e} = 759.478$ with $D_e = 11$. If we consider the upper limit of the plateau in the evaluation of D_2 as $ln(r) \sim 5.0$ we can estimate ρ for the chosen time series as $\rho \sim 760$. In consequence the maximum correlation dimension that is possible to observe with this data set is $D_2^{max} \sim 6$ (see eq. (9.28)), and the number of points needed for evaluation of the Lyapunov exponent will be $N_{min} \sim 1000$ if we consider $D_2 \sim 3$. Similar consideration were made for the other intervals consider in Table 1, in order to check the reliability of the obtained results. In the first interval the convergence of Lyapunov exponent was carefully analyzed due to we used a shorter data set.

In Table 1 we present the obtained values of correlation dimension D_2 and biggest Lyapunov exponent λ_1 for the different EEG intervals values. These two parameters were obtained as an average among the corresponding values obtained for all the considered embedding dimensions. These values are in good agreement with those given in literature concerning to background activity and epileptic seizure $[6 − 7, 17 − 25]$.

From this table we can see that the seizure is characterized by a drop in the corresponding values of the maximum Lyapunov exponent, in comparison with the preseizure values. These Lyapunov exponents are still positive denoting the presence of chaotic attractors. A similar situation is observed for the correlation dimension. From the values showed in Table 1 we can conclude that the dynamic behavior from the start of the seizure to the end are quite similar. They are all different to the preseizure behavior that corresponds to a background electrical activity.

Summarizing, in this contribution we found good evidence that during the epileptic seizure a transition takes place in the dynamic behavior of the neuronal network from a complex to a simple one. Furthermore, it yields insights with respect to the theory of how epileptic seizure occurs.

6   Final Remarks

In this work we reviewed a method that allows to perform the EEG time-frequency analysis in a systematic way. This method lets us give an accurate description of the time evolution of the rhythm defined in the EEG characterization.

We could generate time series that quantifies the dynamic behavior of the brain activity independently from the EEG signal morphology. In particular, the lag correlation among these new time series, gives a good picture about the information transfer process in the epileptiform activity throughout the brain.

We applied this method to intracraneal EEG records of epileptic refractory patients. We can conclude that the epileptic seizure could be characterized by a quasi mono-frequency activity for some of the bands. This characteristic can be used to detect precursors of the seizure, analyze the epileptic seizure and to study the dynamical changes in its time evolution. In particular in this work we found a good evidence that during the epileptic seizure a transition takes place in a dynamic behavior of the neuronal network from a complex to a simpler one.

The use of the present time-frequency analysis together with the clinical patient history and the visual assessment of the EEG, can contribute to the identification of the source of epileptic seizure activity and of its propagation within the brain. Furthermore, it yields new insights with respect to the behavior of the electrical activity during the seizure.

7 Acknowledgements

This work was partially supported by the Consejo Nacional de Investigaciones Científicas y Técnicas (CONICET). Two of us (S. B. and O. A. R.) undertook this work with the support of Fundación Alberto J. Roemmers (Argentina).

8 References

[1] *Handbook of Electroencephalography and Clinical Neurophysiology, Vol. I: Methods of Analysis of Brain Electrical and Magnetic Signals.*
(1987) Gevins A., Rémond A.(eds.), Elsevier, Amsterdam.

[2] *Handbook of Electroencephalography and Clinical Neurophysiology, Vol. II: Clinical Applications of Computer Analysis of EEG and Other Neurophysiological Signals.* (1988) Lopes da Silva F. H., Storm van Leeuwen W., Rémond A. (eds.), Elsevier, Amsterdam.

[3] *Current Practice of Clinical Electroencephalography, 2nd. ed.*
(1990) Daly D. D., Pedley T. A. (eds.), Reven Press Ltd., New York.

[4] Başar E. *Dynamics of Sensory and Cognitive Processing by the Brain. Springer Series in Brain Dynamics, Vol. I.* Springer-Verlag, Berlin, 1988.

[5] Başar E. and Bullock T. H. *Brain Dynamics. Springer Series in Brain Dynamics, Vol. II.* Springer-Verlag, Berlin, 1989.

[6] Başar E. *Chaos in Brain Function.* Springer-Verlag, Berlin, 1990.

[7] West B. J. *Fractal Physiology and Chaos in Medicine. Nonlinear Phenomena in Life Science, Vol 1.* World Scientific, Singapore, 1990.

[8] Gevins A. S. Overview of Computer Analysis. In *Handbook of Electroencephalography and Clinical Neurophysiology, Vol. I: Methods of Analysis of Brain Electrical and Magnetic Signals.* (Ref. [1]), pp. 31 - 83, 1987.

[9] Lopes da Silva F. a) EEG analysis: theory and practice. b) Computer-assisted EEG diagnosis: pattern recognition in EEG analysis, feature extraction and classification. In *Electroencephalography: Basic Principles, Clinical Applications and Related Fields.* Niedermeyer E. and Lopes da Silva F. (eds.), Urban & Schwarzenberg, Baltimore, (a) pp. 685-711, (b) pp. 713-732, 1982.

[10] Gotman J. In *Handbook of Electroencephalography and Clinical Neurophysiology, Vol. II: Clinical Applications of Computer Analysis of EEG and Other Neurophysiological Signals.* (Ref. [2]), pp. 546-597, 1988.

[11] Dumermuth G. and Molinari L. Spectral Analysis of EEG Background Activity. In *Handbook of Electroencephalography and Clinical Neurophysiology, Vol. I: Methods of Analysis of Brain Electrical and Magnetic Signals.* (Ref. [1]), pp. 85-130, 1987.

[12] Gotman J. The use of computers in analysis and display of EEG. In *Current Practice of Clinical Electroencephalography.* (Ref. [3]), pp. 51-83, 1990.

[13] Mars N. J. I. and Lopes da Silva F. H. EEG analysis methods based on information theory. In *Handbook of Electroencephalography and Clinical Neurophysiology, Vol. I: Methods of Analysis of Brain Electrical and Magnetic Signals.* (Ref. [1]), pp. 297-307, 1987.

[14] Pijn J. P. M. *Quantitative Evaluation of EEG Signals in Epilepsy: Nonlinear Associations, Time Delays and Nonlinear Dynamics.* Ph.D. Thesis, Universiteit van Amsterdam, 1990.

[15] Beldhis H. J. A., Suzuki T., Pijn J. P. M., Teisman M. and Lopes da Silva F. H. Propagation of epileptiform activity during development of amygdala kindling in rats: linear and non-linear association between ipsi- and contralateral sites. *Eur. J. of Neuroscience* 5, 944 - 954, 1993.

[16] Pijn J. P. M. and Lopes da Silva F. H. Propagation of electrical activity: nonlinear associations and time delays between EEG signals. In *Basic Mechanism of the EEG.* Zschocke S., Speckmann E. J. (eds.), Bishäuser, Boston, 1993.

[17] Babloyantz A. Evidence of chaotic dynamics of brain activity during the sleep cycle. In *Dimensions and Entropies in Chaotic Systems.* Meyer-Kress G. (eds.), Springer-Verlag, Berlin, pp. 241 - 245, 1986.

[18] Meyer-Kress G. and Layne S. C. Dimensionality of human electroencephalogram. In *Perspectives in Biological Dynamics and Theoretical Medicine.* Koslow S. H., Mandel A. J., Shlesinger M. F. (eds.), Ann. N. Y. Acad. Sci. 504, 1987.

[19] Layne S. C., Meyer-Kress G. and Holzfuss J. Problems associated with dimensional analysis of electroencephalogram data. In *Dimensions and Entropies in Chaotic Systems.* Meyer-Kress G. (eds.), Springer-Verlag, Berlin, pp. 246 - 256, 1986.

[20] Frank G. W., Lookman T., Nerenberg N. A. H., Essex C., Lemieux J. and Blume W. Chaotic time series analysis of epileptic seizures. *Physica D* **46**, 427 - 438, 1990.

[21] Gallez D. and Babloyantz A. Lyapunov exponents for nonuniform attractors. *Phys. Lett.* **161**, 247 - 254, 1991.

[22] Gallez D. and Babloyantz A. Predictability of human EEG: a dynamical approach. *Biol. Cybern.* **64**, 381 - 391, 1991.

[23] Pijn J. P. N., Van Neerven J., Noest A. and Lopes da Silva F. Chaos or noise in EEG signals; dependence on state and brain site. *Electroencephalog. Clin. Neurophysiol.* **79**, 371 - 381, 1991.

[24] Iasemidis L. D., Sackellares J. C., Zaveri H. P. and Williams W. J. Phase space topography and the Lyapunov exponent of electrocorticograms in partial seizures. *Brain Topography* **2**, 187 - 201, 1990.

[25] Lehnetz K. and Elger C. E. Spatio-temporal dynamics of the primary epileptogenic area in temporal lobe epilepsy characterized by neuronal complexity loss. *Elec. and Clinical Neurophys.* **95**, 108 - 117, 1995.

[26] Blanco S., Costa A., Jacovkis P. and Rosso O. A. Characterization of the dynamical evolution of an epileptic seizure. *Proceedings of Primer Coloquio Latinoamericano de Matematica Aplicada a la Industria y la Medicina.* (in press), 1996.

[27] Blanco S., Kochen S., Quian Quiroga R., Rosso O. A. and Salgado P. Characterization of background electrical activity of brain structures by nonlinear dynamical parameters. (to be published), 1996

[28] Gabor D. Theory of communication. *J. Inst. Elec. Eng.* **93**, 429 - 459, 1946.

[29] Blanco S., Quian Quiroga R., Rosso O. A. and Kochen S. Time-frequency analysis of electroencephalogram series. *Phys. Rev. E* **51** 2624 - 2631, 1995.

[30] Blanco S., D'Attellis C., Isaacson S., Rosso O. A. and Sirne R. O. Time-frequency analysis of electroencephalogram series (II): Gabor and Wavelet Transform. *Phys. Rev. E.Physical Rev. E***54**,No. 5, 1996.

[31] Blanco S., Kochen S., Rosso O. A. and Salgado P. Application of the time-frequency analysis to seizure EEG activity in candidates for epileptic surgery. *IEEE Eng. Med. Biol. Mag.* (in press), 1996.

[32] Blanco S., Kochen S., Rosso O. A. and Salgado P. Characterization of epileptic seizure using Gabor Transform. *Proceedings of Primer Coloquio Latinoamericano de Matematica Aplicada a la Industria y la Medicina.* (in press), 1996.

[33] Blanco S., García H., Quian Quiroga R., Romanelli L. and Rosso O. A. Stationarity of the EEG series. *IEEE Eng. Med. Biol. Mag.* **14**, 395 - 399, 1995.

[34] Caputo J. G. Practical remarks on the estimation of dimension and entropy from experimental data. In *Measures of Complexity and Chaos.* Abraham N. B., Albano A. M., Passamante A. and Rapp P. (eds.) NATO, Series B208, pp. 99 - 110, 1989.

[35] Gasser T. General characteristic of the EEG as a signal. In *EEG Informatics: a Didactic Review of Methods and Applications of EEG Data Processing.* Remond A. (eds.), Elsevir, New York, pp. 37 - 55, 1977.

[36] McEwen J. A. and Anderson G. B. Modeling the stationarity and gaussianity of spontaneous electroencephalographic activity. *IEEE Trans. Biomed. Engng.* **22**, 361 - 369, 1975.

[37] Sugimoto H., Ishii N. and Suzumura N. Stationarity and normality test for biomedical data. *Comput. Program. Biomed.* **7**, 293 - 304, 1977.

[38] Jenkins G. M. and Watts D. *Spectral Analysis and Its Applications.* Holden-Day, San Francisco, 1968.

[39] Eckmann J. P. and Ruelle D. Ergodic theory of chaos and strange attractors. *Rev. Mod. Phys.* **57**, 617 - 656, 1985.

[40] Abarbanel H. D. I., Brown R., Sidorowich J. J. and Tsimring L. S. The analysis of observed chaotic data in physical systems *Rev. Mod. Phys.* **65**, 1331 - 1392, 1993.

[41] Peitgen J. O., Jürgens H. and Soupe D. *Chaos and Fractals: New Frontiers of Science.* Spring-Verlag, New York, 1992.

[42] Packard N. H., Crutchfield J. P., Farmer J. D. and Shaw R. S. Geometry from a time series, *Phys. Rev. Lett.* **45**, 712 - 716, 1980.

[43] Mañé R. On the dimension of compact invariant set of certain nonlinear maps. In *Dynamical Systems and Turbulence, Warwick 1980.* Lecture Notes in Mathematics 898, Springer-Verlag, pp. 230 - 242, 1981.

[44] Takens F. Detecting strange attractors in turbulence. In *Dynamical Systems and Turbulence, Warwick 1980.* Lecture Notes in Mathematics 898, Springer-Verlag, pp. 366 - 381, 1981.

[45] Rosenstein M. T., Collins J. J. and De Luca C. J. Reconstruction expansion as a geometry-based framework for choosing proper delay times. *Physica D* **73**, 82 - 98, 1994.

[46] Kennel M. B., Brawn R. and Abarbanel H. D. I. Determining embedding dimension for phase-space reconstruction using a geometrical construction. *Phys. Rev. A* **45**, 3403 - 3411, 1992.

[47] Rosenstein M. T., Collins J. J. and De Luca C. J. A practical method for calculating largest Lyapunov exponents from small data sets. *Physica D* **65**, 117 - 134, 1993.

[48] Grassberger P. and Procaccia P. Characterization of strange attractors. *Phys. Rev. Lett.* **50**, 346 - 349, 1983.

[49] Grassberger P. and Procaccia P. Measuring the stragness of strange attractors. *Phyica D* **9**, 189 - 208, 1983.

[50] Eckmann J. P. and Ruelle D. Fundamental limitations for estimating dimensions and Lyapunov exponents in dynamical systems. *Physica D* **56**, 185 - 187, 1992.

Chapter 10

Characterization of Epileptic EEG Time Series (II): Wavelet Transform and Information Theory

Carlos E. D'Attellis
Lucas G. Gamero
Susana I. Isaacson
Ricardo O. Sirne
María E. Torres

1   Introduction

Records of brain electrical activity from depth and scalp electrodes are used to localize the origin of seizure discharges in epileptic patients who are candidates for surgical removal of the seizure focus. In clinical practice, the epileptogenic loci is infered from visual analysis of the interictal and ictal discharges. Automated systems may be used to detect signal epochs that contain transients, patterns, and characteristic features of abnormal conditions. There are two basic areas of clinical application: 1) an automatic system for data reduction in long-term EEG; or 2) as a short-term detector of epileptic transients. Several techniques have been applied in order to solve the problem of computer assisted detection of epileptiform transients as previously mentioned by Blanco *et al.* in this book and others including template matching [11], parametric [1], mimetic [8] and syntactic [32] methods, neural networks [10], expert systems [12], phase-space topography [14], wavelet transforms [22, 23], and recently, polynomial spline and multiresolution frameworks [21].

As it was recently pointed out [9], the electroencephalography — despite its widespread use— is one of the last routine clinical procedures to be fully automated.

In this work we introduce two new approachs for localizing epileptic events in EEG in a multiresolution framework as introduced by

Unser, *et al.* [31]. These methods were found especially suited for the digital treatment of nonstationary signals, and in particular, for EEG analysis [21].

The first method here presented, based on the energy function, provides a solution to some basic problems in events detection of EEG records:

1. Localization of events in the time-frequency domain.

2. Time localization and characterization of the different types of epileptiform events.

3. Possibility of real-time implementation using fast algorithms.

In order to use a fast multiresolution algorithm for detecting transients, we introduce a simple decomposition of the energy function at each resolution level. In this way, we obtain an event-detector based on digital filters designed from a dyadic scheme.

The second method deals with a particular aspect of the relation between signal analysis and nonlinear dynamics, which is the detection of slight changes in parameters of nonlinear dynamical systems using information measures obtained from experimental data. Physiologists [20, 33] suggest that during the seizure there is a transition from one chaotic system to another of different complexity. Iasemidis [14] proposed a modification of Wolf method [35] for estimating Lyapunov exponents taking into account the nonstationarity of the signal. The results obtained indicate that the largest average Lyapunov exponent can be useful in changes detection. The computational cost of this method is very high due to the algorithm itself and the amount of data required. In order to overcome the already mentioned drawback, we analyze the link between Lyapunov exponents and entropy. In fact, as it was pointed out by Pesin [19], under some hypothesis on the nonlinear system both concepts are related to each other. Thus, complexity changes in a nonstationary signal are related to the Lyapunov exponents and these exponents are connected with the notion of entropy. But the entropy itself is not useful while analyzing nonstationary signals. In fact, the computation of the entropy evolution of the EEG signal does not exhibit the changes introduced in the parameters of a nonlinear model [28].

Looking for an improvement of the entropy performance, we propose a new method based on the wavelet transform [17], generally recognized to be useful for studying nonstationary phenomena and shown to be of value in event detection [6]. A combination of the multiresolution wavelet analysis with the idea of entropy proposed in [28]is here presented. We show that it provides a new tool, called *multiresolution entropy* (MRE), useful for analyzing nonstationary signals and for detecting and localizing slight changes in nonlinear dynamical systems.

The notion of multiresolution entropy has the following advantages: 1) over the entropy, because it is capable to detect changes in a nonstationary signal due to the localization characteristics of the wavelet transform; 2) over the calculation of Lyapunov exponents, because the computational burden is significantly lower since the algorithm involves the entropy computation and uses fast wavelet transforms in a multiresolution framework.

In order to make this chapter self-contained, we present the theoretical background already developed in [6] and [29, 27].

2   Data Collection

Stereo electroencelography (SEEG) is a procedure to measure the electrical activity of the brain in a 3-D way. In each patient, the strategy of the implanted electrodes is planned according to the spatial and temporal organization of the ictal discharges that are simultaneously correlated with clinical symptomatology. A study was carried out on patients using 12 multilead electrodes, 2 cm long and 1.5 mm across. The analysis of interictal and ictal data was accomplished by visual analysis of the SEEG record. Each signal was amplified and filtered using a 1-40 Hz band-pass filter. A four-pole Butterworth filter was used as an anti-aliasing low-pass filter. The SEEG was digitalized at 256 Hz through a 10 bits A/D converter. We took selected EEG signals time intervals from depth electrodes in a patient candidate for surgical treatment.

3   Theoretical Background

3.1   Wavelet Analysis and Filters

Suppose that $\psi(t) \in L^2(\mathbb{R})$ is any basic wavelet, i.e., it verifies [16, 17]

$$\int_{-\infty}^{\infty} \psi(t)dt = 0.$$

The integral wavelet transform $W_\psi f$ of a finite energy signal $f(t)$ is defined by:

$$(W_\psi f)(b, a) = |a|^{-1/2} \int_{-\infty}^{\infty} f(t)\overline{\psi(\frac{t-b}{a})}dt. \qquad (10.1)$$

Where a is the scale parameter, b/a is the shift in time, and the bar indicates complex conjugate.

Using a complex exponential function plus a Hanning window compactly supported as a basic wavelet, Senhadji *et al.* [22, 22] have shown the usefulness of wavelet analysis in the detection of spikes and spikes-and-waves in the EEG. The algorithm proposed by these authors calculates the wavelets coefficients using a discretized version of the integral wavelet transform (10.1). This demands the calculation of an approximate integral for each pair of chosen parameters a, b, so it is a computationally expensive method. To obtain a fast event detector, we propose an algorithm based on digital filters, as it was recently done by Schiff *et al.* [21].

We analyse the event localization problem in the framework given by the polynomial spline wavelet transform [31]. The multiresolution representation obtained from this wavelet transform and the digital filters derived allow us time localization of epileptiform activity in such a way that the requirements specified above are verified.

We choose the multiresolution representation given by [31], using the compact support cubic spline $\phi(t)$ as scaling function

$$\phi(t) = \begin{cases} 1 - |t| + (1/6)|t|^3 - (1/3)(1 - |t|)^3 & \text{if } |t| \leq 1 \\ (2 - |t|)^3/6 & \text{if } 1 \leq |t| \leq 2 \quad (10.2) \\ 0 & \text{if } |t| > 2. \end{cases}$$

The corresponding wavelet function $\psi(t)$ is:

$$\psi(t) = \frac{1}{40320}[-\phi(2t+6) + 124\phi(2t+5) - 1677\phi(2t+4) +$$
$$+ \ 7904\phi(2t+3) - 18482\phi(2t+2) + 24264\phi(2t+1) -$$
$$- \ 18482\phi(2t) + 7904\phi(2t-1) - 1677\phi(2t-2) +$$
$$+ \ 124\phi(2t-3) - \phi(2t-4)].$$

In order to analyze the time-frequency localization properties, we calculate the values of the center and radius of the time and frequency windows and obtain the value of the window's area in the time-frequency plane $area = 2.0013$, i.e. almost the optimal value 2, of the minimun possible window's area [5]. Thus, the selection of this wavelet guarantes a good localization in the time-frequency plane, which was the first of the specifications in the first section. The numerical implementation of the algorithm, is considered below.

The wavelet decomposition for a given EEG signal $s(t)$, initially represented by its polynomial spline coefficients at resolution 0, is:

$$s(t) = \sum_{k=-\infty}^{\infty} c_0(k)\phi(t-k) =$$

$$= \sum_{k=-\infty}^{\infty} c_N(k)\phi(2^{-N}t - k) + \sum_{j=1}^{N}\sum_{k=-\infty}^{\infty} d_j(k)\psi(2^{-j}t - k) \quad (10.3)$$

where the numbers $d_1(k), d_2(k), \ldots, d_N(k)$ are the wavelet coefficients, and the sequence $\{c_N(k)\}$ represents the coarser resolution signal at resolution level N. If this decomposition is carried out over all resolutions levels, the following wavelet expansion is obtained:

$$s(t) = \sum_{j=-\infty}^{\infty}\sum_{k=-\infty}^{\infty} d_i(k)\psi(2^{-j}t - k) \quad (10.4)$$

At each level j, the series in (10.4) has the property of complete oscillation [5], which makes the decomposition useful in applications to event time localization.

At this point, it is convenient to introduce two digital filters that will be used in the algorithm. They are given by the transfer functions:

$$B^{-1}(z) = \frac{6}{z + z^{-1} + 4}$$

$$A^{-1}(z) = \frac{5040}{z^3 + z^{-3} + 2416 + 1191(z + z^{-1}) + 120(z^2 + z^{-2})}.$$

As usual, we indicate with $b^{-1}(k)$ and $a^{-1}(k)$ the impulse responses of these filters.

Following [31], a fast recursive scheme for obtaining the expansion coefficients in (10.3) is given by:

$$c_0(k) = [b^{-1} * s](k) \tag{10.5}$$
$$c_{i+1}(k) = [v^* * c_i]_{\downarrow 2}(k) \tag{10.6}$$
$$d_{i+1}(k) = [w^* * c_i]_{\downarrow 2}(k) \tag{10.7}$$

where the symbol $*$ represents discrete convolution and

$$v^*(k) = (1/2)[[a^{-1}]_{\uparrow 2} * a * u](k) \tag{10.8}$$
$$w^*(k) = (1/2)[[a^{-1}]_{\uparrow 2} * u_s * \delta_1](k) \tag{10.9}$$

$$u_s(k) = (-1)^k u(k), \quad \delta_1 * a(k) = a(k-1)$$

$$[a]_{\uparrow 2}(k) = \begin{cases} a(k/2) & \text{if k is even} \\ 0 & \text{if k is odd} \end{cases}$$

and

$$[a]_{\downarrow 2}(k) = a(2k)$$

Summing up, the coefficients of the expansion (10.3) are calculated with the digital filters of equations (10.8) and (10.9).

3.2 Entropy

Because of the sensitivity of a chaotic system with its initial conditions there is a change in the available information about the states of the system. This change can be thought of as creation of information if we consider that two initial conditions that are different but undistinguishable (with a certain precision) evolve into distinguishable states after a finite time. A notion of entropy was first introduced by Shannon in 1948, in a famous work which originated the information theory.

Given an event with several possible outcomes with probabilities p_j ($j = 1...N$), the information gained by one particular outcome is given by $I_j = Ln(1/p_j)$. The intuitive reasoning for I_j is that the rarer the event, the greater the information gained when it does happen. The information gained is uncertainty lost. The total uncertainty represented by an ensemble of events (for example a time series) defined by Shannon [24] is equal to the weighted sum of individual uncertainties, i.e.,

$$H = -\sum_{i=1}^{N} p_i Ln(p_i) \qquad (10.10)$$

understanding that, $uLn(u) = 0$ if $u = 0$. The entropy H is a measure of the information needed to locate a system in a certain state j^*; it means that H is a measure of our ignorance about the system.

4   Method I: Energy Based Detection Algorithm

With the filters presented in section 3 we build an algorithm for detecting epileptic events in EEG.

When the family $\{\psi_{k,j}(t) = \psi(2^{-j}t - k)\}$ is an *orthonormal* basis in $L^2(\mathbb{R})$, the concept of energy is linked with the usual notion derived from the Fourier theory, and the sum of the square of the coefficients of the series is the energy of the function, i.e.

$$\|s\|^2 = \sum_{k,j} |d_{k,j}|^2 = \sum_{k,j} | < s, \psi_{k,j} > |^2$$

when the wavelet decomposition is given by (10.4). But the wavelets we are using belong to the more general class of *biorthogonal* wavelets [31]. This means that there exists a function $\tilde{\psi}(t)$ such that:

$$< \psi(2^{-i}t - k), \tilde{\psi}(2^{-j}t - l) > = \begin{cases} 2^i & \text{if } i = j \text{ and } k = l \\ 0 & \text{otherwise.} \end{cases}$$

The family $\tilde{\psi}_{j,k} = \tilde{\psi}(2^{-j}t - k)$ is called the *dual basis* of $\psi_{j,k}$. Every signal $s(t)$ can be written as:

$$s(t) = \sum_{j,k} d_j(k)\psi_{j,k}(t) =$$

$$= \sum_{j,k} \tilde{d}_j(k)\tilde{\psi}_{j,k}(t)$$

where

$$d_j(k) = <s, \mathring{\psi}_{j,k}>$$
$$\tilde{d}_j(k) = <s, \psi_{j,k}>$$

In this (biorthogonal) case, the energy of the signal $s(t)$ is given by

$$\|s\|^2 = \sum_{j,k} 2^j d_j(k) \tilde{d}_j(k) \tag{10.11}$$

The dual $\tilde{\psi}(t)$ is in itself a wavelet given by [31]

$$\tilde{\psi}(t) = \sum_k (a * [a_s * a]_{\downarrow 2})^{-1}(k)\psi(t-k),$$

and the corresponding scaling function, $\tilde{\phi}(t)$, is

$$\tilde{\phi}(t) = \sum_k a^{-1}(k)\phi(t-k).$$

The digital filters corresponding to the dual wavelet analysis are:

$$v^\star(k) = (1/2)u(k) \tag{10.12}$$
$$w^\star(k) = (1/2)[a_s * u_s * \delta_1](k) \tag{10.13}$$

and the initialization sequence is

$$\tilde{c}_0(k) = [a * b^{-1} * s](k). \tag{10.14}$$

The equations (10.5)–(10.9) and (10.12)–(10.14), (10.6) and (10.7) allow us to compute recursively the coefficients involved and to obtain the energy function (10.11). Since we are using a dyadic decomposition of the range of frequencies, from a signal of M samples we have $M/2^j$ coefficients at level j. In order to obtain an accurate detection of the events, we uniformily distribute the "atoms" of energy in (10.11) —i.e., the terms $2^j d_j(k)\tilde{d}_j(k)$ along 2^j points [6].

Defining:

$$e_j(r) = d_j(k)\tilde{d}_j(k) \tag{10.15}$$

for integers r in the interval $(k-1)2^j < r \leq k2^j$, the energy at each resolution level $j = 1, \ldots, N$, is

$$E_j = \sum_{r=1}^{M} e_j(r)$$

and the energy in each sampled time $r = 1, \ldots, M$ is:

$$E(r) = \sum_j e_j(r).$$

Different types of epileptic events can be characterized for the values $e_j(r)$ at different resolution levels. The detection is made when the value $e_j(r)$ is greater than a threshold D_j, defined for each level. The thresholds' values were choosen according to a statistical analysis of the EEG series, taken as $D_j = \bar{x} + k_j\sigma$, where $\bar{x}$ is the mean value of the energy e_j at each resolution level j, σ is the standard deviation of the values of e_j, both calculated every eight seconds and k_j is a suitable constant depending on level j. A constant value is assigned when the energy function is greater than the corresponding threshold.

4.1   Results and Comparisons

As mentioned above, different epileptic events appear at different resolution levels. Figure 1 shows eight seconds of signal, levels $j = 3$ and $j = 5$ of energy. It shows that the high and low frequencies corresponding with the spikes and the waves match at level $j = 3$ and at level $j = 5$ respectively

Results obtained with the proposed method in the detection of spike-and-waves are shown in figure 2.

Figure 3 illustrates the results obtained in the detection of a train of spike and an isolated spike assigning a proportional value to the energy function when this function overcomes the threshold.

For comparison, figure 4 shows the results obtained with another signal, detecting sharp transients (bursts). The paroxism begining using the total energy of the signal is detected as shown in figure 5.

In the scheme proposed eight frequency bands (octaves) are analyzed. The characteristic events present in the EEG can be searched for at the proper level. A complete picture of an EEG processed

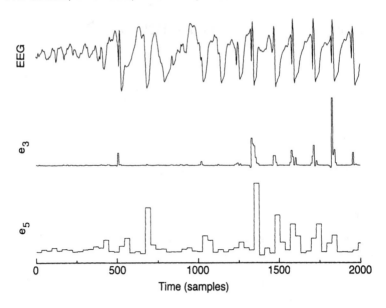

FIGURE 1. Signal, levels $j = 3$ and $j = 5$ of energy (10.15).

with our algorithm are shown in figures 6 and 7. Observing the eight octaves of frequency decomposition and the reconstructions of the signal at each one, we can appreciate that all the features involved in the signal itself are kept in the wavelet analysis. The signal reconstruction using the eight levels is shown in the last row of figure 6 and 7. As it can be seen, there is no difference between the original EEG signal and the reconstructed version.

We compare the performance of the algorithm proposed, based on spline biorthogonal wavelets in a multiresolution scheme, with methods based on:

1. orthogonal wavelets in a multiresolution scheme

2. the complex cubic B-splines modulated, a wavelet introduced [30], outside the multiresolution framework

3. Gabor transform, which is outside the wavelet framework

Thus, we compare our method with other methods at three stages, each one with a change in one characteristic of our analysis. The biorthogonality is changed by orthogonality, keeping the multiresolution scheme. At the second one, the multiresolution analysis is elim-

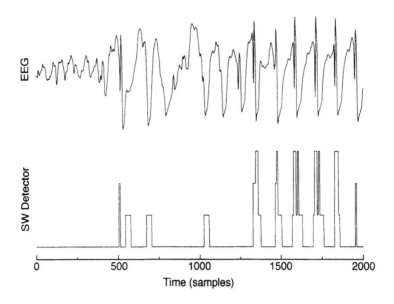

FIGURE 2. Spike-wave (SW) detection, levels $j = 3$ and $j = 5$ with $D_3 = \bar{x}+1/4\sigma$ and $D_5 = \bar{x} + 1/8\sigma$.

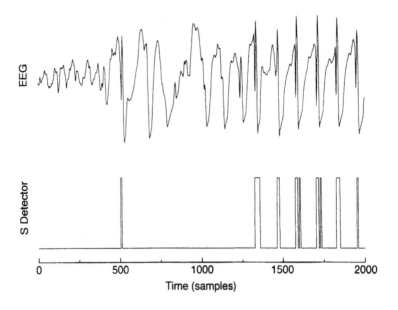

FIGURE 3. Detection of a train of spikes and isolated spike (S detector) with $D_3 = \bar{x} + 1/4\sigma$.

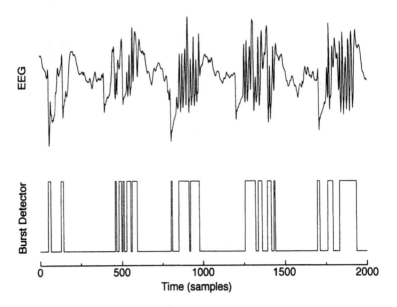

FIGURE 4. Detection of sharp transients (burst) $D_3 = \bar{x} + (1/2)\sigma$ or $D_4 = \bar{x} + (1/2)\sigma$.

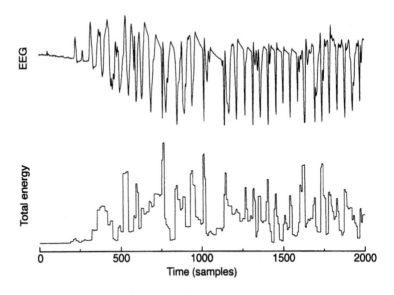

FIGURE 5. Paroxism beginnig analysis using total energy: $E(r) = \sum_j e_j(r)$ in each sample time r.

inated, keeping the wavelet scheme. For comparison with a method based on Gabor transform, see [15, 4].

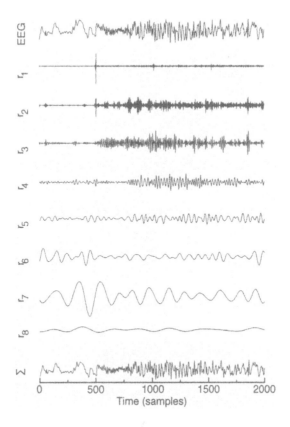

FIGURE 6. EEG record (g_0), the eight multiresolution analysis levels ($r_1, \ldots, r_8$) and reconstruction of EEG from the eight levels ($\sum r_i$).

Comparison 1: Daub4

We have used one of the Daubechies wavelets [7], the simplest and most localized member of the set, often called Daub4. Since this is an orthogonal wavelet, its dual is the same wavelet. As shown in figures 8 and 9, a large number of false positives are given by the algorithm using Daub4, while trying to detect the spikes using the same threshold as in the algorithm proposed in this work.

This comparison shows that despite the fact that different wavelets could be used in our algorithm, the polynomial splines —as it was recently pointed out in [21]— have a number of properties that make them particularly attractive in the present context.

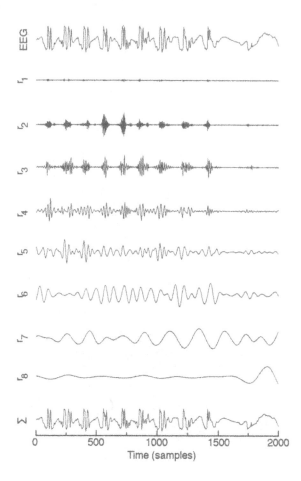

FIGURE 7. EEG record (g_0), the eight multiresolution analysis levels ($r_1, \ldots, r_8$) and reconstruction of EEG from the eight levels ($\sum r_i$).

Comparison 2: Cubic spline modulated wavelet

In order to use a wavelet outside the multiresolution framework, but at the same time use a fast algorithm in its numerical implementation, the complex modulated cubic B-spline, introduced by Unser [30] is considered.

The wavelet transform is given by

$$W f(b, a) = e^{2\pi i (b/a)} [f_a * v_a](b),$$

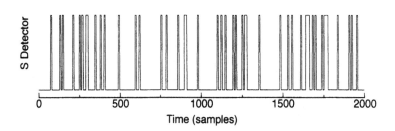

FIGURE 8. The proposed algorithm using Daub4 a) EEG, b) $D_3 = \bar{x} + (1/4)\sigma$.

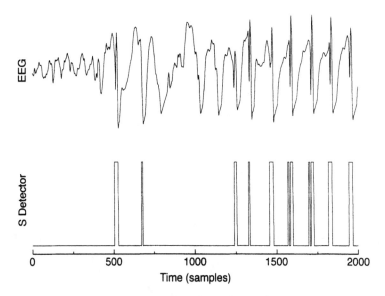

FIGURE 9. The proposed algorithm using Daub4 a) EEG, b) $D_3 = \bar{x} + (1/4)\sigma$.

where

$$f_a(k) = \frac{1}{\sqrt{a}} f(k) e^{-2\pi i(k/a)}$$
$$v_a(k) = v(k/a),$$

$f(k)$ are the samples of EEG signal and $v(t)$ is the cubic B-spline. Choosing integer numbers as scale coefficients a, we can vary the modulation frequency. The computational burden is high if the scale a varies in a large interval. This is the case when we search events of different frequency characteristics. The advantage of the previous analysis over this one is that the same range of frequencies is covered using a small number of levels. As we can see in figures 6 and 7, the analysis of the range from $1/2Hz$ to $128Hz$ is made in eight levels. Of course, if we try to find only events of a specific and well determined frequency, Unser's wavelet is useful. But if the task is to find different types of events, the multiresolution scheme is better.

As an example, we gave the parameter a the value $a = 10$ for detecting spikes, because with this value the frequency-window is centered at the same value than that of the frequency-window of level 3 (where the spikes were detected) of the multiresolution analysis proposed. Figure 10 shows the values $|Wf|$ for $a = 10$, and the results obtained with a detector using $\bar{x} + \sigma$ as threshold, updated each 8 s.

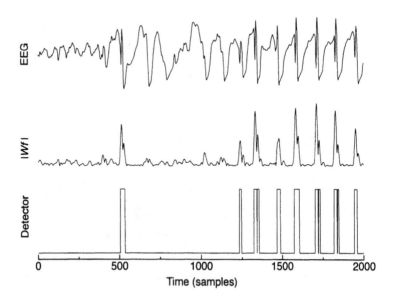

FIGURE 10. a) EEG, b) analysis using Unser's wavelet with $a = 10$, c) detector output with threshold $\bar{x} + \sigma$.

4.2 Discussion

The results presented in this section suggest the capability of the energy function in a multiresolution wavelet analysis for time localization of nonstationary phenomena in EEG signals. The main features of the analysis are:

(a) Different types of epileptogenic events have different frequency localizations, and correspond to different levels in the multiresolution framework.

(b) As figures 6 and 7 clearly show, the complete features contained in the EEG signal can be found in the eight multiresolution levels. In fact, a perfect reconstruction of the EEG signal from these levels is obtained, as shown in the last row of these figures. This fact demonstrates that no information is lost and then, with an adequate use of this information split into eigth levels, the detection of transients is possible.

(c) The energy analysis used allows the detection of characteristic epileptic events. Comparisons with other methods are shown.

(d) An important problem to be addressed in each particular case is the value of the threshold, as it was explained above.

The previous remarks suggest that computational techniques based on the wavelet theory may be incorporated into automatic analysis of EEG signals in order to extract features containing relevant events.

5   Method II: Multiresolution Entropy

As it was previously pointed out, the wavelets have interesting localization properties in the time-frequency plane. The wavelet coefficients corresponding to the level j of the multiresolution analysis are

$$D_j = \{d_j(k), \quad k = 1, \ldots, K_j\}. \tag{10.16}$$

If the total amount of analyzed data is a power of two, i.e. 2^N, the number of coefficients in the level j is 2^{N-j} due to the decimation effect.

Let us assume for simplicity a fix resolution level and its wavelet coefficients

$$D = \{d(k), \quad k = 1, \ldots, K\}. \tag{10.17}$$

They are calculated with the method explained in Section 3.

On this set of wavelets coefficients we define a sliding window depending on two parameters: the width $w \in \mathbb{N}$ (an even number), and the sliding factor $\Delta \in \mathbb{N}$. The definition of the sliding window of data is the following:

$$W(m; w, \Delta) = \{d(k), k = 1 + m\Delta, \ldots, w + m\Delta\}, m = 0, 1, 2, \ldots, M. \tag{10.18}$$

where Δ and w are selected in such a way that $w \leq K$ and $(K - w)/\Delta \in \mathbb{N}_0$. For example, if $w = 10$ and $\Delta = 5$, $S(1; 10, 5) = \{d(1), d(2), \ldots, d(10)\}$, $S(2; 10, 5) = \{d(6), d(7), \ldots, d(15)\}$, etc. The center of the window (10.18) is $d(w/2 + m\Delta)$.

For each window $W(m; w, \Delta)$ we take a partition

$$s_0 = s(m) < s_1 < s_2 < \ldots < s_L = S(m),$$

where

$$s(m) \overset{\Delta}{=} \min\left[W(m; w, \Delta)\right] = \min_k \left[\{d(k), k = 1 + m\Delta, \ldots, w + m\Delta\}\right]$$

$$S(m) \overset{\Delta}{=} \max\left[W(m; w, \Delta)\right] = \max_k \left[\{d(k), k = 1 + m\Delta, \ldots, w + m\Delta\}\right]$$

and consider the set $\{I_l = [s_{l-1}, s_l), l = 1, \ldots, L\}$, of disjoint intervals in such a way that

$$W(m; w, \Delta) = \bigcup_{l=1}^{\overline{L}} I_l. \tag{10.19}$$

We will represent $p^m(I_l)$ as the probability that the wavelet coefficient $d(k) \in W(m; w, \Delta)$ belong to the interval I_l. This probability is the quotient between the number of wavelets coefficients of $W(m; w, \Delta)$ in I_l and the total number of wavelets coefficients in $W(m; w, \Delta)$.

Now we define the *multiresolution entropy* (MRE) as

$$H(m) = -\sum_{l=1}^{L} p^m(I_l) \log(p^m(I_l)), \quad m = 0, 1, \dots, M. \quad (10.20)$$

In this way we obtain the entropy evolution of the wavelet coefficients at the considered resolution level.

In the general case it will be $M = M_j$, $D = D_j$, $K = K_j$, $L = L_j$, $W(m; w, \Delta) = W_j(m; w, \Delta)$, $p^m = p_j^m$ and $H(m) = H_j(m)$.

As it will be shown in the next section, if we plot the MRE (10.20) vs. time at each level, considering the points $\{w/2 + m\Delta, H_j(m)\}$, $m = 1, ..., M_j$, this approach has the localization properties of the wavelets, so it is useful in detecting changes. Moreover, its computational burden is lower than that of the methods based on Lyapunov exponents.

5.1   Results

We will analyze two examples: in the first one, the signal is generated using a known model in which we impose a change in one parameter; the second one corresponds to an electroencephalographic signal from an epileptic patient. Further examples corresponding to signals obtained for different models are analysed in [27].

Case 1: Signals generated by a discrete model

First, we present a simulation performed using the quadratic map model

$$x_{n+1} = a_n\, x_n\, (1 - x_n) \quad (10.21)$$

Consider parameter a_n evolving in order to produce a change at the dynamic behavior as

$$a_n = \begin{cases} a_1 & if & n < n_1 \\ a_1 + [(n - n_1)(a_2 - a_1)/(n_2 - n_1)] & if & n_1 \le n \le n_2 \\ a_2 & if & n > n_2 \end{cases}$$

$$(10.22)$$

Figure 11 shows the temporal evolution of the quadratic map with a linear variation of the parameter a_n at a band where a chaotic attractor is present [26], with $a_1 = 3.57$, $a_2 = 3.60$, $n_1 = 1024$ and

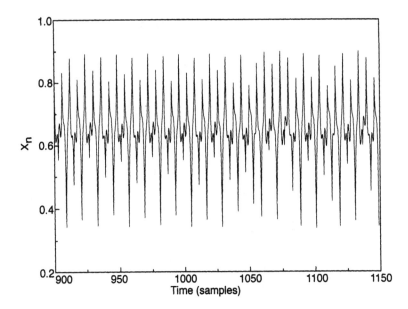

FIGURE 11. Temporal evolution of the quadratic map (eq. 10.21), with linear variation of the parameter a_n as in (eq. 10.22), with $a_1 = 3.57$, $a_2 = 3.60$, $n_1 = 1024$, $n_2 = 1048$, and initial condition $x_1 = 3.60$.

$n_2 = 1048$. The entropy does not exhibit any change worth noticing in its evolution (figure 12). However, the MRE evolution shows an important change at levels 1,2 and 3 (figure 13).

We assume a linear perturbation *at the output* x_n of the system (10.21) with constant parameter a_n. As a first perturbation case, we perform a multiplication by a factor b_n and we consider:

$$y_n = b_n \, x_n; \; where \; b_n = \begin{cases} 1 & if \quad n \leq n_o \\ b & if \quad n > n_o \end{cases} \tag{10.23}$$

We obtain the behavior shown in figure 14, for a_n constant, $a_n = 3.60$, $b = 1.1$, $n_o = 1024$, and initial condition $x_1 = 3.60$. In figures 15 and 16 it can be appreciated that the entropy presents a smooth modification, like a lobe, around n_o, but the MRE does not present any *jump* at n_o at its three first levels.

As a third case, we add a value c_n to the output:

$$y_{n+1} = x_{n+1} + c_n; \; where \; c_n = \begin{cases} 0 & if \quad n \leq n_o \\ c & if \quad n > n_o \end{cases} \tag{10.24}$$

In figures 17, 18 , 19 we show y_n evolution and its entropy and

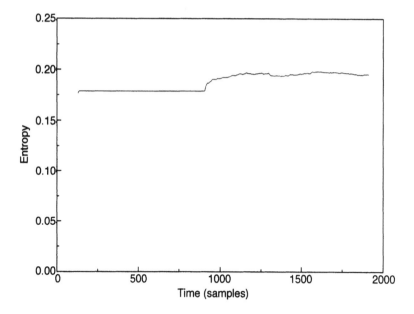

FIGURE 12. Entropy temporal evolution of the quadratic map signal (figure 11). It does not exhibit any change worth noticing in its evolution.

MRE evolutions respectively, for a_n, n_o and x_1 as it was mentioned before and $c = 0.1$ (> 10 of $Max[x_n]$). Again the entropy presents a lobe around n_o, and the MRE keeps without any perceptible *jump* at its three first levels, even if something appears at the fourth one

Case 2: EEG signal

Some physiological systems behave in a nonlinear chaotic way, and different methods have been developed to determine the presence of an attractor, its dimension, and the values of the Lyapunov exponents. In this case, we present an example of EEG signals corresponding to epileptic patients. Some authors have shown that the variability of the EEG signals does not represent noise but an attractor [2, 3, 18]. As it was pointed out by Iasemidis *et al.* [13], electrocorticograms of partial epilepsy of temporal lobe origin indicate that the epileptogenic focus also generates signals characteristic of a nonlinear dynamic system. Iasemidis *et al.* [14] established that during the preictal period, signals from each electrode exhibit positive values of the first Lyapunov exponent L_1, with multiple transient drops; at the onset of the seizure, signals show a simultaneous drop in L_1 to its lowest value.

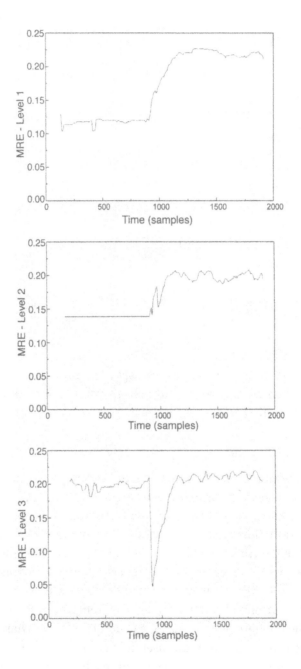

FIGURE 13. MRE evolution of the quadratic map signal (figure 11). It shows an important change at levels 1, 2 and 3.

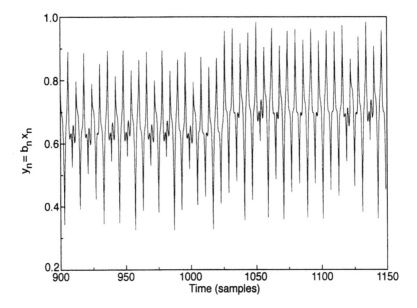

FIGURE 14. Temporal evolution of the quadratic map (eq. 10.21), with linear perturbation of the output x_n as in (eq. 10.23), with $a_n = 3.60, b = 1.1, n_0 = 1024$ and initial condition $x_1 = 3.60$.

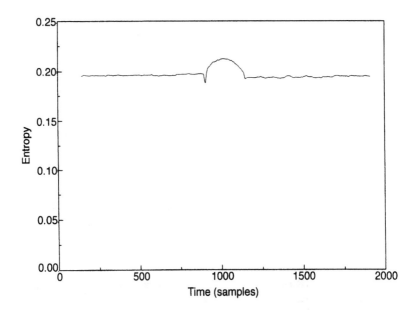

FIGURE 15. Entropy temporal evolution of the quadratic map signal(figure 14). It presents a smooth modification, like a lobe, around n_0.

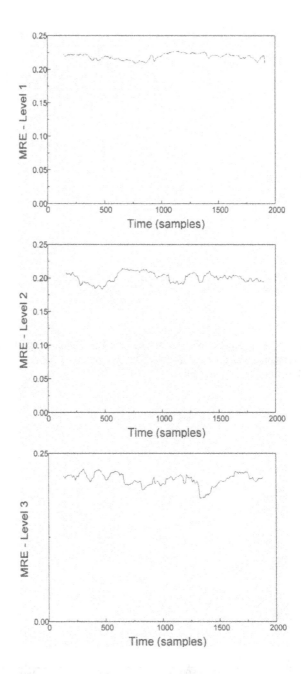

FIGURE 16. MRE time evolution of the quadratic map signal in figure 14. The MRE does not present any *jump* at n_0.

In preliminar papers [28, 29] we have shown that the MRE offers an alternative method for the analysis of EEG signals, corresponding to

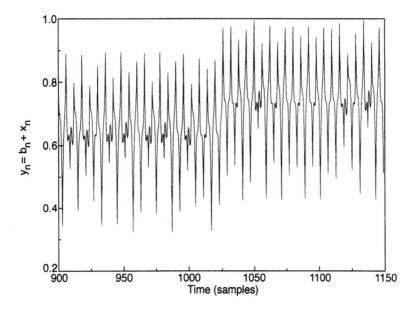

FIGURE 17. Temporal evolution of the quadratic map (eq. 10.21), with linear perturbation of the output x_n as in (eq. 10.24), for a_n, n_o and x_1 as it was mentioned before and $c = 0.1$.

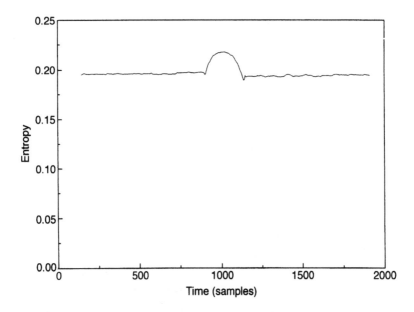

FIGURE 18. Entropy temporal evolution of the quadratic map signal at figure 17. Again the entropy presents a lobe around n_o.

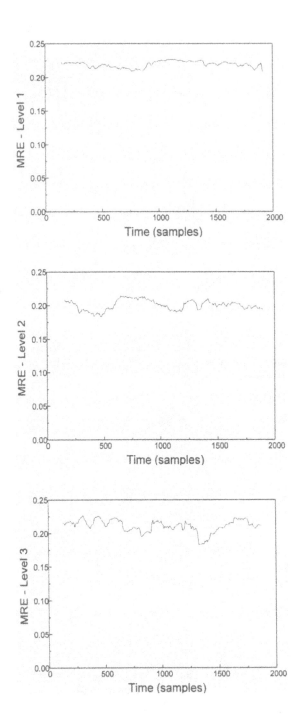

FIGURE 19. MRE evolution of the quadratic map signal in figure 17. The MRE keeps without any perceptible *jump* at its three first levels.

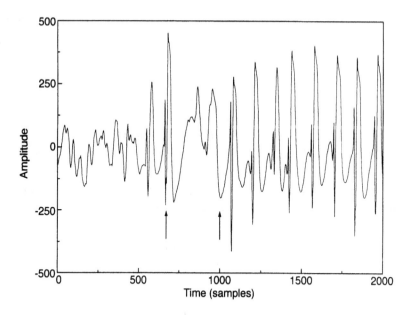

FIGURE 20. EEG showing an epileptic seizure beginning by a spike paroxysm followed by a spike-wave paroxysm.

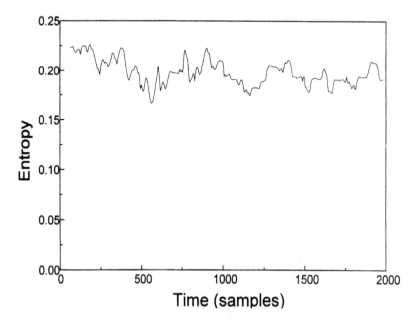

FIGURE 21. Entropy time evolution of EEG signal (figure 20).

epileptic patients. Figure 20 shows one channel of an epileptic seizure beginning with a spike followed by a spike-wave paroxysm. The mark was made by the physiologist and corresponds to the starting point of the seizure activity. The frequency distribution of this signal suggests the presence of chaos which is confirmed by the presence of a strange attractor at the phase space diagram obtained by the time delay method [25]. The real time data is assumed to be embedded in a $p = 7$ phase space. Even if the signal entropy time evolution shows several drops (figure 21), it is difficult to obtain information about the seizure onset localization.

Figure 22 shows the MRE time evolution of the signal shown in figure 20 at the three first resolution levels, obtained with $w = 128$ and $\Delta = 5$. At the seizure onset it suffers a sudden drop to its minimum value at the first level. This slope is preceded by similar ones of less magnitude at the three first levels which corresponds to the spike paroxysm shown by the first arrow. The first drop has been observed at other channels of the EEG at which only the spike paroxysm is present. It can be observed that the MRE increases its value again, oscillating between the same values while the EEG signal is arranged at a spike-wave paroxysm as the seizure evolves. Iasemidis [14] established a similar behavior using Lyapunov exponents. Comparing the EEG shown in figure 20 with the MRE time evolution drop (figure 22), it is clear the agreement between them at the onset of the epileptic seizure.

5.2   Discussion

In the simulated case, we discussed the behavior of the MRE and the entropy of signals obtained for constant parameters values, under a linear modification of one of the output states of the system. As it was shown, the MRE presents a clear change at the three first resolution levels in most of the simulated models. Even if the entropy shows a similar behavior, it is not so clear in some cases. In almost all the cases, the MRE keeps invariant in face to a linear change of the output signal, that does not affect the original signal's graphic in an obvious way.

Using a multiresolution analysis approach we have developed a method that has the wavelets localization properties, so it is useful for detecting changes. The results obtained with nonlinear discrete and continuous models at chaotic parameters range suggest that the

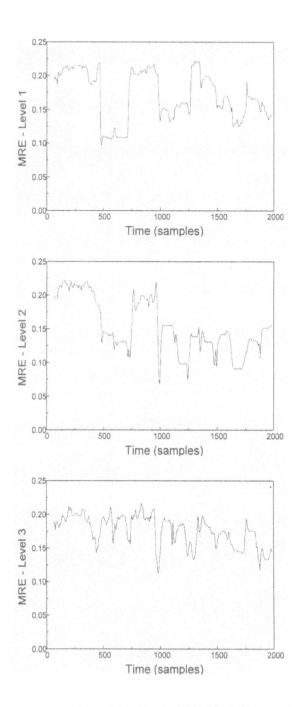

FIGURE 22. MRE time evolution of EEG signal (figure 20).

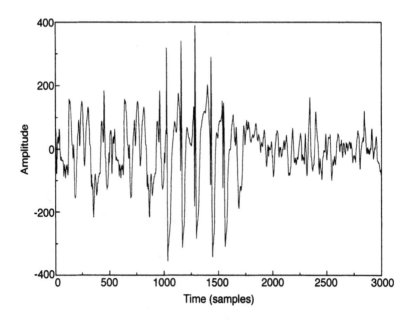

FIGURE 23. EEG showing an epileptic seizure beginning by a spike paroxysm.

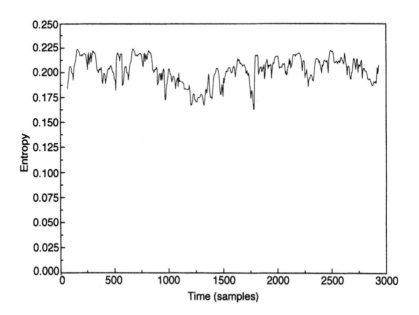

FIGURE 24. Entropy time evolution of EEG signal (figure 23).

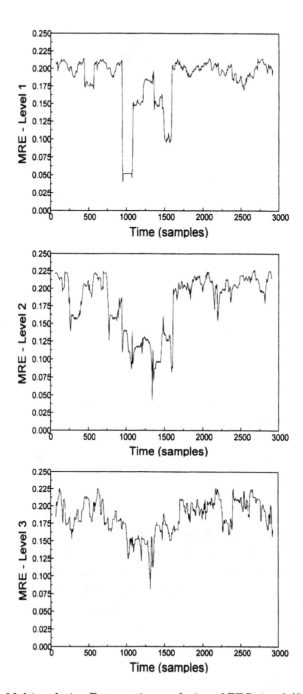

FIGURE 25. Multiresolution Entropy time evolution of EEG signal (figure 23).

MRE method must be considered as a complement or as an alternative for detecting changes in the parameters of the system.

Both, the entropy and the MRE have shown to be useful for detecting changes in simulated systems. However, the studied cases show that the proposed method allows to distinguish between changes in one parameter and linear perturbations at the output of the system.

In the case of EEG signals, we could establish a suitable agreement between the seizure onset localization obtained by visual inspection, the Lyapunov exponent method and the MRE method proposed in this paper. Furthermore, the sub-band decomposition of the signal should allow to establish a differentiation between frequency levels, and thus at which ones the transition —suggested by Pijn [20] and West [33, 34]— from one chaotic system to another of different complexity level is dominant.

The main advantage of the entropy and the MRE method, over the Iasemidis' method for evaluating the Lyapunov exponents, with the goal of detecting changes in the signal, is that it allows the analysis with short signal segments at less computational cost. This computational cost is very high due to the algorithm complexity and the fact of requiring a register segment of around 10^D to 30^D points, where D is the attractor dimension. For example, in the EEG case presented here, D is between 2 and 3 at the ictal case [13]. So the record length must be around 20 min., starting 10 min. before the seizure onset. With the method here proposed, 8 s analysis windows were used. The algorithms have been implemented in *Mathematica 2.2* and run on a PC–486 DX2 (66 MHz). For an EEG segment of 3180 samples, the CPU time for the evaluation of the temporal evolution of L has been of 61706s. The corresponding CPU time for the MRE has been of 246 s. The former has been evaluated with Iasemidis algorithm [14] with a temporal evolution of 20 samples. For most of the MRE evaluations here presented the parameters have been $\Delta = 5$, $w_1 = 128$ and $M = 10$. A modification in the values of w_1 and Δ produce a zoom-in or zoom-out effect in the results observed at the MRE evolution.

6   Conclusions

In this work we have introduced two new approaches for localizing epileptic events in EEG in a multiresolution framework, especially suited for the digital treatment of nonstationary signals, and in particular, for EEG analysis. The first method presented, based on the energy function, provides a solution to some basic problems in events detection of EEG records. A real-time implementation of it is possible using fast algorithms. As different types of epileptogenic events have different frequency localizations, it is possible to detect and characterize them at different levels in the multiresolution framework. We have shown that, from this decomposition, perfect reconstruction of the EEG signal is obtained. This fact demonstrates that no feature is lost and then, the detection of transients is possible. On the other hand, the energy analysis using biorthogonal spline wavelets allows the detection of characteristic epileptic events, with a better performance in regard to the energy based in other methods, as considered in this work.

The second method hereby presented deals with a particular aspect of the relation between signal analysis and nonlinear dynamics, which is the detection of slight changes in parameters of nonlinear dynamical systems using information measures obtained from experimental data. Taking into account that the EEG underlying dynamics corresponds to a non linear system and because of the sensitivity of a chaotic system with its initial conditions, there is a change in the available information about the states of the system. This change, thought of as creation of information, is not recognized by the entropy of the signal itself, but we have shown that it is possible to identify it by means of the MRE.

The previous remarks suggest that computational techniques based on the wavelet theory may be incorporated into the automatic analysis of EEG signals in order to extract features containing relevant events, and that when properly used with techniques provided by the Information Theory, they may give a further insight on the underlying non linear dynamics.

7 References

[1] K. Arakawa, D. Fender, H. Harashima, H. Miyakawa, and Y. Saitoh. Separation of a nonstationary component from the EEG by a nonlinear digital filter. *IEEE Trans. Biomed. Eng.*, 33:724–726, 1986.

[2] A. Babloyantz. Evidence of chaotic dynamics of brain activity during the sleep cycle. In Meyer-Kress, editor, *Dimensions and Entropies in Chaotics Systems*, pages 241–245, Springer Verlag, Berlin, 1986.

[3] E. Basar. *Chaos in Brain Function*. Springer Verlag, Berlin, 1990.

[4] S. Blanco, C. D'Attellis, S. Isaacson, O. Rosso, and R. Sirne. Time-frequency analysis of EEG series (ii): comparison between methods based on gabor and wavelets transforms. Physical Rev. E, Vol. 54, No. 5, 1996.

[5] C. K. Chui. *An Introduction to Wavelets*. Academic Press, San Diego, 1992.

[6] C. E. D'Attellis, S. I. Isaacson, and R. O. Sirme. Detection of epileptic events in EEG using wavelet theory. To appear in Annals of Biomedical Engineering.

[7] I. Daubechies. *Ten Lectures on Wavelets*. Siam, Philadelphia, USA, 1992.

[8] P. Guedes de Oliveira, C. Queiroz, and F. Lopes de Silva. Spike detection based on a pattern recognition approach using a microcomputer. *Electroenceph. Clin. Neorophysiol*, 56:97–103, 1983.

[9] A. Dingle, R.D.Jones, G.J.Carroll, and W.R.Fright. A multi-stage system to detect epileptiform activity in the EEG. *IEEE Trans. Biomed. Eng.*, 40:1260–1268, 1993.

[10] R.C. Eberhart, R.W. Dobbins, and W.R. Webber. Eeg waveform analysis using case-net. *Proc. Conf. IEEE Eng. Med. Biol. Soc.*, 2046–2047, 1989.

[11] G. Fischer, N.J.I.Mars, and F. Lopes da Silva. Pattern recognition of epileptiform transients in the electroencephalogram. *Inst. Med. Physi*, Rep. 7, Utrecht, 1980.

[12] J.R. Glover, D.N. Varmazis, and P.Y. Ktonas. Continued development of a knowledge-based system to detect epileptogenic sharp transients in the EEG. *Proc. Conf. IEEE Eng. Med. Biol. Soc.*, 1374–1375, 1990.

[13] L.D. Iasemidis, J. C. Sackellares, H.P. Zaveri, and W.J. Williams. Modelling of ecog in temporal lobe epilepsy. *25th Ann. Rocky Mountain Bioing. Symposium*, 1201–1203, 1988.

[14] L.D. Iasemidis, J. C. Sackellares, H.P. Zaveri, and W.J. Williams. Phase space topography and the lyapunov exponent of electrocorticograms in partial seizures. *Brain Topogr.*, 2(3):187–201, 1990.

[15] S. Kochen, C. D'Attellis, R. Sirne S. Isaacson, P. Salgado, O. Rosso, C. Parpaglione, and L. Riquelme. Seizure localization using gabor transform, wavelet analysis and neural networks in depth EEG. *Annual Meeting American Epilepsy Soc. New Orleans*, 4–6, 1994.

[16] Y. Meyer. *Ondelettes*. Hermann Ed, Paris, France, 1990.

[17] Y. Meyer. *Wavelets, Algorithms and Applications*. SIAM, Philadelphia, USA, 1993.

[18] G. Meyer-Kress and S.C. Layne. Dimensionality of human electroencephalogram. In M. F. Shlesinger S. H. Koslow, A. J. Mandel, editor, *Perspectives in Biologycal Dynamics and Theoretical Medicine*, Ann. N.Y. Acad. Sci 504, New York, USA, 1987.

[19] Y. B. Pesin. Characterisctic lyapunov exponents and smooth ergodic theory. *Russ. Math. Surv*, 32(4):55–114, 1977.

[20] J. P. Pijn and F. H. Lopez da Silva. *Propagation of electrical activity: nonlinear associations and time delays between EEG signals*. Birkhäuser, Boston, 1993.

[21] J. F. Schiff, A. Aldroubi, M. Unser, and S. Sato. Fast wavelet transformation of EEG. *Electroencelography and Clinical Neurophysiology*, 91:442–455, 1994.

[22] L. Senhadji, G. Carrault, and J. J. Bellanger. *Detection et cartographie multi-echelles en EEG*, pages 609–614. Editions Frontieres, Paris, 1993.

[23] L. Senhadji, G. Carrault, J.J. Bellanger, and G. Passarello. Quelques nouvelles applications de la transformee en ondelettes. *Innov. Tech. Biol. Med.*, 14:389–403, 1993.

[24] C. E. Shannon. A mathematical theory of comunication. *The Bell System Technical Journal*, 27(3):379–423, 1948.

[25] F. Takens. *Detecting strange attractors in turbulence*, pages 366 – 381. Berlin, 1981.

[26] J. M. Thompson and Stewart. *Non linear Dynamics and Chaos*. Wiley and Sons, 1993.

[27] M. Torres, L. Gamero, and E. D'Attellis. *Detection of changes in Nonlinear dynamical systems using Multirresolution Entropy*. Technical Report 2812, INRIA-Rapport de recherche, 1996.

[28] M. Torres, L. Gamero, and E. D'Attellis. A multirresolution entropy approach to detect epileptic form activity in the EEG. In I. Pitas, editor, *IEEE Workshop on non linear signal and image processing*, pages 791–794, 1995.

[29] M. Torres, L. Gamero, and E. D'Attellis. Pattern detection in EEG using multirresolution entropy. *Latin American Applied Research*, (53):53–57, 1995.

[30] M. Unser. Fast Gabor-like windowed Fourier and continuous wavelet transform. *IEEE Signal Process. Letters*, (1):76–79, 1994.

[31] M. Unser, A. Aldroubi, and M. Eden. A family of polynomial spline wavelet transforms. *Signal Process.*, 30:141–162, 1993.

[32] R. Walters, J. Principe, and S. Park. Spike detection using a syntactic pattern recognition approach. In *IEEE Eng. Med. Biol. Soc.*, pages 1810–1811, 1989.

[33] B. J. West. *Fractal Physiology and Chaos in Medicine, Nonlinear Phenomena in Life Science*. Volume I, World Scientific, Singapore, 1990.

[34] B. J. West and Patterns. *Information and Chaos in Neuronal Systems, Nonlinear Phenomena in Life Science*, chapter II. World Scientific, Singapore, 1993.

[35] A. Wolf, J. B. Swift, H. L. Swinney, and J. A. Vastano. Determining Lyapunov exponents from a time series. *Physica D*, 16:285 – 317, 1985.

Part III

Applications in Physical Sciences

Chapter 11

Wavelet Networks for Modelling Nonlinear Processes

N. Roqueiro
E. L. Lima

1 Introduction

The objective of engineering study is basically, the design and operation of useful systems for humankind development. Historically, this work has been done by trial-and-error procedures. The development of "theory" has always been related to "experiments" that deny or confirm hypothesis. The incorporation of the acquired knowledge to the design of systems, generally using mathematical language, leads to "models" that approach real situations. With these models, a group of equations that relates system variables, it is possible to analyze system behavior and to plan operational changes. In order to conceive a model it is necessary to define its application. In the present work models that emulate a system dynamic behavior that could be used in the design and in the operation of controllers, will be developed. This objective imposes some desirable characteristics to this model, such as: system representation in a wide range of operation, easy implementation and fast prediction of variables of interest.

The modelling of a system can be achieved using basic principles of nature (conservation laws, for example) or empirically, based on processed data. The identification problem can be divided into two principal tasks: the determination of its structure and the parameters estimation. The determination of the structure means the definition of equations that better describe the system, their number, their class (differential equations, algebraic, and etc.) and the choice of involved variables. The estimation of parameters means the determination of the constants used in the model equations. Determination of model structure is often a hard task, generally based on previous knowledge of the system or trial-and-error method. The present work intends to present a new structure for non-linear model based on a single layer neural network that uses wavelets as activation functions. Some particularities of adjusting recurrent network parameters (by the least square method) are analyzed through the influence of the over-estimation and the incorporation of steady-state information in the signals used for estimation. Furthermore it is studied the ill-posedness problem of the estimation arising from using delayed output signals (reccurrence) and from the activation functions employed (compactly supported). A solution for this problem is then proposed. Through simulation, the ability of the model to represent important characteristics of non-linear systems, as non-linear gain, multiplicity of the steady state and chaotic behavior is investigated. Finally, problems related to the new structure, that have not yet been solved, are pointed out, and some lines of research and development for applications in process control are proposed. Section *2* presents a brief summary of non-linear models employed, specially those generated by series of functions, aiming at placing single layer networks in this context. A single layer neural network that uses scalar activation functions that can represent multivariable systems is presented in section *3*, the main contribution of the present work. The simulation studies are presented in section *4*.

2   Models for non-linear system identification

In the last four decades, several models have been developed for identification of non-linear systems. Haber and Unbehauen [1] review identification methods of non-linear systems, presenting the following chronology of development:

In the sixties:

Volterra's series

Block oriented models

In the seventies:

Cascade models

Hierarchical models

In the eighties:

Non linear models, linear in the parameters

It could be added to this sequence the use of neural networks and the fuzzy logic models.

The description of non-linear models presented below is not exhaustive and aims to show the evolution of non linear models, linear in the parameters. For major details, refer to the works of Haber and Unbehauen [1], Hsia [3], Eykhoff [4], Chen and Billings [5], Wu et.al. [6].

2.1 Volterra's series

A linear system, invariant in time and causal, with an input $u(t)$ and an output $y(t)$ can be described by the convolution integral:

$$y(t) = \int_0^\infty g(t - \tau)u(\tau)d\tau$$

where $g(t)$ is the weight function or impulse response of the system.

The weight function is generally unknown, and the problem of identification consists in its determination from the signals $u(t)$ and $y(t)$ already known.

One extension of this model for non-linear models is through Volterra's series [4], which, for systems with one input and one output, can be described as:

$$
\begin{aligned}
y(t) \;=\;& \int_0^\infty g_1\left(\tau\right) u\left(t - \tau\right) d\tau \\
&+ \int_0^\infty \int_0^\infty g_2\left(\tau_1, \tau_2\right) u\left(t - \tau_1\right) u\left(t - \tau_2\right) d\tau_1 d\tau_2 \\
&+ \ldots + \int_0^\infty g_n\left(\tau_1, \ldots, \tau_2\right) \prod_{i=1}^{n} u\left(t - \tau_i\right) d\tau_i + \ldots
\end{aligned}
$$

where the kernel of Volterra $g_n(\tau_1, \ldots, \tau_n)$ represents a weighting function of degree n. Assuming that the signals $u(t)$ and $y(t)$ are sampled with a period T, it is possible to approximate Volterra's series by:

$$y(k) = \sum_{i=0}^{p} h(i)u(k-i) + \sum_{i=0}^{p}\sum_{j=0}^{p} h(i,j)u(k-i)u(k-j)$$
$$+ \sum_{i=0}^{p}\sum_{j=0}^{p}\sum_{m=0}^{p} h(i,j,m)u(k-i)u(k-j)u(k-m) + \cdots$$

$$(11.1)$$

for $k > p$. The simplified notation employed in equation (11.1) is a consequence of the following considerations:

1. The settling time is pT

2. The sampling times are kT

3. The weight sequences in the sampling time are the same as in the weight functions (i.e. $h(i) = g_1(\tau = i\,T\,)$)

4. The sampling time is implicit $k = kT$, $p = pT$, etc.

The Volterra's series represents a system linear in its parameters (h) whose non-linearities involves exclusively products of inputs. The parameters can be estimated by the least square method. In general, the models that will be presented below can also be adjusted minimizing a quadratic perfomance index, linear in the parameters. There are other non-linear models in which the parameters do not appear linearly, as the multi-layer neural networks, whose convergence problems to local minimum and delayed computational time to reach the minumum are known [14], [2], [8]. In this work, the development of models linear in the parameters was treated.

2.2 Block oriented models

The block oriented models have in common the fact that they are combinations of linear sub-systems and non-linear gains. It is possible to define three basic models:

1) The Wiener model, showed in Figure 1, is constructed by two blocks in series. The first block, associated to the input, is formed by a dynamical linear model. Then, operating over the output of

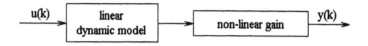

FIGURE 1. Wiener's model.

this first block a second block formed by non-linear constant gain is imposed.

2) The Hammerstein's model, shown in Figure 2, uses the same two blocks in an opposite order

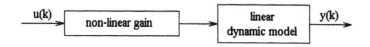

FIGURE 2. Hammerstein's model.

3) The general model, shown in Figure 3, combines the two previous models, arranging three blocks in series. The central block is the non-linear gain and the others are the linear models [3].

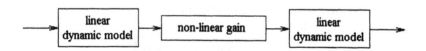

FIGURE 3. General model.

A great number of block oriented models can be defined blending linear dynamic models and non linear constant gains [1]. In order to illustrate this group of models, the model of Hammerstein is presented for a system with one input and one output. The dynamic linear block is represented by ARX model (Auto Regressive with eXogenous variable).

$$A(q^{-1})y(k) = B(q^{-1})x(k),$$

where

$$A(q^{-1}) = 1 + a_1q^{-1} + \cdots + a_nq^{-n},$$
$$B(q^{-1}) = b_0 + b_1q^{-1} + \cdots + b_nq^{-n}.$$

The non-linear gain is represented by a polynomial of order p

$$x(k) = \gamma_1 u(k) + \gamma_2 u^2(k) + \cdots + \gamma_p u^p(k)$$

and q^{-1} is the shift operator, so that $q^{-1}y(k) = y(k-1)$.

In this case, the identification problem consists of the estimation of parameters a_i, b_i and γ_i , for chosen values of n and p, from known values (experimental) of the signals $u(k)$ and $y(k)$.

Applications of this model can be found in [9], [1].

As this kind of model has linear dynamics and non-linear gains in polynomial form, it is not appropiate for the identification of systems that exhibit multiplicity of steady state or complex dynamics as chaos.

2.3   Non-linear models linear in the parameters

The non-linear models linear in the parameters can be described, in general, as models based in time series of non-linear functions, and it is possible to define several realizations depending on the choice of these functions ([1], [5] and [6]). The single layer neural networks can be characterized as a particular case of this model where the activation functions stand for non linear functions of the series and the weights are the corresponding parameters. It is easy to verify that a series of functions

$$y(k) = \sum_i c_i f_i(u(k))$$

can be represented as a single layer network (Figure 4), where each circle corresponds to a function f_i.

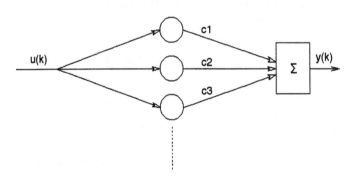

FIGURE 4. Single layer neuralnetwork model model.

Bakshi and Stephanopoulos [2] used single layer neural networks with families of functions called **wavelets** as activation functions. These networks can be considered a particular case of radial basis function networks (RBFN)[8].

A single layer neural network with wavelets as activation functions, or a time series of wavelets, or even though the expansion of a function in a basis formed by families of wavelets, are different insights for the same model.

We presented diferent realizations of non-linear dynamical models linear in the parameters. They show the evolution of the time series in the representation of non-linear dynamical systems and its relation to single layer neural networks.

Realizations of NARMAX model

The NARMAX models (Nonlinear Auto Regressive Moving Average with eXogenous variable) can describe a large number of non-linear dynamical systems, according to equation (11.2).

$$y(k) = f(y(k-1), \cdots, y(k-n_y), u(k-1), \cdots, u(k-n_u),$$
$$e(k-1), \cdots, e(k-n_e)) + e(k) \qquad (11.2)$$

where $y(k)$ is the output variable, $u(k)$ is the input variable, $e(k)$ is the white noise and n_y, n_u and n_e are maximum delays of each variable. The function f, in general unknown, can be approximated by time series of non-linear functions linear in the parameters. There are several proposed regressors [6], [1], In order to simplify the notation, the models will be presented without noise.

Polynomial models are particular cases of NARMAX models:

Linear models

$$y_s(k) = a_{s,0} + \sum_{i=1}^{n_y} a_{s,i}^T y(k-i) + \sum_{i=1}^{n_u} b_{s,i}^T u(k-i).$$

Linear models with time variant parameters

$$X(k) = [y(k), ..., y(k-n_y+1), u(k), ..., u(k-n_u+1)]$$

$$y(k) = a_0\left(X(k-1)\right) + \sum_{i=1}^{n_y} a_i^T\left(X(k-1)\right) y(k-i)$$
$$+ \sum_{i=1}^{n_u} b_i^T\left(X(k-1)\right) u(k-i),$$

and bilinear models

$$y(k) = a_0 + \sum_{i=1}^{n_y} a_i^T y(k-i) + \sum_{i=1}^{n_u} b_i^T u(k-i) + \sum_{i=1}^{n_y}\sum_{j=1}^{n_u} a_i y(k-i) * b_j u(k-j).$$

All of these models can be considered particular cases of the GMAPS model (Generalized Adaptive Polynomial Synthesis) [6], that is a polynomial model with constant parameters which represents all possible combinations of input and output variables. It is possible to generalize this model by adding other types of non-linear functions such as: sin, cos, tan, sinh, cosh, tanh, exp.

Threshold-NARMAX model (TNARMAX)

If the system to be represented has different behaviors in the sub-domain of the input-output region, it is possible to define one model for each region. TNARMAX models are built by a set of models NARMAX defined for different regions of the process domain. A general expression for these models is:

$$y(k) = f^i\left(X(k-1)\right)$$
$$X(k) = [y(k),\ldots,y(k-n_y+1), u(k),\ldots,u(k-n_u+1)]$$
$$X(k-1) \in \mathbb{R}e^i$$

where $\mathbb{R}e^i$ are regions of the process domain. This model allows the use of different regressors. The MARS model is a realization of the TNARMAX model that uses bases of spline functions uniformly distributed on the input-output space [6]. For multivariable systems, the basis is obtained as composition of spline functions, and the TNARMAX model becomes similar to radial basis function networks (RBFN) with only one layer. In those networks activation functions of each neuron are defined from a center and a radius that will generate the action domain of the neuron. The center location and the determination of the action radius are made by a heuristic method. This method is based in the data density in a determined region of the input-output space.

Radial basis function networks (RBFN)

There are several families of functions being used as activation function of the neural network, or as regressors in a time series. The most used functions are polynomials, gaussian functions, and spline functions. Even though polynomial models allow accurate approximations of a system for practical purposes, their structures (number of series elements) grow fast with the number of independent variables. In these models, it must be considered the monomial that represents a linear system and also all the products of two or more dependent variables (eq.(11.3)). The radial basis function networks are an alternative to the polynomial series.

A RBFN can be define as:

$$
\begin{aligned}
X(k) &= [y(k), \ldots, y(k - n_y + 1), u(k), \ldots, u(k - n_u + 1)] \\
y(k) &= a_0 + \sum_{i=1}^{M} a_i \vartheta \left(\| X(k-1) - c_i \| \right)
\end{aligned}
\tag{11.3}
$$

where the number of functions (M), and thus the centers c_i, should be smaller than the number of available points. In Pottmann and Seborg [8] it is presented a network with such characteristics. It was concluded that the use of functions that decay inversely to the difference $\| X(k-1) - c_i \|$, improves the stability of the identified models. The function utilized by Pottmann and Seborg was:

$$
\begin{aligned}
\vartheta(r) &= (r^2 + \beta)^{-\frac{1}{2}}, \quad \beta \geq 0 \\
r &= \| X(k-1) - c_i \|.
\end{aligned}
$$

The solution to the estimation problem for the parameters a_i demand an orthogonalization process in order to became stable. Determination of the position and the number of centers c_i is made by statistic methods.

Wavelet networks

Wavelet networks are, essentially, radial basis function networks (RBFN). The difference between them lies in the definition of the functions ϑ.

The change of scale in Pottmann and Seborg for RBFN [8] would be represented by the change of β. In wavelets, the change of scale is associated to the choice of the family, i.e., once defined the system variable domain and the family of wavelets, then the center and radius of all functions for every scale is immediately defined. As wavelets are orthogonal among them, the relative value of the coefficients associated to each function of the series permits to select those terms that contribute the most to the global approximation.

Wavelets

A multivariable non-linear dynamical system can be described through NARMAX model (eq.(11.2)). It has an input-output relation characterized by the function f, that is usually known. The identification of a process consists in selecting a model that approximates satisfactorily f. Projection of f over a subspace allows the determination of a model, that will be a linear combination of the basis elements. From a set of experimental data (input-output pairs), parameters of the linear combination are fitted to give an output model as similar as possible to the experimental output. The use of orthogonal bases simplifies fitting since it is possible to determine the coefficient of each basis element in a independent way. When experimental data are not uniformly distributed in the input-output space, fitting is improved using of localized functions and approximating in different resolution levels.

The model proposed here uses linear combinations of wavelets. A full description can be found in the works of Daubechies [11], Strang [10] and Bakshi and Stephanopoulos [2].

3   Wavelet networks

Networks that use wavelets as activation functions are new in the literature and so, few results were published on their use in process identification. In the present work, the use of single layer neural networks to avoid convergence problems of multilayer networks [14], [2], [8], is suggested. The practical problems of implementation (creation and calculation of a great number of activation functions) of this new kind of networks [2] for the identification of multivariable processes, leads to the proposition of a multivariable model based on a new single layer network that uses single variable wavelets.

Although there is not, yet, any theoretical result about the representation of this new structure, ease of implementation and nice simulation results indicate this approach as a good model for non-linear systems. In what follows we will intoduce a brief description of the wavenet suggested in [2]. Then we show the advantages offered by our proposal to deal with the multivariable functional problem. To simplify the explanation only inputs and outputs up to time $k-1$ (used as network inputs) will be considered. The extension for data before time $k-1$ is a trivial problem.

3.1   Multivariable wavelet networks

Single layer neural networks based on the wavelet multiresolution approach, the wavenets, were recently introduced [2]. In these networks the activation function of each neuron is a wavelet or a scale function. The output signal is the result of the activation function evaluation on the signal and an adjustable parameter.

Mathematical model for a SISO system, using a single layer feedforward network, can represent dynamic or static behavior. In the last case, shown in Figure 5, it represents a non-linear gain.

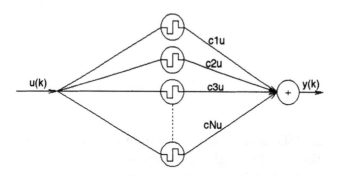

FIGURE 5. Schematic representation of a static net.

The map, in this case, can be described as:

$$y(k) = c_u^T . f_u(u(k))$$

with

$$f_u = [f1_u, f2_u, f3_u, k, fN_u] \ .$$

A net that represents a dynamical system, with an external input and feedback outputs, is represented in Figure 6.

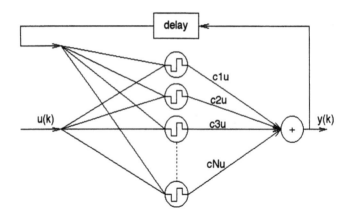

FIGURE 6. Schematic representation of a dynamic recurrent net.

This kind of net, also called recurrent, can be described as:

$$y(k) = c^T . f(u(k), y(k-1)).$$

In this case, the activation functions are multivariable.

Using multivariable wavelets it is necessary to evaluate a large number of functions. This problem leads us to suggest another network structure with the same dynamic characteristics and input-output relations, using single layer wavelets. This new structure will be called "parallel structure".

3.2 Parallel structure of single variable wavelets

SISO model

For the SISO model the new structure, indicated in Figure 7, proposes the use of a single layer wavelet network for each input, in a way that the parallel structure output is the direct sum of the outputs of each network.

The map represented by this structure can be described as:

$$y(k) = c_y^T f_y(y(k-1)) + c_u^T f_u(u(k))$$

In this structure, the dynamic characteristics is represented by the feedback of the output on the first group of neurons, with the external input acting directly on the creation of this output, through the second neuron group.

An advantage of this arrangement is that it is not necessary to use

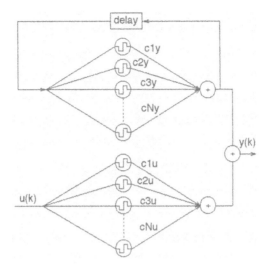

FIGURE 7. Schematic representation of a SISO dynamic net with parallel structure.

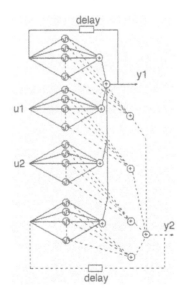

FIGURE 8. Schematic representation of a dynamic network with parallel structure for 2*2 system.

composed wavelet functions, simplifying the construction of neural network.

MIMO model

The MIMO model uses a sub-net for each output and another one for each input of the system. So, each neuron of any sub-net will have as many outputs as outputs the MIMO system has.

Figure 8 represents a network with two inputs and two outputs, being the general multivariable case, for m inputs and n outputs, an extension of this one.

The mathematical model associated to this diagram is:

$$
\begin{aligned}
y_1(k) = \; & c_{y11}^T f_{y1}(y_1(k-1)) + c_{u11}^T f_{u1}(u_1(k-1)) \\
& + c_{u12}^T f_{u2}(u_2(k-1)) + c_{y12}^T f_{y2}(y_2(k-1))
\end{aligned}
$$

$$
\begin{aligned}
y_2(k) = \; & c_{y21}^T \cdot f_{y1}(y_1(k-1) + c_{u21}^T f_{u2}(u_1(k-1)) \\
& + c_{u22}^T f_{u2}(u_2(k-1)) + c_{y22}^T f_{y2}(y_2(k-1))
\end{aligned}
$$

That is, the number of parameters vectors **c** for a system with n outputs and m inputs will be $n \times n + n \times m = n \times (n+m)$. In words, the number of parallel vectors grows linearly with the number of inputs and quadratically with the number of outputs. As the wavelets are single variable, the number of different functions used for this representation is two, a scale function and a wavelet.

Implementation

Since the proposal is the use of single variable wavelet families for each sub-net, a methodology must be defined for the creation and growth of each of these sub-nets. The methodology consists in the use of the same function for all sub-nets and the estimation of the parameters associated to each function using experimental data, in one step. The next step, is done in the same way, including another wavelet family for each sub-net. To estimate the parameters it is used the difference between the last step approximation error and the output of the new included wavelet family. The last step approximation error is the difference between the process outputs and the fitted network outputs. Although, for the model proposed, there is not evidence of stability, convergence and representation, several identifications and simulations of non-linear processes have been done, and satisfactory results have been observed.

To illustrate the described method, some steps to obtain a SISO model are presented below. Let:

$$u = [u(k), u(k-1), \ldots, u(k-p)]$$
$$y = [y(k), y(k-1), \ldots, y(k-p)]$$

be experimental data to fit the model.

Initially the same function for each sub-net is used (Figure 7); that is, $f_{y1} = f_{u1}$ and the minimization problem

$$\min \left\{ \underbrace{\begin{bmatrix} y(k-1) \\ y(k-2) \\ \\ y(k-p-1) \end{bmatrix}}_{r} - \underbrace{\left(\begin{bmatrix} c_{y1} f_{y1}(y(k-1)) \\ c_{y1} f_{y1}(y(k-2)) \\ \\ c_{y1} f_{y1}(y(k-p-1)) \end{bmatrix} + \begin{bmatrix} c_{u1} f_{u1}(u(k)) \\ c_{u1} f_{u1}(u(k-1)) \\ \\ c_{u1} f_{u1}(u(k-p)) \end{bmatrix} \right)}_{a}^{2} \right\}$$

is solved for c_{y1} and c_{u1}. Then, another function is used, $f_{y2} = f_{u2}$, and another minimization problem:

$$\min \left\{ [r-a] - \left(\begin{bmatrix} c_{y2} f_{y2}(y(k-1)) \\ c_{y2} f_{y2}(y(k-2)) \\ \\ c_{y2} f_{y2}(y(k-p-1)) \end{bmatrix} + \begin{bmatrix} c_{u2} f_{u2}(u(k)) \\ c_{u2} f_{u2}(u(k-1)) \\ \\ c_{u2} f_{u2}(u(k-p)) \end{bmatrix} \right) \right\}^{2}$$

is solved for c_{y2} and c_{u2}, and so on. The fitting finds one parameter a time and the network growth is checked in relation to the approximation error. The same methodology is used for multivariable systems.

3.3   Conclusions

In the last section, a new single layer network structure, with mono-variable activation functions, that allows to represent multivariable systems has been presented. The construction of this network, its conception and the developement of a methodology, were motivated by the need of a mathematical model for process control. Therefore the model should have simple structure and a fast fitting methodology. The proposed network does not allow the representation of functions to any arbitrary degree of precision.

Meanwhile, we belive that it allows the representation of a large number of systems and be useful for process simulation and control.

4   Examples

This section shows examples of process identification using the proposed model on two non-linear systems. The first one is a SISO chaotic system. The other one is a chemical process involving non linear and multivariable characteristics. At first, some particularities of non-linear system identification problems will be presented.

4.1   The non-linear identification problem

In the last fifty years important advances have been done in the theoretical analysis and synthesis of dynamical systems, mainly for linear systems, in the time or the frequency domains.

The starting point of a dynamical system analysis is the determination of a mathematical model. The natural way to obtain this kind of models is by the use of physical and chemical principles.

These models are called phenomenological and their accuracy relies on adopting simplified hypothesis. Sometimes it is not possible to obtain a phenomenological model, or even to solve the equations that describe the model. In such cases the solution is to use empirical black box models.

This problem was studied using the theory of linear dynamical systems. Otherwise, they are inaccurate in order to predict the behavior of strong non-linear systems, indicating the necesity to develop black box non-linear mathematical models. This is an open research problem [1]. In the last fifteen years several non-linear models have been developed, but none has obtained the name of "universal model".

Nowadays, the black box models based on artificial intelligence are widely used, especially those using syntactic rules (fuzzy logic models) and mathematical functions (neural nets).

Two systems have been identified in this work, and for better evaluation of the proposed network only one wavelet family was used as activation function. The selected family is based on cubic splines. A simulation using a Haar basis is also presented, to compare two models based on two different wavelet families. The choice of the wavelet family has no rules, depending only on the user experience. Comparing models based on the two families (Haar and cubic splines), we observed that a good approach depends on the wavelet family used.

When wavelets obtained by numerical methods are used, the models are not continuous functions. The interpolation between two points is done by constant functions. Consequently, the obtained model shows a discretization effect, even when using spline functions as initial elements, but this effect may be smooth by using interpolation function or increasing the number of points that represent the wavelet functions.

The inputs and outputs were normalized between 0 and 1 defining the problem domain. The first step is to define n scales and n wavelet functions into the domain $(0,1)$, forming the first layer of wavelets. The next step is to define $n*2$ wavelet functions with half the support of first layer wavelets, forming the second layer of wavelets, and so on. This methodology is for one sub-net, and must be repeated for the other sub-nets. Every new neuron incorporated represents a new approximation.

The neural network model proposed consists of sub-nets with their outputs added to give the global model output. Each sub-net output is the sum of the neuron outputs weighted with constant coefficients. The least-square method fits those coefficients by minimization of the error square between the net output and the process output. The perturbation signals should be "persistently exciting" for satisfactory results when using least square methods with linear models [7]. In network model, the inputs are delayed external input sequences and delayed output feedbacks.

For a good identification using least square methods, the input signals must "excite" every neuron. As it was indicated, these input signals may be external inputs or feedback outputs. If one of these signals has all its values out of the domain of a neuron, the output of this neuron will be zero. So the least square method will not fit neural network parameters because the matrix involved does not have full rank. Zero output is an extreme case and sometimes the output of one neuron may be near zero. The output near zero means the rank is full but the problem is ill conditioned. The solution proposed is designed to eliminate neurons until the problem becomes well conditioned. The delayed values of external inputs and feedback outputs, used as inputs of the neural net, and the number of neurons in each sub-net (number of wavelet layers) defines the structure of the proposed neural net.

4.2   Identification of a chaotic attractor

The identification of a chaotic attractor is used as the first performance test for the proposed network (4) and shows the representation capability of a non-linear dynamic system.

The system to be identified is based on the quadratic map:

$$y(k) = ay(k-1)(1 - y(k-1))$$

whose behavior for $a > 3.7$ is chaotic [2].

In order to incorporate an external input without introducing modifications in the main attractor characteristics the parameter a was modulated by a sin function,

$$a = 3.9 * u(k)$$

with

$$u(k) = .9 + 0.07 * sin(k/20) * (1 + 0.1 * rand(1)),$$

where *rand* indicates random number between 0 and 1, with uniform distribution. Basically, this process has chaotic behavior like the quadratic map with $a = 3.9$, when the signal u is at its maximum value, and oscillatory behavior when u is at its minimum value, as can be seen in Figures 9 and 10.

In this process identification problem, a wavelet network model with two sub-nets, considering an external input and a feedback output, was used. Each sub-net was created using one scale function level and two wavelet levels, resulting in a total of 16 parameters to fit. The model can be represented by the equation:

$$y(k) = f(y(k-1)) + g(u(k-1))$$

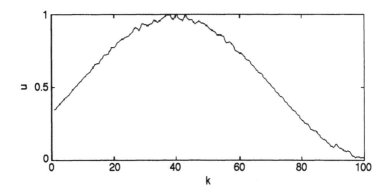

FIGURE 9. Input signal u.

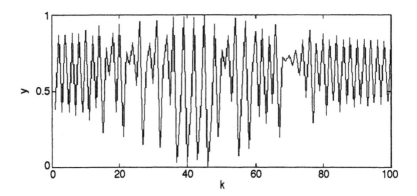

FIGURE 10. Output of chaotic map.

To fit the model, a set of 100 input and output process values was used (Figure 9). For the simulation of the resulting model, constant inputs, at values $u = 0.84$ and $u = 0.97$ (Figure 4.2 and Figure 13), were used to evaluate representation characteristics at oscillating and chaotic behavior. Two independent simulations were realized to analyze the representation characteristics, with the same fitted coefficients and input values, but modifying the feedback output. One simulation used the process output and the other used the model output. The process outputs for inputs $u = 0.84$ and $u = 0.97$ can be observed in Figure 4.2.

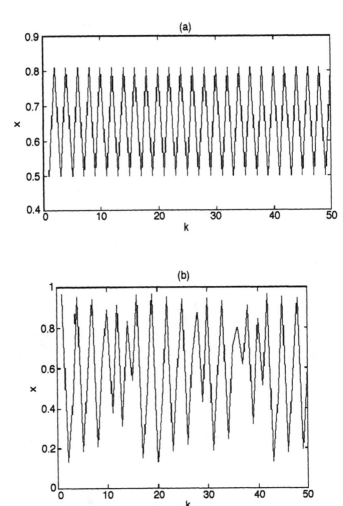

FIGURE 11. Process outputs for constant inputs

For a better visualization of the model representation characteristics when it presents chaotic behavior, Figure 12 shows the relation $y(k+1)$ vs. $y(k)$.

The first simulation was done using process outputs as feedback. The model outputs in this case, for inputs $u = 0.84$ and $u = 0.97$, are shown in Figure 13

An oscillatory behavior is observed for $u = 0.84$, and the time sequence with input $u = 0.97$ seems to be chaotic. In Figure 14 it is

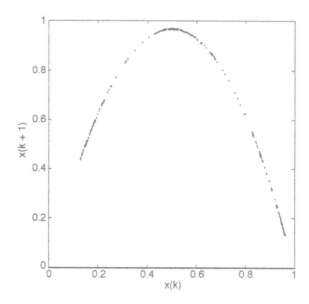

FIGURE 12. Process outputs for input $u = 0.97$

observed that the identified map is quadratic, although $y(k+1)$ vs. $y(k)$ is not a dense curve as in Figure 12.

This difference comes from the used identification model. The parallel model does not represent variable multiplications that are present in the process equation, introducing a model error. The second simulation was done using the model outputs as feedback. Results are shown in Figure 15

For $u = 0.84$ the oscillatory behavior was reproduced with some differences in amplitude. Meanwhile for $u = 0.97$ it is observed an output pattern repetition for every 20 points. The map that relates output in $k+1$ with the output in k retains the quadratic characteristic, though it is obviously discontinuous (Figure 16) and this can, probably, explain the repeated pattern observed.

The experimental results were satisfactory since it was possible to identify a non-linear dynamic process. The model does not use variable or function multiplication, being exclusively a sum of two single variable functions. Other identifications using greater number of points and wavelet levels have not presented better results.

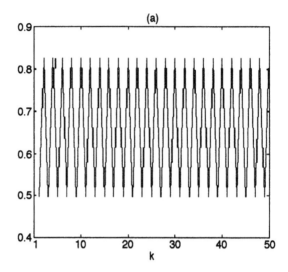

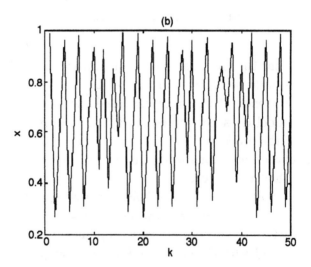

FIGURE 13. Model outputs for external constant inputs using process outputs as feedback.

In the process identification problem, the proposed model uses information from process outputs [14], decreasing the modeling error. So the next simulations use exclusively process outputs instead of model outputs as feedback.

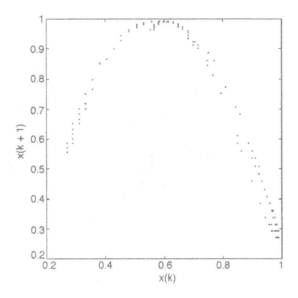

FIGURE 14. Process outputs for input $u = 0.97$

4.3   Continuous stirred tank reactor identification.

The continuous stirred tank reactor (CSTR) is a benchmark in control and system identification [8], so it was chosen to be identified by the proposed wavelet net. The equations for the continuous stirred tank reactor were taken from Embirucu et.al. [13], with same parameters, allowing the evaluation of the proposed model and comparing with other published results. The process is a jacketed CSTR with an exothermic first order chemical reaction A $\rightarrow$ B, represented by a three differential equation system:

$$\frac{dx_1}{d\tau} = -\phi * k * x_1 + q * (x_{1f} - x_1)$$

$$\frac{dx_2}{d\tau} = \beta * \phi * k * x_1 - (q + \delta) * x_2 + \delta * x_3 + q * x_{2f}$$

$$\frac{dx_3}{d\tau} = \frac{q_c * (x_{3f} - x_3)}{\delta_1} + \frac{\delta * (x_2 - x_3)}{\delta_1 * \delta_2}$$

$$k = e^{\left\{\frac{x_2}{1 + x_2/\gamma}\right\}}$$

It is described by the following dimensionless variables:

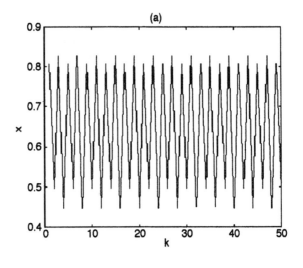

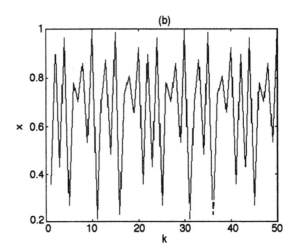

FIGURE 15. Model outputs for external constant inputs using its own outputs as feedback.

$$\beta = \frac{-\Delta H * C_f * \gamma}{\rho * C_p * T_{f0}} \qquad \phi = \frac{V * k_0 * e^{-\gamma}}{Q_0} \qquad x_2 = \frac{(T - T_{f0}) * \gamma}{T_{f0}}$$

$$\gamma = \frac{E}{R * T_{f0}} \qquad \tau = \frac{Q_0 * t}{V} \qquad x_{2f} = \frac{(T_f - T_{f0}) * \gamma}{T_{f0}}$$

$$\delta = \frac{U * A}{\rho * C_p * Q_0} \qquad q = \frac{Q}{Q_0} \qquad x_3 = \frac{(T_c - T_{f0}) * \gamma}{T_{f0}}$$

$$\delta_1 = \frac{V_c}{V} \qquad x_1 = \frac{C}{C_{f0}} \qquad x_{3f} = \frac{(T_{cf} - T_{f0}) * \gamma}{T_{f0}}$$

$$\delta_2 = \frac{\rho_c * C_{pc}}{\rho * C_p} \qquad x_{1f} = \frac{C_f}{C_{f0}}$$

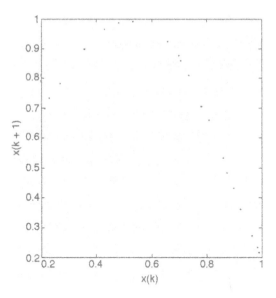

FIGURE 16. Process outputs for inputs $u = 0.97$

The thermal load m is defined as:

$$m = q * x_{2f}$$

and the used parameter values are:

β	=	8.0	x_{1f}	=	1.0	ϕ	=	0.072	x_{2f}	=	0.0
δ	=	0.3	q	=	1.0	γ	=	20.0	x_3	=	0.0
δ_1	=	0.1	δ_2	=	0.5	x_{3f}	=	-1.0	q_c	=	1.65102

The stable steady states are $(x_1 = 0.8933, x_2 = 0.5193, x_3 = -0.5950)$ and $(x_1 = 0.1890, x_2 = 5.1373, x_3 = 0.6359)$.

The simulation was done by solving the differential equation system near a stable steady state, using a Runge-Kutta method. The process inputs were uniformly distributed random signals. These inputs and the simulation outputs are the identification data set. To validate the model another data set with input defined as step functions was used.

A MISO problem.

In this section a MISO model, with one output (x_1) and two inputs (q) and (m) will be fitted. For the process identification different networks were tested and the best results were obtained for a network with four external inputs, two output feedbacks and 10 neurons for each sub-net, with cubic spline activation functions. The model was validated for pulse external inputs (Figure 17).

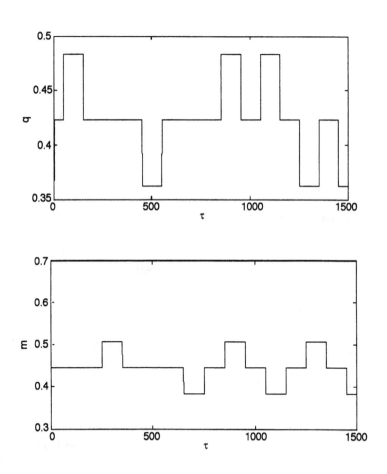

FIGURE 17. External inputs to validate MISO model.

The pulse functions on the external inputs have enough amplitude to modify the steady state, that is, after perturbation, the process becomes stabilized on any of the two steady states. This simulation confirms the identification of the process steady state multiplicity. The model and process output are seen in Figure 18. The difference between them is measured by a quadratic error $\left(\sum_{i=1}^{n} e_i^2/n \right)$ that is equal to 1.0109.

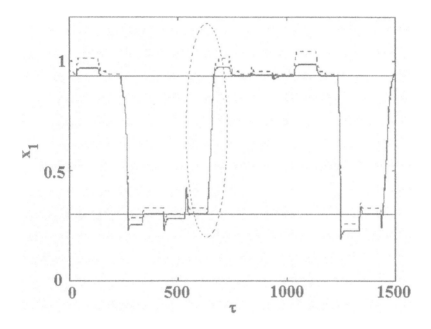

FIGURE 18. Process and model output for the step perturbation.

The figure enlargement in regions where transition of steady state happens (for example between points 630 and 639 in Figure 18) demonstrates that these dynamics were well identified (Figure 19).

The cubic splines were replaced by Haar functions as activation functions in the same neural network and the identified model behavior is shown in Figure 20.

It can be seen in this figure that the model identifies the steady state multiplicity, though the output presents worse steady state transition and worse process representation for perturbations near the steady state. The quadratic error is 2.5311.

Using Haar basis, algorithms become simpler, since these functions can be constructed from logic sequences ("if"), instead of mathemat-

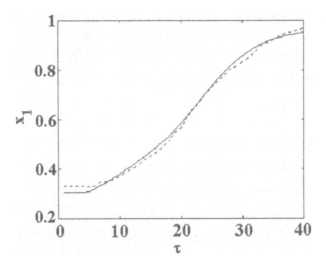

FIGURE 19. Enlargement of Figure V.15 between points 630 and 639.

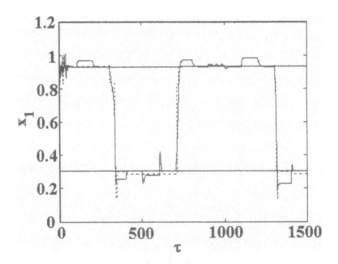

FIGURE 20. Process and model output for step perturbation (Haar basis).

ical calculus using floating point operations. This is the only advantage of Haar basis.

A process as a CSTR with a strong non-linear characteristic and multiple steady state, can not be identified by linear models. This problem can be observed in Figure 21 were results obtained by [13] are presented.

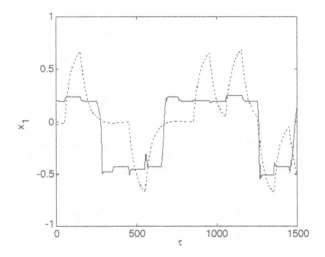

FIGURE 21. Simulation with linear model (deviation variables).

The models identified predict one step ahead process output. For applications in predictive control it is necessary to know the process outputs several steps ahead so the model should have one output for each prediction step. Models for several steps ahead prediction were identified using the same structure and experimental data. A prediction two steps ahead can be observed in Figure 22 and another three steps ahead in Figure 23.

It is possible to see that the model behavior becomes worse as the number of prediction steps increases. The steady state transition and the dynamic behavior remain well identified. The quadratic error grows to 2.9766 in the two steps ahead prediction and to 3.3512 in the three steps ahead prediction.

Through those experiences it was possible to infer that a multiple steady state process can be represented and its behavior predicted. The success of several step ahead prediction depends on appropriate structure selection.

MIMO problem

The same procedure used for MISO system was applied to identify a system with two inputs (q, m) and two outputs (x_1, x_2). The process inputs were identical to the signal used in the MISO problem, but the model structure was different, with four external inputs, four output

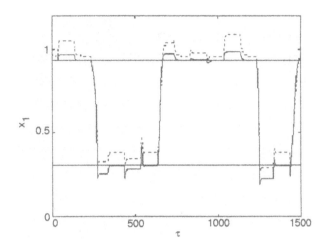

FIGURE 22. Two steps ahead prediction.

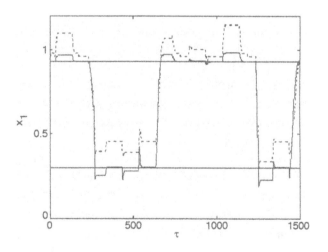

FIGURE 23. Three steps ahead prediction.

feedbacks (two for each output) and two wavelet levels for each sub-
net. This selection allows the comparison between MIMO and MISO
models. In Figure 24 and Figure 25 it is presented a simulation of
one step ahead prediction.

Comparing concentration outputs in Figures 18 and 24 better performance can be observed for the MIMO case, due to more information incorporated (temperature output). The quadratic errors are

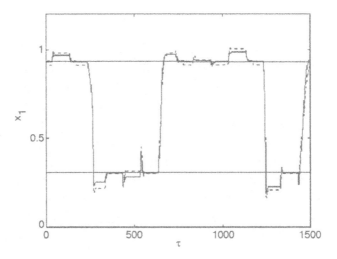

FIGURE 24. MIMO model concentration output.

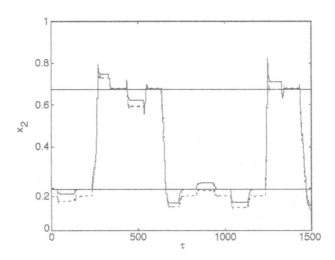

FIGURE 25. MIMO model temperature output.

0.7850 for concentration output and 1.1277 for temperature output.

4.4 Conclusions

In this section were presented identification results of two processes, each one with important non-linear characteristics. The first one, a simple SISO dynamical system, presents chaotic behavior for some inputs. This particular behavior was observed in polymerization reactors. The chaotic attractor identification can be considered successful. The identified model had not optimized structure and the wavelets used as activation functions were organized as a table (and so discontinuous). The discontinuous effect is important because chaos depends on a continuous quadratic map. The interpolation between two points was done by a constant function. Other interpolation functions may give a smooth discretization effect and better global model performance.

Another important characteristic of dynamical non-linear systems is steady state multiplicity. The second identified process (CSTR) presents two stable steady states and an unstable one. The perturbation used for identification and simulation, makes the process variables to move through the space near the two steady states. The process simulation was done with one, two and three prediction steps, and the results were satisfactory.

5   General conclusion

The analysis of complex dynamic systems and the project of efficient controllers needs accurate process model. Complex processes should be represented by non-linear models. Haber and Unbehauen [1] noticed that since the sixties non-linear models have been developed with low mathematical complexity and valuable representative characteristic. Nowadays, neural network model are frequently used for representing complex systems. The use of neural networks for process modelling has been a useful tool for analysis and control.

The most used neural networks are multilayer, with backpropagation training methods. These networks, in general, use sigmoidal functions. In other cases radial basis functions are used, as for example, gaussian functions. Both kinds of neural networks are universal approximators, that is, they are empirical models for general use. The radial basis function, specially those with compact support, allows the definition of disjoined regions on input and output system spaces. Therefore, the parameters fitting is strongly influenced by data close to the support, that is, additional data implies only the new calculation of some parameters. Otherwise, on neural networks with sigmoidal activation functions, new data implies the calculation of every net coefficient. Another characteristic of multilayer neural network is that numerical methods used for parameter fitting, in general steepest descent, presents local minimum stagnation. The parameter fitting of a single layer net is a linear optimization problem with just a global minimum. From the two main ideas, compactly supported activation functions and single layer neural network, Bakshi and Stephanopoulos [2] proposed single layer network with wavelet activation functions.

A system representation through single layer neural network can be described as projections on a subspace defined by wavelets. Multivariable neural network models use multivariable wavelets defined as combination of single variable wavelets. The number of wavelets needed to define a space basis grows rapidly with the number of variables.

A single layer neural network formed by the combination of neural networks with single variable wavelets, proposed in section 4, was used to represent complex dynamical systems. It allowed a satisfactory system representation and to decrease the total calculation time comparing with Bakshi e Stephanopoulos networks [2]. The characteristics of the new neural network such as convergence and stability have not been analyzed mathematically, though some results on identification and system simulation encourage their use as empirical models. There is an important relationship between the identification problem using single layer neural networks and projection on function spaces, allowing the use of functional analysis principles in the development of new neural networks. Also, it can be observed that fuzzy set modelling is strongly related with single layer networks, because transformations between syntactic and mathematical rules frequently used the mexican hat function, that is a kind of wavelet. Finally, it should be said that identification with single layer neural networks depends strongly on developments in functional analysis area.

6 References

[1] Harber, R. e Unbehauen, H. *Structure identification of nonlinear dynamic systems.* Automatica , 16, p.651, 1990.

[2] Bakshi, B. e Stephanopoulos WaveNet: A multiresolution hierarchical neural network with localized learning. G. AIChE.J., 39, p.57, 1993.

[3] Hsia, T.C. *System identification : Least-Squares Methods.* Lexington Books, 1977.

[4] Eykhoff, P. *System identification: Parameter and State Estimation* John Wiley & Sons, 1974.

[5] Chen, S. e Billings, S.A. *Modelling and analysis of non-linear time series.* Int. J. Control, vol.50, N 6, p.2151, 1989.

[6] Wu, X - DeCicco, J e Llinar, A. *Empirical dynamic models for linear and nonlinear multivariavel process.* Proceedings of PSE, p.1323, 1994.

[7] Ljung L. *System identification - Theory for the user.* Prentice Hall. 1987.

[8] Pottmann, M. e Seborg, D.E. *Identification of non-linear process using reciprocal multiquadratic functions.* J.Proc.Contr., vol.2, 4, 189, 1992.

[9] Zhu, X e Seborg, D.E. *Nonlinear predictive control based on Hammerstein models.* Proceedings of PSE, p.995, 1994.

[10] Strang G. *Wavelets and dilation equations: A brief introduction.* Siam Review, Vol 31, 4, 1989.

[11] Daubechies, I. *Orthonormal bases of compactly supported wavelets.* Comm. on Pure and Appl. Math. , vol.XLI, 909, 1988.

[12] Mallat, S.G. *A theory for multiresolution signal descomposition: The wavelet representation.* IEEE Trans. Patt. An. Mach. Int., 11, 7, p.674, 1989.

[13] Embirucu, M., Tauhata, T. Wanderley, A. e Lima, E. L. *Model Predictive Control of a Nonlinear Process.* Latin American Applied Research, 25: 177-180 (1995).

[14] Bezerra de Souza Junior, Mauricio *Redes Neuronais Multicamada Aplicadas a Modelagem e Controle de Processos Quimicos.* Tese de Doutorado, PEQ/COPPE/UFRJ, 1993.

[15] Meyer, Y. and Ryan, R.D. *Wavelets* Siam, 1993.

Nomenclature

ΔH	Reaction heat
C	Reaction composition
C_f	Feed composition
C_{f_0}	Reference feed composition
ρ	Process liquid density
ρ_c	Coolant density
C_p	Process liquid heat capacity
C_{pc}	Coolant heat capacity
T	Reactor temperature
T_f	Feed temperature
T_{f_0}	Reference feed temperature
T_c	Coolant temperature
T_{c_0}	Feed coolant temperature
E	Activation energy
R	Gas-law constant
U	Heat transfer coefficient
A	Heat transfer are
V	Reactor volume
V_c	Cooling jacket volume
k_0	Frequency factor
Q	Feed flow
Q_0	Reference feed flow

Chapter 12

Higher order asymptotic boundary conditions for an oxide region in a semiconductor device

Irene M. Gamba

1   Introduction

When modeling steady potential flow problems in polygonal non–convex domains, it is expected to find singularities being develop at the corners. The asymptotic behavior of these singularities in reentering corners depends on the boundary data, on the corner angle and on the permittivity constants associated with the potential equation when modeling inhomogeneous media.

Here we focus on a problem that arises when modeling the electrostatic potential that corresponds to a 2-dimensional electronic semiconducting device which combine a semiconducting region and an oxide region with a common interface. We assume that device is formed of two homogeneous materials device. (See Markowich [6], Markowich at al. [7] and Selberherr [9] as references on the modeling of semiconductor devices.)

This problem is motivated by the large scale computations that are required when modeling these devices (see, as an example, the discussion in chapter 5 of Selberherr [9] about the boundary conditions on MOSFET devices).

We propose here a further approximation to the reduction from computing the electrostatic potential of a full a boundary value problem in a nonconvex domain with an interface, to compute the electrostatic potential of a boundary value problem in a convex domain with a weak boundary condition at the interface with an error of fourth order in the thickness of the device, all under geometrical assumptions on the shape of the device.

Since the space charge is neutral in the oxide region, we use there the explicit solution in terms of their Fourier series representation combine with the fact that the solution of the full boundary value problem has enough regularity in the interface so that it has a Fourier series representation along the interface. The higher order approximating problem is obtained by estimating the Fourier coefficients in terms of the thickness of the oxide region and using the regularity of the solution to the full problem there.

The domain Λ under consideration consists in the union of two disjoint rectangles. The larger one, denoted by Ω, is the semiconducting region, and the smaller one, Φ_δ, is the oxide region with thickness δ (see figure 1). One side of the oxide Φ_δ is completely included in a side of Ω.

We had already analized in [1] and [2] a boundary value problem for a Drift-Diffusion model for transport of charged-particles in a semiconducting device. A complete study of regularity and asymptotic behavior at the boundary was presented there. In addition, if the oxide region is of small thickness, then [3] analyses asymptotically the approximating boundary value problem on the semiconducting region. Thus in [3] we proved that solutions of the full Poisson problem in the non-convex geometry converge to different asymptotic limit according to the ratio $\frac{\delta}{\varepsilon_1}$ as the area of the oxide region converges to zero, where δ is the oxide width or thickness and ε_1 the oxide permittivity constant. There, we provide the correct boundary value approximation to the full problem by given a relationship between the Fourier series coefficients of the solution on the interface region between Φ_δ and Ω, away from the endpoints of the interface, where the asymptotic behavior described in [1] and [3] takes over.

Hence, we concentrate on the thickness of the oxide of relatively small size and we consider the case where the quotient $\frac{\delta}{\varepsilon_1}$ is of order of 1. It was shown in [3] that for this case, the limiting problem is the one that replaces the oxide region by a mixed condition (or Robin conditions) weakly in $H^{-1/2}(I)$, with I is the interface. The boundary condition depends on the ratio of thickness–permitivity.

We present here a higher order correction to the interface–boundary condition. We shall obtain a weak $H^{-1/2}(I)$ interface–boundary condition that involves the normal and double tangential derivatives to the boundary and it is an approximation of order (δ^4) to the original boundary value problem for the full geometry, when δ relatively small. The limiting boundary value problem is a nice elliptic problem that has a variational formulation.

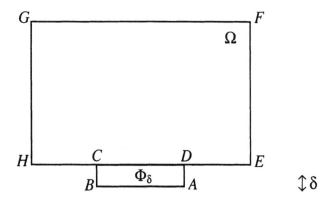

FIGURE 1.   The oxide–semiconductor device geometry

2   The full problem

Let $\Omega \subset I\!\!R^2$ be a rectangle and Φ_δ a smaller disjoint rectangle such that $I = \partial\Omega \cap \partial\Phi_\delta$ is one of the sides of Φ_δ contained in $\partial\Omega_I$, with the endpoints of $I = \{x_1, x_2\}$ away from the endpoints of $\partial\Omega_I$. The length of the sides of $\partial\Phi_\delta$ adjacent to I are equal to δ (see fig.1).

We consider the following boundary value problem in the set Λ, that denotes the interior of the closure of $\Omega \cup \Phi_\delta$.

Let $u = \{u_1, u_2\}$ the electrostatic potential that solves the boundary value problem

$$\Delta u_2 = q\,\varepsilon_2^{-1}(f + C) \quad \text{in} \quad \Omega \qquad (12.1)$$
$$\Delta u_1 = 0 \qquad\qquad \text{in} \quad \Phi_\delta \qquad (12.2)$$

with in interface transfer conditions

$$u_1 = u_2 \quad \text{and} \quad \varepsilon_2(u_2)_\nu = \varepsilon_1(u_1)_\nu \quad \text{on} \quad I = \overline{CD}$$

and boundary conditions

$$u_1 = U_0 \quad \text{on} \quad \overline{AB}, \qquad u_2 = \begin{cases} U_1 & \text{on } \overline{HC} \\ U_2 & \text{on } \overline{DE} \\ U_0 & \text{on } \overline{GF} \end{cases} \qquad (12.3)$$

and

$$(u_1)_\nu = (u_2)_\nu = 0 \qquad\qquad \text{on} \quad \partial\mathcal{N} = \partial\Lambda \setminus \partial\mathcal{D}$$

where the horizontal wall sections, denoted by $\partial D = \overline{\mathcal{BA}} \cup \overline{\mathcal{HC}} \cup \overline{\mathcal{DE}} \cup \overline{\mathcal{GF}}$, represent contacts (i.e. Dirichlet boundary conditions), and the vertical wall sections, denoted by $\partial \mathcal{N}$ represent insulating boundary conditions (i.e., homogeneous Newman data).

The parameters ε_1 and ε_2 are positive constants (effective permittivity) in Φ_δ and Ω respectively and q a constant (the electric charge).

The function $C(x)$ is the given predefined doping profile and may have a jump discontinuity surface (the pn–junctions), denoted by $J_{p,n}$, contained in $\overline{\Omega}$ that intersects I in two points away from the endpoints $x_1 = C$ and $x_2 = D$.

The function $f \in C^\alpha(\overline{\Lambda})$ denotes carrier densities.

2.1   Regularity of the full problem

From classical potential theory (see [4] and [5] and references therein) it is known that given f bounded and measurable there exists a unique solution $u = \{u_1, u_2\}$ of problem (12.1) belonging to $H^1(\Lambda) \cap C^\alpha(\Lambda)$ and the $H^1(\Lambda)$–norm of u depends on the numbers $\varepsilon_1, \varepsilon_2$ and δ. Hence, $u_1 \in H^{1/2}(\partial \Phi_\delta)$ and $u_2 \in H^{1/2}(\partial \Omega)$.

Let $B_i = B(x_i, R_i)$ denote the ball centered at x_i with radious R_i for $i = 1, 2$. Let Λ_1 denote $\Lambda_1 = \Lambda \setminus \{B_1 \cup B_2\}$. Then from [1] we know that

$$u \in W^{2,p}\left(\Lambda_1 \cap \overline{\Phi}_\delta\right), \cap W^{2,p}\left(\Lambda_1 \cap \overline{\Omega}\right), \quad 1 < p < \infty \qquad (12.4)$$

as shown there for the solutions of the full drift–diffusion system of semiconductor models; and the following estimate holds

$$\|u_1\|_{W^{2,p}(\Lambda_1 \cap \Phi_\delta)} + \|u_2\|_{W^{2,p}(\Lambda_1 \cap \Omega)} \leq$$
$$C\left(\|u_1\|_{L^p(\Phi_\delta)} + \|u_2\|_{L^p(\Omega)} + \|f - C\|_{L^p(\overline{\Omega})}\right) \qquad (12.5)$$

where C depends on Ω, δ, ε_1, ε_2, and q.

Moreover, in [3], the following estimate was derived on the H^1–norm of the solution $u = \{u_1, u1, u_2\}$ of the boundary value problem (12.1)

$$\|u\|_{H^1(\Lambda)}^2 \leq \|u_1\|_{H^1(\Phi_\delta)}^2 + \|u_2\|_{H^1(\Omega)}^2 \leq 1 + \log^+\left(\frac{\varepsilon_1}{\delta}\right). \qquad (12.6)$$

In case that the function $C(x)$ has a jump discontinuity along J_{np}, let R_i be small enough such that $B_i \cap J_{np}$ is empty, $i = 1, 2$. Then, it has been shown also in [1] that the solutions $u = \{u_1, u_2\}$ of the boundary value problem 12.1 are controlled from above and below by

$$-W_{1,i} \leq u_1 \leq W_{1,i} \quad \text{in} \quad B_i \cap \Phi_\delta \quad \text{and}$$
$$-W_{2,i} \leq u_2 \leq W_{2,i} \quad \text{in} \quad B_i \cap \Omega, \qquad (12.7)$$

where the functions $W_k\, i$ and $-W_k\, i$ are upper and lower barrier functions for u_k, $k = 1, 2$ solutions of the full boundary value problem, respectively, at the points x_i, $i = 1, 2$.

These barrier functions were explicitly constructed in [1] and they are

$$W_{k,i}(x) = (A\, v_{k,i} + w)(x), \qquad k = 1, 2. \qquad (12.8)$$

The functions $v_{k,i}$ are

$$v_{1,i}(\tau_i,\, \theta_i(\theta)) = \tau_i^{\alpha_\varepsilon} \frac{\cos \alpha_\varepsilon(\frac{3\pi}{2} - \theta)}{\cos \alpha_\varepsilon \pi/2} + U_i, \qquad \begin{matrix} \pi \leq \theta \leq \frac{3\pi}{2} \\ 0 \leq \tau_i \leq R_i \end{matrix} \qquad (12.9)$$

$$v_{2,i}(\tau_i,\, \theta_i(\theta)) = \tau_i^{\alpha_\varepsilon} \frac{\sin \alpha_\varepsilon \theta}{\sin \alpha_\varepsilon \pi} + U_i \qquad \begin{matrix} 0 \leq \theta \leq \pi \\ 0 \leq \tau_i \leq R_i \end{matrix} \qquad (12.10)$$

with $(\tau_i, \theta_i, (\theta))$ are the polar coordinates centered at x_i in $B_i(x_i, R_i)$, defined as $\tau_i = |x - x_i|$ and

$$\theta_i = \begin{cases} \theta_i = \theta + \pi \\ \\ \theta_2 = -\theta \end{cases} \quad 0 \leq \theta \leq \frac{3\pi}{2} \qquad (12.11)$$

The parameter denoted by α_ε is positive and satisfies the relationship

$$\frac{\varepsilon_2}{\varepsilon_1} \alpha_\varepsilon \cot \alpha_\varepsilon \pi = \alpha_\varepsilon \tan \alpha_\varepsilon \frac{\pi}{2} \qquad (12.12)$$

Remark. If $\frac{\varepsilon_2}{\varepsilon_1} \ll 1$, asymptotically $0 < \alpha_\varepsilon \approx \frac{2}{\pi} \frac{\varepsilon_2}{\varepsilon_1}$.

The function w from (12.8) is given by

$$w(x) = w(x, y) = -\frac{B\, \varepsilon_1}{2\varepsilon_2} (y - y_i)^2 \quad \text{in} \quad (B_1 \cup B_2) \cap \Lambda, \quad (12.13)$$

for $B = \sup_{\Omega} |f - C|$ and $y_1 = y_2$. The A is a constant in (12.8) depends on B, $\varepsilon_2/\varepsilon_1$, δ and $\sup_{\Omega} |u_2|$, therefore, estimate (12.8) implies

$$|u_i(\tau_i, \theta_i)| \leq C \tau_i^{\alpha_\varepsilon} \quad \text{in} \quad B_i(x_i, R_i), \qquad (12.14)$$

Moreover, we showed [1] that

$$|D_i\, u(x)| \leq C_1 |x - x_k|^{\alpha_\varepsilon - 1} \quad \text{and} \quad |D_{ij}\, u(x)| \leq C_2 |x - x_k|^{\alpha_\varepsilon - 2}$$
$$(12.15)$$

in $B(x_k, R_k) \cap \Lambda$, $\quad k = 1, 2$.

Therefore, since $\alpha_\varepsilon > 0$ the solutions $u_k \in H^{1/2}(I)$ and $\nabla u_k \in L^1(I)$ and hence, each u_k and $(u_k)_\nu$, $k = 1, 2$ has a Fourier series representation along I.

Furthermore, since the second pure derivative in the tangential direction of u_2 (i.e. $(u_2)_{ss} = (u_2)_{xx}$) is bounded along I away from the endpoints $x_i, i = 1, 2$ (see [2]), then $(u_2)_{ss}$ has a Fourier series representation in $I \setminus \{x_1, x_2\}$. Here $u_s = \partial u / \partial s$ denotes the partial differentition with respect to the tangential direction to the boundary.

2.2 Equivalent problem formulated in terms of the Fourier representation

As anticipated in the introduction the solution u_1 corresponding to the electrostatic potential with neutral space charge and boundary-interface conditions as in (12.3) can be solved with a closed form that is computed using separation of variables in a periodic δ-wide infinite strip obtained from infinitely many even reflections across the insulating homogeneous data of the oxide region and the harmonic operator. These explicit solutions are constructed as the solutions of the Dirichlet problem for a circular annulus of an even function with prescribed constant on the edge of the annulus and they are given in Fourier series representation (see Mikhlin [5, Ch. 13, section 2].)

Then, it was shown in [3] that, problem (12.1) is equivalent to the following one: Find a $u_{\varepsilon_1, \delta} \in H^1(\Omega)$ solution to the problem

$$\Delta u_{\varepsilon_1, \delta} = q\, \varepsilon_2^{-1}\, (f + c) \quad \text{in} \quad \Omega, \qquad (12.16)$$

with the boundary conditions

$$u_{\varepsilon_1, \delta} = \begin{cases} U_1 & \text{on } \overline{HC} \\ U_2 & \text{on } \overline{DE} \\ U_3 & \text{on } \overline{GF} \end{cases}$$

$$(u_{\varepsilon_1, \delta})_\nu = 0 \quad \text{on} \quad \partial \mathcal{N} = \partial \Omega \setminus (\partial \mathcal{D} \cup I),$$

and the following boundary condition on I

$$(u_{\varepsilon_1, \delta})_\nu (x) = \frac{1}{\varepsilon_2} \frac{\varepsilon_1}{\delta} \left[(a_{0,2} - U_0) + \sum_{n \geq 1} a_{n,2} \, 2\pi n\delta \cos 2\pi nx \, \frac{\cosh 2\pi n\delta}{\sinh 2\pi n\delta} \right]$$

$$(12.17)$$

where $u_{\varepsilon_1, \delta}(x) = \sum_{n \geq 0} a_{n,k} \cos 2\pi nx$ on I has a Fourier series representation. In addition we showed that the boundary condition (12.17) on I is equivalent to

$$(\widehat{u}_{\varepsilon_1, \delta})_{\nu, 0} = \frac{1}{2\pi} \frac{\varepsilon_1}{\varepsilon_2} \frac{\varepsilon_1}{\delta} \left[(\widehat{u}_{\varepsilon_1, \delta})_0 - U_0 \right] (\widehat{u}_{\varepsilon_1, \delta})_{\nu, n} = \mathcal{F}_{\delta, n} \cdot (\widehat{u}_{\varepsilon_1, \delta})_n \quad , \quad n > 1$$

$$(12.18)$$

where $(\widehat{\cdot})_{\nu, n}$ denotes the n-th Fourier coefficient of the normal derivative of $u_{\varepsilon_1, \delta}$ and $\mathcal{F}_{\delta, n}$ is defined as

$$\mathcal{F}_{\delta, n} = \frac{1}{\varepsilon_2} \frac{\varepsilon_1}{\delta} 2\pi n\delta \frac{\cosh : 2\pi n\delta}{\sinh : 2\pi n\delta} \quad n > 1. \qquad (12.19)$$

In addition, it was shown that, if $\frac{\varepsilon_1}{\delta} \to C_0$ as $\delta \to 0$ (i.e. the relevant physical case), then the limiting problem to (12.1) becomes

$$\Delta u = q \, \varepsilon_2^{-1} \, (f + C) \quad \text{in} \quad \Omega, \qquad (12.20)$$

with the boundary conditions

$$u = \begin{cases} U_1 & \text{on } \overline{HC} \\ U_2 & \text{on } \overline{DE} \\ U_3 & \text{on } \overline{GF} \end{cases}$$

$$(u)_\nu = 0 \quad \text{on} \quad \partial \mathcal{N} = \partial \Omega \setminus (\partial \mathcal{D} \cup I),$$

and the following boundary condition on I

$$u_\nu = \frac{1}{\varepsilon_2} C_0 (u - U_0) \quad \text{weakly in } H^{-1/2}(I). \qquad (12.21)$$

2.3 Higher order asymptotic behavior of the Oxide boundary condition

Due to the regularity of the solution of the full problem in the non-convex domain the above limiting problem can be improved in order to get higher accuracy on the interface condition when δ is relatively small.

Thus we show here that, if δ is sufficiently small, standard asymptotic analysis on the Fourier coefficients of the term (12.18) yields an approximate local boundary condition weakly in $H^{-1/2}(\tilde{I})$, with $\tilde{I}$ compactly contained in I. This condition provides a higher order correction to the limiting problem (12.20 - 12.21) , and proves that for small delta, the solution of problem (12.20 - 12.21) provides a quite accurate approximation to the full problem.

Lemma 2.1 *Let $0 < \alpha_\varepsilon$ and $\frac{\varepsilon_1}{\delta} \to C_0$ as $\delta \to 0$, then the limiting problem to (12.1) becomes*

$$\Delta u = q\,\varepsilon_2^{-1}\,(f + C) \quad in \quad \Omega, \qquad (12.22)$$

with the boundary conditions

$$u = \begin{cases} U_1 & on \ \overline{HC} \\ U_2 & on \ \overline{DE} \\ U_3 & on \ \overline{GF} \end{cases}$$

$$(u)_\nu = 0 \quad on \quad \partial \mathcal{N} = \partial\Omega \setminus (\partial D \cup I),$$

and the following boundary condition on I

$$u_\nu = \frac{1}{\varepsilon_2}\,C_0(u - U_0 - \frac{\delta^2}{3}u_{ss}) \quad weakly \ in \ H^{-1/2}(\tilde{I}). \qquad (12.23)$$

for u_{ss} denoting the double tangential derivative to the boundary and $\tilde{I} = (I \cap \Lambda_1) \setminus (I \cap J_{p,n})$, where $\Lambda_1 = \Lambda \setminus (B_1(x_1,\ R_1) \cup B_2(x_2,\ R_2))$, and if $C(x)$ is a smooth function, then the boundary condition is valid for $\tilde{I} = I \cap \Lambda_1$. Moreover from (12.13), u behaves as $\tau_i^{\alpha\varepsilon}$ and its first and second derivatives as in (12.15), in $B_i(x_i,\ R_i) \cap \Omega$, $i = 1, 2$, respectively.

Proof: The proof of this lemma contains two parts. The first one is the asymptotic analysis on the Fourier coefficients. The second one studies the convergence of the interface condition in terms of the relationship between the Fourier coefficients of the normal derivatives into a local boundary condition that holds weakly in $H^{-1/2}(\tilde{I})$. This technique was used in [3] in order to show the approximation to the limiting problem (12.20 - 12.21).

From estimate (12.17) $u_{\varepsilon_1,\delta}$ remains uniformly in $H^1(\Omega)$ and hence $u_{\varepsilon_1,\delta}|_{\partial\Omega}$ is uniformly in $H^{1/2}(\partial\Omega)$.

Therefore $\sum_{n\geq 1} n^{1/2}(\hat{u}_{\varepsilon_1,\delta})_n \leq K < \infty$

Next we estimate the term $\mathcal{F}_{\delta,n}$ from (13.2) for $2\pi n\delta < 1$. Writing

$$\cosh 2\pi n\delta = \sum_{k\geq 0} \frac{(2\pi n\delta)^{2k}}{k!} \quad \text{and} \quad \sinh 2\pi n\delta = \sum_{k\geq 0} \frac{(2\pi n\delta)^{2k+1}}{(2k+1)!} .$$

$$(12.24)$$

Then, for $n\delta < \frac{1}{2\pi}$

$$\frac{\cosh 2\pi n\delta}{\sinh 2\pi n\delta} \asymp \frac{1 + \frac{(2\pi n\delta)^2}{2!} + \frac{(2\pi n\delta)^4}{4!} + \mathcal{O}(n\delta)^6}{2\pi n\delta + \frac{(2\pi n\delta)^3}{6} + \frac{(2\pi n\delta)^5}{5!} + \mathcal{O}(n\delta)^7}$$

$$(12.25)$$

The right-hand side of (12.25) is equals to

$$
\begin{aligned}
&\frac{1}{2\pi n\delta} \cdot \frac{1 + \frac{(2\pi n\delta)^2}{2!} + \mathcal{O}(n\delta)^4}{1 + \frac{(2\pi n\delta)^2}{6} + \mathcal{O}(n\delta)^4} \\
= \ &\frac{1}{2\pi n\delta} \cdot \frac{\left(1 + \frac{(2\pi n\delta)^2}{2!} + \mathcal{O}(n\delta)^4\right)\left(1 - \frac{(2\pi n\delta)^2}{6!} - \mathcal{O}(n\delta)^4\right)}{1 - \frac{(2\pi n\delta)^4}{6^2} + \mathcal{O}(n\delta)^6} \\
= \ &\frac{1}{2\pi n\delta} \cdot \left(1 + \frac{(2\pi n\delta)^2}{3} + \mathcal{O}(n\delta)^4\right)\left(1 + \mathcal{O}(n\delta)^4\right)
\end{aligned}
$$

Hence, asymptotically, for $n\delta < \frac{1}{2\pi}$,

$$\frac{\cosh 2\pi n\delta}{\sinh 2\pi n\delta} \asymp \frac{1}{2\pi n\delta}[1 + \frac{(2\pi n\delta)^2}{3} + \mathcal{O}(n\delta)^4]. \qquad (12.26)$$

Therefore, the term $\mathcal{F}_{\delta,n}$ is approximated by

$$\mathcal{F}_{\delta,n} \asymp \frac{1}{\varepsilon_2} \frac{\varepsilon_1}{\delta}\left(1 + \frac{(2\pi n\delta)^2}{3} + \mathcal{O}(n\delta)^4\right), \quad \text{for} \quad 2\pi n\delta < 1, \quad (12.27)$$

so that the relationship given in (12.18) becames

$$(\widehat{u}_{\varepsilon_1,\delta})_{\nu,n} \asymp \frac{1}{\varepsilon_2}\frac{\varepsilon_1}{\delta}\left(1 + \frac{(2\pi n\delta)^2}{3} + \mathcal{O}(n\delta)^4\right)a_{n,k} \quad \text{for} \quad 1 < n < \frac{1}{2\pi\delta},$$
(12.28)

and $a_{n,k}$ the n-th Fourier coefficient of $u_{\varepsilon_1,\delta}$.

Therefore the boundary condition on I defined in (12.17) can be expressed asymptotically for $2\pi n\delta < 1$ as

$$(u_{\varepsilon_1,\delta})_{\nu}(x) \asymp \frac{1}{\varepsilon_2}\frac{\varepsilon_1}{\delta}[(a_{0,2} - U_0) +$$

$$+ \sum_{1<n\leq\frac{1}{2\pi\delta}} a_{n,2}\left(1 + \frac{(2\pi n\delta)^2}{3} + \mathcal{O}(n\delta)^4\right)\cos 2\pi nx$$

$$+ \sum_{n\geq\frac{1}{2\pi\delta}} a_{n,2}\,2\pi n\delta\,\cos 2\pi nx\frac{\cosh 2\pi n\delta}{\sinh 2\pi n\delta}\Big]$$
(12.29)

Recall that $u(\cdot,b)$ is regular in the x direction along the interface $I = \{y = b\}$ with the exception of the points x_i, $i = 1, 2$ and on the points of intersection of I with the pn-junction $J_{p,n}$ whenever the doping profile function $C(x)$ is discontinuos along $J_{p,n}$. However, even if $C(x)$ is discontinuos along $J_{p,n}$ and the curve $J_{p,n}$ is orthogonal to I at their intersecting points, still $u_{xx} = u_{ss}$ is bounded, for $\partial/\partial s$ denoting the partial differentition with respect to the tangential direction to the boundary, as shown in [2]. Then $D_j u(x,b), j = 0,1,2$ have a Fourier series representation given by

$$u(x,b) = \sum_{n\geq 0}\widehat{u}_n \cos 2\pi nx \quad \text{on} \quad A = I \setminus [\{x_1,x_2\} \cup (I \cap J_{n,p})] \quad (12.30)$$

and

$$u_x(x,b) = \sum_{n\geq 1}\widehat{u}_n\,(2\pi n)\sin 2\pi nx \quad (12.31)$$

and

$$u_{xx}(x,b) = -\sum_{n\geq 1}\widehat{u}_n\,(2\pi n)^2\cos 2\pi nx. \quad (12.32)$$

In order to show (12.23), given $w \in H^{1/2}(\tilde{I})$ we need to study the limit (we drop the ε_1, δ indexes) of

$$\hat{w}_0(\hat{u}_\nu)_0 - \frac{\hat{w}_0}{\varepsilon_2} C_0(\hat{u}_0 - U_0) + \sum_{n \geq 1} \hat{w}_n \left[(\hat{u}_\nu)_n - \frac{C_0}{vp_2} \left(1 - \frac{\delta^2}{3}(2\pi n)^2 \right) \hat{u}_\nu \right]$$

$$(12.33)$$

for $\varepsilon_1/\delta \to C_0$ as $\delta \to 0$.

Susbstitute each $(\hat{u}_\nu)_n$ by using the coefficients of the Fourier Series representation in (12.29) for the approximation of u_ν where $\hat{u}_n = a_{n,2}$, then (12.33) becomes

$$\frac{\hat{w}_0}{\varepsilon_2} \left[(\frac{\varepsilon_1}{\delta} - C_0)(\hat{u}_0 - U_0) \right] + \sum_{n \geq \frac{1}{2\pi\delta}} \frac{C_0}{vp_2} \hat{w}_n \, \hat{u}_n (2\pi n\delta \coth 2\pi n\delta) + \mathcal{O}(n\delta)^4$$

$$(12.34)$$

on $\tilde{I}$.

Now, if $t = 2\pi n\delta > 1$, then $|2\pi n\delta \coth 2\pi n\delta| < 4\pi n$, so that the second term of (12.34) is bounded by

$$\sum_{n \geq \frac{1}{2\pi\delta}} \frac{C_0}{\varepsilon_2} n\hat{w}_n \, \hat{u}_n \leq 4\pi \frac{C_0}{\varepsilon_2} \left(\sum_{n \geq \frac{1}{2\pi\delta}} n\hat{w}_n^2 \right)^{1/2} \left(\sum_{n \geq \frac{1}{2\pi\delta}} n\hat{u}_n^2 \right)^{1/2}.$$

$$(12.35)$$

Since w and u are in $H^{1/2}(I)$ uniformly in δ, then, for β arbitrary, there is a $N_0 = N_0(\beta, C_0/\varepsilon_2, |u|_{H^{1/2}(I)})$ such that (12.34) is estimated from above by

$$\frac{\hat{w}_0}{\varepsilon_2} (\frac{vp_1}{\delta} - C_0)(\hat{u}_0 - U_0) + \beta + \mathcal{O}(N_0\delta)^4$$

$$(12.36)$$

Clearly, if $\varepsilon_1/\delta \to 0$ as $\delta \to 0$, then (12.36) converges uniformly to zero as $\delta \to 0$.

Therefore, the solution to the full problem in the non-convex domain and interface condition (12.17) converges to the solution of problem (12.22 - 12.23) where the condition (12.23) on the interface subregion $\tilde{I}$ holds weakly in $H^{-1/2}$. See that, if $C(x)$ is a smooth function, then the boundary condition is valid for $\tilde{I} = I \cap \Lambda_1$.

2.4 Conclusions

The approximating boundary value problem corresponds to a nice elliptic problem that has a variational formulation. Thus, it may be used in order to optimize computational methods for potential equations in geometries as the ones described above, where the oxide region (or smaller rectangular section), corresponding to neutral space charge, is replaced by a boundary conditions on the interface. In addition, most of these techniques and results can be extended to three dimensional geometries corresponding to boundary value problems for semiconductor devices. However, in the three dimensional case the singularities at the points x_1 and x_2 would become singularities along a curve corresponding to the boundary of the interface, which now is a two dimensional rectangular surface. The regularity and asymptotic behavior at the boundary has not been studied for the three dimensional boundary value problem yet. These regularity results are needed in order to prove an equivalent result for a three dimensional approximating model.

Therefore, in the present framework , the asymptotic problem to the full problem 12.1, can then be computed using the approximating problem (12.22 - 12.23).

Consider the rectangle $\Omega = GFEH$, where $I = \overline{CD}$ (as in figure 1), and solve the following boundary value problem numerically

$$\Delta u = q\,\varepsilon_2^{-1}\,(f + C(x)) \quad \text{in} \quad \Omega \setminus (B_1(C,\ R_1) \cup B_2(D,\ R_2)) \quad (12.37)$$

with boundary conditions

$$u = \begin{cases} U_1 & \text{on } \overline{HC} \\ U_2 & \text{on } \overline{DE} \\ U_3 & \text{on } \overline{GF} \end{cases} \qquad u_\nu = 0 \quad \text{on} \quad \overline{GH} \cup \overline{FE}$$

$$(12.38)$$

and on the interface $I = \overline{CD}$ set first

$$u_\nu = \frac{1}{2\pi\,\varepsilon_2}\,\frac{\varepsilon_1}{\delta}\left(u - \frac{\delta^2}{3}\,u_{ss} - U_0\right) \quad \text{on } \tilde{I} \qquad (12.39)$$

for $\tilde{I} = (x_1 + R_C,\ J_1) \cup (J_1,\ J_2) \cup (J_2,\ x_2 - R_D)$, where u_{ss} is bounded even if $C(x)$ has discontinuities on $\{J_1, J_2\} = J_{np} \cap \overline{CD}$. If $C(x)$ is smooth in I, u_{ss} is smooth on I away from the endpoints, then (26.3) holds in $\check{I} = (x_1 + R_C,\ x_2 - R_D) \subset I$.

In the rest of I, for α_ε given by (12.12), set

$$u(R_1, \theta + \pi) = R_1^{\alpha_\varepsilon} \frac{\sin \alpha_\varepsilon \theta}{\sin \alpha_\varepsilon \pi} + U_1, \qquad 0 \leq \theta \leq \pi \mathrm{on} \Omega \cap \partial B(C, R_1)$$
$$(12.40)$$

near $x_1 = C$, and set

$$u(R_2, -\theta) = R_2^{\alpha_\varepsilon} \frac{\sin \alpha_\varepsilon \theta}{\sin \alpha_\varepsilon \pi} + U_2 \qquad 0 \leq \theta \leq \pi \text{ on } \Omega \cap \partial B(D, R_2)$$
$$(12.41)$$

near $x_1 = C$.

Therefore, the computational complexity of a potential equation in this non-convex domain Λ with a connected separating interface with standard transmission conditions have been transfered into a mixed-higher order local condition that depends on the size of quotient of the permittivity and thickness of the geometrical source of non-convexity and under the assumption that the thickness parameter δ is small with an accuracy of order δ^4 with respect to the problem in the full non–convex domain.

3   References

[1] Gamba, I.M. *Asymptotic behavior at the boundary of a semi-conductor device in two space dimensions* Based on the thesis dissertation. Istituto di Analisi Numerica del C.N.R. Pub.N.740. Pavia, Italy (1990). Ann. di Mat. Pura App. (IV) CLXIII, 1993, pp 43–91.

[2] Gamba, I.M. *Behavior of the potential at the pn-Junction for a model in semiconductor theory* Appl. Math. Lett. vol. 3, 1990, pp 59–63.

[3] Gamba, I.M. *Asymptotic boundary conditions for an oxide region in a semiconductor device* , Asymptotic Analysis Journal, vol. 7, 1993, pp 37–48.

[4] Gilbarg, D. and Trudinger, N. S. *Elliptic Partial Differential Equations of Second Order*, Springer-Verlag, New York, 1983.

[5] Ladyženskaja, O. A. and Uralt'ceva, N. N. *Équations aux dérivées partielles du type elliptique* Dunod, Paris, 1968.

[6] Markowich, P. *The stationary Semiconductor Device Equations,* Springer, Vienna, 1986.

[7] Markovich, P. , Ringhofer, C.A., and Schmeiser, C., *Semiconductor Equations,* Springer, Wien-New York, 1989.

[8] Mikhlin, S.G. *Mathematical Physics, An Advanced Course,* North–Holland, Amsterdam, 1970.

[9] Selberherr, S. *Analysis and Simulations of Semiconductor Devices,* Springer, Vienna, 1984.

Chapter 13

Estimation of the complex plain–wave modulus in viscoelastic media

E. M. Fernández-Berdaguer
J. E. Santos

1 Introduction

In this work we treat an estimation problem arising from the study of wave propagation in solids. Before dealing with the estimation problem we pay special attention to the physical model, formulated in the space–frequency domain.

The paper is presented in order to show the important points and the difficulties of the model and the inverse problem. No proofs are presented. The proofs, which are a generalization of those in [1] for the finite–dimensional case, will appear elsewhere.

First we describe the physics related to the direct problem, which concerns the modelization of wave propagation in attenuating media.

It is known that waves suffer attenuation and dispersion when travelling in rocks and other solid materials. The quality factor q, measuring the fraction of energy loss per cycle of a harmonic oscillation of angular frequency ω is known to be essentially independent of frequency over a wide range of frequencies and with the attenuation increasing significantly with increasing frequency. Since attenuation effects yield information about rock properties like lithology, saturation levels and porosity distribution, it is important to include such effects in the direct model to be used in the parameter estimation procedures.

Several models have been proposed in the literature to represent attenuation and dispersion effects in rocks and other solid materials. Here we will employ a viscoelastic model presented by [10] which derives stress–strain relations in the space–frequency domain using the Boltzmann after–effect equation for a continuous superposition of standard linear solids [21]. This formulation yields a causal model consistent with the behaviour of rocks and other solid materials at very low and high frequencies [11].

Although our work is intended for applications in geophysical exploration, the analysis and algorithms presented here are also of interest in other fields such as non–destructive material testing [4], polymer physics [8] and ocean acoustics [13], [14].

Since attenuation and dispersion effects are described stating the stress–strain relations in the space–frequency domain, and the absorbing boundary conditions used at the artificial boundaries lead to a pseudo–differential formulation of the propagation phenomena, it is natural to formulate and solve the problem in the space–frequency domain rather than in the space–time domain as opposite to the classic formulation and analysis of wave propagation in dispersive solid materials. As an additional and very important advantage from the computational point of view, this approach avoids the necessity of storing the history of the system and yields naturally parallelizable algorithms.

The parameters to be estimated, which are functions of time and the spatial variable, will be the quality factor and the relaxed plane wave modulus (i.e., the plane wave modulus at zero frequency) characterizing the viscoelastic behavior of the material. These parameters appear nonlinearly in the differential and boundary equations of the direct model.

Due to realistic considerations (for example the model does not match exactly the physical phenomena, the measurements are usually inaccurate) the observed data is not in the range of the solution operator of the equations modelling the problem. Since we do not expect to exactly match the observed data to the data of the model we approach the estimation problem in the least squares sense, that is, as an optimization one. It consists on the minimization of a quadratic functional involving the observed data and the traces of the model.

For a general treatment of parameter estimation using the Output Least Squares Method, we refer to the book by T. Banks and K. Kunisch [3].

Minimization problems where the direct model is stated as the solution of a partial differential equation can be approached by discretizing the differential problem first and then using optimization algorithms to solve the discretized version, or applying optimization algorithms to the continuous problem; i.e., minimizing the infinite dimensional model and performing the discretization only as a final step. Our estimation algorithm is of the second type and therefore it is formulated without specifiying the type of discretization employed in the corresponding numerical implementation.

This estimation procedures appeared earlier in [2] and were employed by A. Tarantola ([15], [16], [17], [18]) to solve inverse problems in geophysics.

An important feature of this approach is that any point in the space is a candidate for a change of the elastic properties (i.e. an interface). This is in contrast with our previous work in this subject in [1] we assumed the interface value positions on each layer known while searching for the quality factor and the plane wave modulus and in [7], in the elastic case, we estimated the location of the interfaces assuming the velocities known.

The highlights of the method are the sensitivity equations and the adjoint integral that allow to calculate the derivatives of the solution with respect to the parameters by solving a differential problem similar to that of the direct model but with modified source and boundary conditions. Those features make the algorithm very fast, which is needed when there is a large number of parameters to be estimated, as it is the case in the geophysical applications in which we are interested.

Questions when dealing with parameter estimation are those of existence, uniqueness and stability of solutions of the inverse problem. Conditions for the existence of solutions to the estimation problem are known only for particular sets of parameters, mainly in the case where the parameter belongs to a finite dimensional space. In general, inverse problems are ill-posed since the solutions are very sensitive to changes in the input data. We overcome the unstability problem using a standard Thikhonov regularization. At the present there is a great deal of work about the regularization problem. Among them we mention only a few publications on the subject: the book by Groetsch about regularization for linear (integral) equations [9], more recently we cite [5], [12], [19], [20], the interested reader can consult the bibliography therein.

1.1 Description of the Viscoelastic Model

Let $u(x,t)$ denote the displacement at $x \in \Omega = (0,1)$ and $t \in [0,\infty)$. The parameters characterizing the viscoelastic behaviour at $x \in \Omega$ will be chosen to be the relaxed plane wave modulus $m = m(x)$ and the mean quality factor $q = q(x)$. Let $\rho = \rho(x)$ be the mass density at x. We use the usual notation $\hat{f}(\omega)$ for the Fourier transform of a function $f(t)$. Denote by Υ the parameter function $\Upsilon(x) = (m(x), q(x))$. Then the differential model for the direct problem is the following: find $\hat{u}$ such that

$$-\rho\omega^2 \hat{u}(\Upsilon, x, \omega) - \frac{\partial}{\partial x}\left(f(\Upsilon)\frac{\partial \hat{u}(\Upsilon, x, \omega)}{\partial x}\right) = \hat{S}(x, \omega), \quad x \in \Omega, \quad (13.1)$$

with boundary conditions

$$\frac{\partial \hat{u}(\Upsilon, x, \omega)}{\partial \nu} + i\alpha\, \hat{u}(\Upsilon, x, \omega) = 0, \quad x \in \Gamma \equiv \partial\Omega. \qquad (13.2)$$

The source term $\hat{S}(x, \omega)$ is chosen such that its inverse Fourier transform $S(x,t)$ is real and $S(x,t) = 0$ for $t \leq 0$.

The condition (13.2) is a perfectly absorbing boundary condition which makes the boundary Γ transparent for outgoing waves. The complex coefficient α in the boundary condition (13.2) satisfies the dispersion relation

$$- \rho\omega^2 + \alpha^2\Upsilon = 0. \qquad (13.3)$$

The parametric function $f(\Upsilon)$ is given by

$$f(\Upsilon, \omega, x) = \frac{m(x)}{\beta(\omega, q(x)) - i\gamma(\omega, q(x))} \qquad (13.4)$$

with

$$\beta(\omega, q) = 1 - \frac{1}{\pi q} \ln\frac{1 + \omega^2\tau_1^2}{1 + \omega^2\tau_2^2}, \quad \gamma(\omega, q) = \frac{2}{\pi q} \tan^{-1}\frac{\omega(\tau_1 - \tau_2)}{1 + \omega^2\tau_1\tau_2}. \qquad (13.5)$$

The functions $\rho(x)$, $m(x)$ and $q(x)$ will be assumed to be bounded above and below by positive constants ρ_*, ρ^*, m_*, m^*, q_*, and q^*, respectively.

Realistic values for $q(x)$ in rocks lie in the range 50 to 1000, with low values of q representing highly dissipative materials and high values of q "almost" elastic materials. The constants τ_1 and τ_2, having units of time, will be assumed to be given and such that the quality factor $q^f(x,\omega) = (\operatorname{Re} f(\Upsilon)/\operatorname{Im} f(\Upsilon))(x,\omega)$ is approximately equal to $q(x)$ in the frecuency range $\tau_1^{-1} \leq \omega \leq \tau_2^{-1}$. The choice of the function $f(\Upsilon, \omega, x)$ as given in (13.4)–(13.5) corresponds to the viscoelastic behavior described in [10] associated with a continuous distribution of relaxation times.

A weak form of (13.1)–(13.2) is given by seeking $\widehat{u}(\Upsilon, x, \omega) \in H^1(\Omega)$ such that

$$- \omega^2(\rho\widehat{u}, v) + (f(\Upsilon)\nabla\widehat{u}, \nabla v) + \langle i\alpha f(\Upsilon)\,\widehat{u}, v\rangle = (\widehat{S}, v), \quad v \in H^1(\Omega). \tag{13.6}$$

The following existence and uniqueness result is an immediate generalization of that given in [11].

Theorem 1.1 *For $\omega \neq 0$, there exists a unique solution of (13.6) for any $\widehat{S}(\cdot, \omega) \in L^2(\Omega)$. Also, for $\omega = 0$, (13.1)–(13.2) has a solution (unique up to an additive constant) provided the condition*

$$\int_0^1 \widehat{S}(x, 0)dx = \int_0^1 \int_0^\infty S(x, t)dt\, dx = 0$$

is satisfied.

In order to have optimal a priori error estimates for the finite element procedures to be used in the discretization of the differential problems associated with the direct model and the derivatives of the solution with respect to the parameters, we need to know the behaviour of the elliptic regularity constants as functions of ω as $\omega \to 0$ or ∞, [6], [11]. This asymptotic behaviour is also needed to show the invertibility of the Fourier transform to obtain the solution $u(x, t)$. The result is stated in the following theorem is an extension of that proved in [1].

Theorem 1.2 *Let $\widehat{S}(\cdot, \omega) \in L^2(\Omega)$ be given. Then, the solution of (13.1)–(13.2) satisfies the estimates*

$$\|\widehat{u}(\cdot, \omega)\|_0 \leq \frac{C_1(\omega, \Upsilon)}{\omega}\|\widehat{S}(\cdot, \omega)\|_0, \tag{13.7}$$

$$\|\nabla\widehat{u}(\cdot, \omega)\|_0 \leq C_2(\omega, \Upsilon)\|\widehat{S}(\cdot, \omega)\|_0 \tag{13.8}$$

with $C_\ell(\omega, \Upsilon) = O(1)$ as $\omega \to 0$ and $C_\ell(\omega, \Upsilon)$ bounded as $\omega \to \infty$, $\ell = 1, 2$.

1.2   Formulation of the Estimation Problem

We denote by $\widehat{u}^{obs}(x, \omega)$ the measured values of the displacement at recording points $x = x_{r_\ell}$, $1 \leq \ell \leq K$, inside Ω and for $\omega \in \mathbf{R}$. We want to infer from the measurement vector $\widehat{u}^{obs}(\omega) = (\widehat{u}^{obs}(x_{r_\ell}, \omega))_{1 \leq \ell \leq K}$ the actual value of the parameters m and q. The set of admissible parameters is

$$\mathcal{P} = \{\Upsilon = \Upsilon(m(x), q(x)) : m \text{ and } q \text{ measurable}, \qquad (13.9)$$
$$m_* \leq m(x) \leq m^*, \; q_* \leq q(x) \leq q^*\}. \qquad (13.10)$$

Then, for $\widehat{u}^{obs}(\omega) \in L^2(\mathbf{R}, \mathbf{R}^K)$, the least–square estimation problem is:

minimize $J(\Upsilon)$ over $\mathcal{P}$, where $J(\Upsilon) = J(\Upsilon(m, q))$ is given by

$$J(\Upsilon) = \frac{1}{2} \int_{-\infty}^{+\infty} \sum_{\ell=1}^{K} |\widehat{u}(\Upsilon, x_{r_\ell}, \omega) - \widehat{u}^{obs}(x_{r_\ell}, \omega)|^2 \, d\omega. \qquad (13.11)$$

Our parameters are functions of x with no special representation, and consequently the problem is infinite dimensional in contrast to those were we prescribe a special representation for $m(x)$ and $q(x)$ with a finite number of parameters; for example if $m(x), q(x)$ are in a space of polynomials of finite dimension or are represented by spline functions. It is known that for finite dimensional problems as the number of parameters increases the inverse problem becomes unstable. One way of overcoming this problem is by using regularization techniques where the original (unstable) problem is approached by a sequence of stable ones whose solutions converge to the solution of the original one. Thus we consider the regularized problems: minimize $J_\eta(\Upsilon)$ over $\mathcal{P}$, where

$$J_\eta(\Upsilon) = J(\Upsilon) + \eta\|\Upsilon - \Upsilon_0\|^2, \qquad (13.12)$$

where η is chosen so that the minimization of J_η is stable. Chosing suitable sequences of η's becomes crucial. It shouldn't be too small or the above problem becomes unstable and it should be small enough so that the solution of (1.12) approaches that of (1.11).

Next we state a result that guarantees that if the parameter belongs to compact sets of $L^2(\Omega)$ the minimization problem has at least a solution. Assume that the source function $\widehat{S}(x, \omega)$ is such that

$C_1(\omega,\Upsilon)\widehat{S}(x,\omega)\omega \in L^2(\mathbf{R},L^2(\Omega))$ where $C_1(\omega,\Upsilon)$ is the constant in theorem 1.2. Then the mapping

$$(\mathcal{P},\|\cdot\|_0) \;\longrightarrow\; H^1(\Omega), \tag{13.13}$$
$$\Upsilon \;\longrightarrow\; \widehat{u}(\Upsilon,\cdot,\omega), \tag{13.14}$$

that associates the corresponding solution of (13.6) to each element $\Upsilon \in \mathcal{P}$ is continuous. Thus, the continuity of the mapping $J : \mathcal{P} \to \mathbf{R}^+$ follows by combining Sobolev's inequality with the above result and we conclude that the minimization problem (13.11) has a solution for the parameter in compact sets.

1.3   The Gateaux Derivatives

Most iterative schemes to minimize the functional J require the calculation of the derivative of J with respect to the parameter $\Upsilon = (m,q)$. In this section we will present the equations satisfied by the Gateaux derivative which with some other hypotheses on the source is enough for the convergence of the algorithms.

Let $N > 2$. We consider as the space of perturbations of the parameters the vector space

$$\widetilde{\mathcal{P}} = \{\delta\Upsilon = \delta\Upsilon(\delta m(x),\delta q(x)) : \quad \delta m,\delta q \in L^N(\Omega)\}. \tag{13.15}$$

Note: the choice of $L^N(\Omega)$ as the space of perturbations is due to technical hypotheses needed in the proofs.

We denote the Gateaux derivative of $\widehat{u}$ with respect to Υ by $D\widehat{u}(\Upsilon)$. That is, for $\Upsilon \in \mathcal{P}$, $\delta\Upsilon \in \widetilde{\mathcal{P}}$ and λ real we have that

$$\lim_{\lambda\to 0} (1/\lambda)\|\widehat{u}(\Upsilon+\lambda\delta\Upsilon,\cdot,\omega) - \widehat{u}(\Upsilon,\cdot,\omega) - \lambda(D\widehat{u}(\Upsilon)\delta\Upsilon)(\cdot,\omega)\| = 0 \tag{13.16}$$

where $D\widehat{u}(\Upsilon)$ is the linear operator defined from the space $\widetilde{\mathcal{P}}$ of all possible perturbations $\delta\Upsilon$ of Υ into $H^1(\Omega)$. Moreover, for $\delta\Upsilon = (\delta m,\delta q)$ belonging to the vector space of perturbations $\widetilde{\mathcal{P}}$ the function $D\widehat{u}(\Upsilon)\delta\Upsilon$ can be obtained as the solution of the following differential problem:

$$-\rho\omega^2(D\widehat{u}(\Upsilon)\delta\Upsilon) - \frac{\partial}{\partial x}\left(f(\Upsilon)\frac{\partial(D\widehat{u}(\Upsilon)\delta\Upsilon)}{\partial x}\right)$$
$$= \frac{\partial}{\partial x}\left(Df(\Upsilon)\delta\Upsilon\frac{\partial\widehat{u}}{\partial x}\right), \qquad x \in \Omega \tag{13.17}$$

$$\frac{\partial(D\hat{u}(\Upsilon)\delta\Upsilon)}{\partial\nu} + i\alpha(D\hat{u}(\Upsilon)\delta\Upsilon) = -iD\alpha(\Upsilon)\delta\Upsilon\hat{u}(\Upsilon), \quad x \in \Gamma.$$

$$(13.18)$$

where the function $Df(\Upsilon)\delta\Upsilon$ is defined by

$$(Df(\Upsilon)\delta\Upsilon)(x,\omega) = Dm(\Upsilon,x,\omega)\delta m(x) + Dq(\Upsilon,x,\omega)\delta q(x) \quad (13.19)$$

with

$$Dm(\Upsilon,x,\omega) = \frac{1}{\beta(q(x),\omega) - i\gamma(q(x),\omega)} \quad (13.20)$$

and

$$Dq(\Upsilon,x,\omega) = \frac{(1 - \beta(q(x),\omega) + i\gamma(q(x),\omega))m(x)}{(\beta(q(x),\omega) - i\gamma(q(x),\omega))^2}. \quad (13.21)$$

The function $D\alpha(\Upsilon)\delta\Upsilon$ in 13.18 is given by

$$D\alpha(\Upsilon)\delta\Upsilon = -\frac{Df(\Upsilon)\delta\Upsilon\alpha(\Upsilon)}{2f(\Upsilon)} \quad (13.22)$$

A weak form for problem 13.17–1.19 can be stated as follows: find $v \in H^1(\Omega)$ such that

$$-\omega^2(\rho D\hat{u}(\Upsilon)\delta\Upsilon, v) + (f(\Upsilon)\frac{\partial D\hat{u}(\Upsilon)\delta\Upsilon}{\partial x}, \frac{\partial v}{\partial x}) + \langle i\alpha f(\Upsilon)D\hat{u}(\Upsilon)\delta\Upsilon, v\rangle$$
$$= -(Df(\Upsilon)\delta\Upsilon\frac{\partial\hat{u}(\Upsilon)}{\partial x}, \frac{\partial v}{\partial x}) - \langle\frac{Df(\Upsilon)\delta\Upsilon\alpha(\Upsilon)}{2}, v\rangle, \quad v \in H^1(\Omega) (13.23)$$

Now the functional J has a Gateaux derivative with respect to the parameter Υ given by

$$J'(\Upsilon)\delta\Upsilon = \int_0^\infty \text{Re } (\delta(x - x_r)(D\hat{u}(\Upsilon)\delta\Upsilon(\cdot,\omega)), (\hat{u} - \hat{u}^o bs)(x_r,\omega)) \, d\omega.$$

$$(13.24)$$

Using this directional derivative in the following section we will discuss a steepest descent method to search for the parameters.

1.4 The Algorithms

Equations 13.17 and 13.18 allow us to compute the directional derivatives with respect to the parameter.

The problem of minimizing $J(\Upsilon)$ can be approached using iterative algorithms. Among them we worked the present problem using two of the most common; Newton-type algorithms and steepest descent.

We used the first one in [1] and it proved to be a very accurate but computationally expensive method.

Next we describe the implementation of the steepest descent method for our problem. It can be summarized as follows;

1. Pick an initial guess Υ^0. Set $n = 0$.
2. Calculate the directional derivative

$$d^n(x) = d(\Upsilon^n, x) = 2 \int_0^\infty \text{Re}\,(D\hat{u}(\Upsilon^n)^*(\delta(x - x_r)\hat{u}(x_r, \omega) - u^{obs}(\omega)))d\omega \tag{13.25}$$

3. Calculate the length of the step $\mu^n = \mu(\Upsilon^n)$

$$\mu^n = \frac{\displaystyle\int_\Omega |d^n(x)|^2\, dx}{\displaystyle\int_{-\infty}^\infty |D\hat{u}(\Upsilon^n)d^n|^2(x_r, \omega)\, d\omega + \eta\|d^n\|^2} \tag{13.26}$$

4. $\Upsilon^{n+1} = \Upsilon^n - \mu^n d^n$
5. If $\|\hat{u}^{n+1}(x_r, \cdot) - \hat{u}^{obs}\|^2 / \|\hat{u}^n(x_r, \cdot) - \hat{u}^{obs}\|^2 < \epsilon$ STOP. Otherwise set n=n+1, go to step 2.

1.5   The Adjoint Problem

Note that in step 2 of the algorithm we need to compute the adjoint of the operator $D\hat{u}(\Upsilon)$ applied to $\delta(x - x_r)(\hat{u}(x_r, \omega) - u^{obs}(\omega))$. This calculation can be carried out by solving an equation similar to that of the direct model as follows: Let $H(\cdot, \omega) \in L^2(\Omega)$ and W be the solution of

$$-\rho\omega^2 W(x, \omega) - \frac{\partial}{\partial x}\left(\overline{f(\Upsilon)}\frac{\partial W(x, \omega)}{\partial x}\right) = H(x, \omega) \quad x \in \Omega, \tag{13.27}$$

with boundary conditions

$$\frac{\partial\hat{u}(\Upsilon, x, \omega)}{\partial\nu} - i\bar{\alpha}\,W(x, \omega) = 0, \quad x \in \Gamma. \tag{13.28}$$

Then,

$$(D\hat{u}(\Upsilon))^* H \quad = -\int_\Omega (Df(\Upsilon))^* \frac{\partial\hat{u}}{\partial x}\frac{\overline{\partial W}}{\partial x}\, dx \tag{13.29}$$

$$-\frac{1}{2}\int_\Gamma i((Df(\Upsilon))^*\alpha\hat{u}\overline{W})(\sigma, \omega)d\sigma. \tag{13.30}$$

Now the function under the integral sign in 13.24 can be written as

$$(D\hat{u}(\Upsilon))^* \left(\delta(x - x_r)(\hat{u} - \hat{u}^{obs})(x_r, \omega) \right) \qquad (13.31)$$

$$= -(Df(\Upsilon))^* \frac{\partial \hat{u}}{\partial x} \frac{\overline{\partial W}}{\partial x}(x, \omega) - \frac{1}{2}i((Df(\Upsilon))^* \alpha \hat{u}\overline{W})(x, \omega)\chi_\Gamma(\text{13}.32)$$

where χ_Γ denotes the characteristic function of the set Γ and W is the solution of 13.27,13.28 for

$$H(x, \omega) = \delta(x - x_r)(\hat{u} - \hat{u}^{obs})(x, \omega). \qquad (13.33)$$

1.6 Numerical Experiments

The estimation procedure presented here was applied to solve a problem arising in inversion of seismic data.

The continuous estimation algorithm described in the previous section was discretized in space using an standard finite element procedure, and the integrations in frequency were computed over a finite range, as explained in [1]. The data was chosen as follows. The source function $\hat{S}(x, \omega)$ in (13.1) is the Fourier transform (on time) of $S(x, t)$ given by

$$S(x, t) = \delta(x - x_s)(-2\xi(t - t_s)e^{-\xi(t - t_s)^2}), \quad t \geq 0,$$

with $\xi = 8f_0^2$, $t_s = 5/4f_0$ and f_0 being the characteristic source frequency, chosen to be $25Hz$. The domain Ω consists of five layers as described in Table 1, with the source and the receiver location equal to $5m$. The mass density ρ was chosen to be $1gr/cm^3$.

Table 1 describes both the "true" layered model and the initial guess used in the example.

Figure 1 displays values of the relaxed plane wave modulus m_j for the true layered model, the initial guess and several stages of the algorithm.

Depth range(m)	True Model		Initial Guess	
	$m_j(10^{10}dynes/cm^2)$	q_j	$m_j(10^{10}dynes/cm^2)$	q_j
0–60	4	148	4	148
60–82.5	6	148	4	148
82.5–120	4	148	4	148
120–142.5	3	148	4	148
142.5–165	4	148	4	148

Table 1

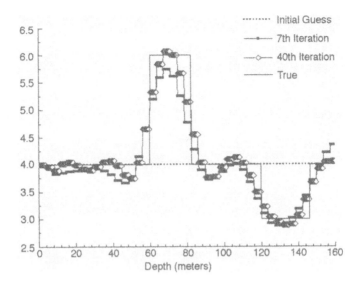

FIGURE 1. Values of the relaxed modulus at several steps of the estimation algorithm

2   References

[1] Armentano M.G., Fernández–Berdaguer, E.M., Santos J.E. (1995) A frequency domain parameter estimation procedure in viscoelastic layered media *Comp. Appl. Mat., 14* , pp. 191–216.

[2] Banks H. T., Groome G. M. Jr. (1973) Convergence theorems for parameter estimation by quasilinearization. *J. Math. Anal. Appl. Vol. 42* , pp 91–109.

[3] Banks H. T., Kunish K. (1989) *Estimation techniques for distributed parameter systems* Birkhauser, Boston.

[4] Burk J. J. and Weiss V. (1979) Non destructive evaluation of materials, Plenum Press.

[5] Chavent G., Kunisch K. (1994) Convergence of Tikhonov regularization for constrained ill-posed inverse problems *Inverse Problems Vol. 10*, pp. 63-76.

[6] Douglas J. Jr., Santos J. E., Sheen D., Bennethum L. S. (1993) Frequency domain treatment of one–dimensional scalar waves *Mathematical Models and Methods in Applied Sciences Vol. 3*, pp. 171–194.

[7] Fernández–Berdaguer E. M., Santos J. E.(1996) On the solution of an inverse scattering problem in one-dimensional acoustic media. *Computer Methods in Applied Mechanics and Engineering.* *vol 129*, pp.95–101.

[8] Ferry J. D. (1970) Viscoelastic properties of polymers John Wiley.

[9] Groetsch C. W. (1994) The theory of Tikhonov regularization for Fredholm equations of the first kind *Pitman, London.*

[10] Liu H. P., Anderson D. L., Kanamori H. (1976) Velocity dispersion due to anelasticity; implications for seismology and mantle composition. *Geophys. J. R. astr. Soc. Vol. 47* , pp. 41–58.

[11] Ravazzoli C. L., Santos J. E. (1995) *Consistency analysis for a model for wave propagation in anelastic media Latin American Applied Research Vol. 25* pp. 141–151.

[12] Scherzer O., Engl H. W., Kunisch K. (1993) Optimal a posteriori parameter choice for Thikhonov regularization for solving nonlinear ill–posed problems *SIAM J. Numer. Anal. Vol 30*, pp. 1796–1838.

[13] Stoll R. D. (1977) Acoustic waves in ocean sediments *Geophysics Vol. 42*, pp. 715-725.

[14] Stoll R. D. and Bryan G. M. (1969) Wave attenuation in saturated sediments *J. Acoust. Soc. Am. Vol. 47*, pp. 1440-1447.

[15] Tarantola A. (1984) Linearized inversion of seismic reflection data. *Geophysical Prospecting Vol. 32* , pp. 998–1015.

[16] Tarantola A.(1984) Inversion of seismic reflection data in the acoustic approximation. *Geophysics Vol. 49* , pp. 1259–1266.

[17] Tarantola A. (1986) A strategy for nonlinear elastic inversion of seismic reflection data *Geophysics vol. 51*, pp. 1893–1903.

[18] Tarantola A. (1987) Inverse problem theory - methods for data fitting and model parameter estimation Elseivier, New York .

[19] Tautenhahan U. (1994) Error estimates for regularized solutions of nonlinear ill–posed problems *Inverse Problems Vol. 10*, pp. 485–499.

[20] Tautenhahan U. (1994) On the asymptotical regularization of nonlinear ill–posed problems *Inverse Problems Vol. 10*, pp. 1405–1418.

[21] Zener C.(1948) Elasticity and anelasticity of metals . The University of Chicago Press.

Chapter 14

Numerical Modelling of Maxwell's Equations with Applications to Magnetotellurics

Luis Guarracino
Juan Enrique Santos

1   The Differential Model and the Iterative Hybrid Finite Element Domain Decomposition Algorithm

The magnetotelluric method consists on inferring the earth's electric conductivity distribution from measurements of natural electric and magnetic fields on the earth's surface. Information about characteristics of the earth's interior may be inferred from the conductivity models by exploiting relationships between electrical conductivity and physical properties of rocks such as composition, texture, temperature, porosity, and fluid content. The magnetotelluric method has been employed in oil exploration in regions where the seismic reflection procedure technique is very expensive or not possible to perform, as in volcanic regions. In the high-frequency range, it has been used to detect groundwater reservoirs and mineral deposits.

The natural sources of electromagnetic waves are the electric storms for frequencies higher than 1 Hz. Those of frequencies lower than 1 Hz are due to solar wind fluctuations which generate hydromagnetic waves in the magnetosphere. When those waves reach the ionosphere they are transformed into electromagnetic waves. The downgoing waves are assumed to be plane waves and normally incident upon the earth surface. Most of the incident energy is reflected, but a small fraction is transmitted into the earth, generating electric currents known as *telluric* currents. The amplitude, phase, and directional relations between electric and magnetic fields on the earth's surface will depend on the earth's conductivity profile.

Our objective is to present a domain decomposition procedure to determine the electromagnetic field induced inside the earth when a plane monocromatic electromagnetic wave arrives normally to the earth surface. It will be assumed that the earth can be represented by a two–dimensional body.

Numerical methods to solve the direct method in magnetotellurics have been presented by several authors. See, for example, [17], [20], [21], [22] for descriptions of global finite element procedures to solve the direct problem in magnetotellurics. The iterative hybridized finite element domain decomposition procedure presented here is related to that described by Deprés in [4], [5], and [6]. It is also related to the procedure presented by Douglas, et al. in [9] for the approximate solution of second–order elliptic problems. For general references on domain decomposition procedures for mixed finite element approximations we refer to [3], [12], [14], [15], [18], [19].

The organization of the paper is as follows. In Subsection 1.1 we describe the physical problem and the differential equation and boundary conditions employed for its mathematical description. In Subsection 1.2 we consider the formulation of a nonoverlapping domain decomposition procedure at the differential level, which is used to motivate the discrete domain decomposition technique. In Subsection 1.3 we present the iterative hybridized finite element domain decomposition procedure and prove some convergence results. Finally in Subsection 1.4 we present the results of experimental calculations.

1.1   The Differential Model

Let $\mathbf{E}$ and $\mathbf{H}$ denote the electric and magnetic fields, respectively. Then the time–harmonic Maxwell's equations state that

$$
\begin{aligned}
&\text{i)} \quad \nabla \times \mathbf{H} = (\sigma + i\omega\epsilon)\mathbf{E}, \\
&\text{ii)} \quad \nabla \times \mathbf{E} = -i\omega\mu\mathbf{H},
\end{aligned}
\tag{14.1}
$$

where ϵ is the electric permitivity, μ the magnetic permeability and σ the electric conductivity of the medium.

The boundary conditions at an interface Γ_{12} between two media Ω_1 and Ω_2 having different electromagnetic properties are

$$
\begin{aligned}
&\text{i)} \quad \mathbf{E}^{(1)} \cdot \tau_1 + \mathbf{E}^{(2)} \cdot \tau_2 = 0, \\
&\text{ii)} \quad \mathbf{H}^{(1)} \cdot \tau_1 + \mathbf{H}^{(2)} \cdot \tau_2 = 0, \\
&\text{iii)} \quad \sigma_1\mathbf{E}^{(1)} \cdot \nu_1 + \sigma_2\mathbf{E}^{(2)} \cdot \nu_2 = 0, \\
&\text{iv)} \quad \mu_1\mathbf{H}^{(1)} \cdot \nu_1 + \mu_2\mathbf{H}^{(2)} \cdot \nu_2 = 0,
\end{aligned}
\tag{14.2}
$$

where τ_i, ν_i are the unit tangent and outer unit normal, respectively, to the interface as seen from Ω_i. Equations (1.2) express the continuity of the tangential and normal components of the electric and magnetic fields, and the continuity of the normal component of the current density and magnetic flux at the interface Γ_{12}.

The terms $\sigma\mathbf{E}$ and $i\omega\epsilon\mathbf{E}$ in (1.1) represent conduction and displacement currents, respectively. The ratio between these two types of currents is

$$\frac{|iw\epsilon\mathbf{E}|}{|\sigma\mathbf{E}|} = \frac{w\epsilon}{\sigma} = \frac{2\pi f \frac{1}{36\pi}10^{-9}}{\sigma} \approx \frac{1}{2}\frac{f}{\sigma}10^{-10},$$

where f denotes temporal frequency. The range of frequencies used in magnetotellurics sounding do not exceed 100 Hz, and conductivities normally encountered in the earth lie in the range 0.001 to 1 $(\text{ohm.m})^{-1}$. Therefore, we will neglect the displacement currents. Also, the magnetic permeability μ will be taken to be constant and equal to the magnetic permeability μ_0 of the free space $(\mu_0 = 4\pi \cdot 10^{-7}$ Henry/m$)$. These are standard assumptions used in magnetotelluric modelling (see for example [20]). Consequently, from (1.1) we obtain

$$\begin{aligned} &\text{i)} \quad \nabla \times \mathbf{H} = \sigma\mathbf{E}, \\ &\text{ii)} \quad \nabla \times \mathbf{E} = -iw\mu_0\mathbf{H}. \end{aligned} \tag{14.3}$$

We are concerned with electromagnetic induction in a two-dimensional conductivity model consisting of horizontally–layered model Ω_p identifie with $R_+^2 = \{(x, z) \in R^2 : z > 0\}$ containing a cylindrical inhomogeneity Ω_s. The top layer Ω_0, associated with the air region, will be assigned a very low but positive conductivity σ_0.

The associated electric conductivity $\sigma(x, z)$, $z > 0$ will be assumed to be bounded below and above by positive constants and will be taken to be of the form

$$\sigma(x, z) = \begin{cases} \sigma_p(z) & \text{in } \Omega_p, \\ \sigma_p(z) + \sigma_s(x, z) & \text{in } \Omega_s. \end{cases}$$

The function $\sigma_s(x, z)$ represents the conductivity anomaly.

Let us assume a plane electromagnetic field $\mathbf{E}_0 = (E_{0x}, E_{0y}, 0)$, $\mathbf{H}_0 = (H_{0x}, H_{0y}, 0)$, of normal incidence upon the top surface of Ω_p. The electromagnetic response may be expressed as a linear combination of two uncoupled modes. We consider separately the electromagnetic field induced in the earth by the components across-strike

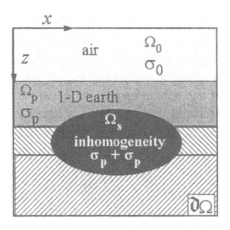

FIGURE 1. The conductivity model

E_{0x} and along-strike E_{0y} of the incident electric field. The E_{0x}-component generates an electric field $\mathbf{E} = (E_x, 0, E_z)$ and a magnetic field $\mathbf{H} = (0, H_y, 0)$, exciting the transverse magnetic mode (TM–mode). On the other hand, the E_{0y}-component induces an electric field $\mathbf{E} = (0, E_y, 0)$ and a magnetic field $\mathbf{H} = (H_x, 0, H_z)$, exciting the transverse electric mode (TE–mode). This paper deals specifically with the TE–mode.

Replacing the electromagnetic field components for the TE–mode in (1.3) we obtain the following scalar equations in R_+^2:

$$\begin{aligned}
\text{i)} \quad & \frac{\partial H_x}{\partial z} - \frac{\partial H_z}{\partial x} = \sigma(x, z) E_y, \\
\text{ii)} \quad & \frac{\partial E_y}{\partial x} = -i\omega\mu_0 H_z, \\
\text{iii)} \quad & \frac{\partial E_y}{\partial z} = i\omega\mu_0 H_x.
\end{aligned} \tag{14.4}$$

From equations (1.4) we obtain the following second–order TE–equation:

$$\nabla^2 E_y = \frac{\partial^2 E_y}{\partial x^2} + \frac{\partial^2 E_y}{\partial z^2} = i\omega\mu_0\sigma E_y \qquad \text{in } R_+^2. \tag{14.5}$$

The forward problem will be formulated in terms of the secondary or scattered field following the idea of Coggon, [2]. The total field consists of a *primary* part which can be calculated analytically and a *secondary* part that is defined as the difference between the total and the primary fields. This formulation allows for an increase in accuracy in the numerical calculations.

The secondary electric field u is defined by

$$u = E_y - E_{yp}, \qquad (14.6)$$

where $E_{yp}(z)$ is the physically meaningful solution (determined up to a constant) of (1.5) for the one-dimensional conductivity model $\sigma = \sigma_p(z)$. The electric field E_{yp} has a simple form that can be calculated explicitly [16].

Next let $\Omega \subset R_+^2$ be a rectangular domain containing the inhomogeneity Ω_s, big enough so that $\Gamma \equiv \partial\Omega$ is far away from Ω_s and the secondary field is negligible on $\partial\Omega$ and outside Ω. Thus we can formulate the inhomogeneous boundary value problem for the secondary electric field u as follows:

$$
\begin{aligned}
&\text{i)} \quad \nabla^2 u - iau = f \quad \text{in } \Omega,\\
&\text{ii)} \quad u = 0 \quad\quad \text{on } \partial\Omega,
\end{aligned}
\qquad (14.7)
$$

where $a(x,z) = \omega\mu_0\sigma(x,z)$ and $f(x,z) = i\omega\mu_0\sigma_s(x,z)E_{yp}(z)$.

Similarly, the secondary magnetic field $\mathbf{v}$ is defined by

$$\mathbf{v} = \mathbf{H}(x,z) - \mathbf{H}_{xp}(z), \qquad (14.8)$$

where $\mathbf{H}_{xp}(z)$ is the primary magnetic field (determined up to a constant) corresponding to the conductivity model $\sigma_p(z)$, whose analytical expression can be computed analytically.

Using (1.6), (1.8), (1.4.ii) and (1.4.iii) we obtain the following expressions for the components of the secondary magnetic field $\mathbf{v}$:

$$
\begin{aligned}
&\text{i)} \quad \tfrac{\partial u}{\partial x} = -i\omega\mu_0 v_z,\\
&\text{ii)} \quad \tfrac{\partial u}{\partial z} = i\omega\mu_0 v_x.
\end{aligned}
\qquad (14.9)
$$

This completes the formulation of the problem in terms of the secondary electromagnetic fields. In the next section we will present a paralelizable numerical procedure for the approximate solution of problem (1.7).

1.2   A Differential Domain Decomposition Formulation

The domain decomposition method consists of transforming the original differential problem (1.7) in a set of formally equivalent differential problems stated over a partition $\tau_h^{n_x,n_z}$ of Ω. This procedure leads in a natural way to the formulation of an iterative algorithm whose solution is approximated by a hybridized mixed finite element method. Applications of the domain decomposition technique to flow in porous media modelling and simulation of waves in dispersive media can be found in [8], [9], [10], and [11].

Let us consider a nonoverlapping partition $\tau_h^{n_x,n_z}$ of Ω into rectangles Ω_{jk}:

$$\Omega = \bigcup_{j,k=1}^{n_x,n_z} \Omega_{jk} \; ; \quad \Omega_{jk} \cap \Omega_{lm} = \emptyset \quad (j,k) \neq (l,m).$$

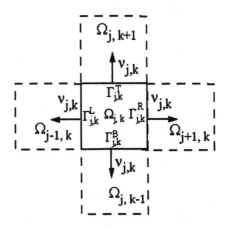

FIGURE 2. Mesh description

Let $\partial\Omega_{jk}$ denote the boundary of Ω_{jk}, which is decomposed in the form

$$\partial\Omega_{jk} = \bigcup_{s=L,R,T,B} \Gamma_{jk}^s,$$

where Γ_{jk}^L, Γ_{jk}^R, Γ_{jk}^T, Γ_{jk}^B denote the left, right, top, and bottom boundaries, respectively (see Figure 2). Also, ν_{jk} denotes the outer unit normal on Γ_{jk}^s.

Then, a formally equivalent domain decomposition formulation of (1.7) can be stated as follows: find u_{jk} such that

i) $\nabla^2 u_{jk} - ia u_{jk} = f, \quad$ in Ω_{jk},

ii) $u_{jk} = 0, \quad$ on $\partial\Omega$,

iii) $\dfrac{\partial u_{jk}}{\partial \nu_{jk}} + i\beta_{jk}^s u_{jk} = -\dfrac{\partial u_{j^*k^*}}{\partial \nu_{j^*k^*}} + i\beta_{jk}^s u_{j^*k^*}, \quad$ on Γ_{jk}^s, $\quad \Gamma_{jk}^s \cap \partial\Omega = \emptyset$,

$$(14.10)$$

where

$$
\begin{array}{llll}
j^* = j - 1, & k^* = k, & \text{on } \Gamma^L_{jk}, \\
j^* = j + 1, & k^* = k, & \text{on } \Gamma^R_{jk}, \\
j^* = j, & k^* = k - 1, & \text{on } \Gamma^B_{jk}, \\
j^* = j, & k^* = k + 1, & \text{on } \Gamma^T_{jk}.
\end{array}
$$

Note that the natural consistency conditions imposing continuity of u_{jk} and $\frac{\partial u_{jk}}{\partial \nu_{jk}}$ across $\partial \Omega_{jk}$ have been replaced by the Robin transmission boundary condition (1.10.iii), with $\beta^s_{jk} = \beta^s_{j^*k^*}$ being a real positive constant.

Also, since uniqueness holds for problem (1.10), the weak form for the associated Galerkin approximation to (1.10) will be well-defined. Existence for problem (1.10) cannot be assured because a characteristic function is not a multiplier for $H^s(\partial \Omega_{jk})$ for $s = \frac{1}{2}$ or $s = -\frac{1}{2}$. Consequently, the differential domain decomposition (1.10) and its hybridization to be defined below must be considered as a motivation for the iterative numerical procedure.

To define a hybridized form of (1.10) we introduce Lagrange multipliers ([1], [13])

$$
\lambda^s_{jk} \sim -\frac{\partial u_{jk}}{\partial \nu_{jk}}, \quad \text{on } \Gamma^s_{jk}, \ s = L, R, T, B, \tag{14.11}
$$

for all boundaries Γ^s_{jk} such that $\Gamma^s_{jk} \cap \partial \Omega = \emptyset$. Thus, we can state a hybrid formulation of (1.10) as follows: find (u_{jk}, λ_{jk}) such that

i) $\nabla^2 u_{jk} - i a u_{jk} = f, \quad$ in Ω_{jk},

ii) $u_{jk} = 0, \quad$ on $\partial \Omega$,

iii) $-\lambda^s_{jk} + i\beta^s_{jk} u_{jk} = \lambda^{s^*}_{j^*k^*} + i\beta^s_{jk} u_{j^*k^*}, \quad$ on $\Gamma^s_{jk}, \quad \Gamma^s_{jk} \cap \partial \Omega = \emptyset$, (14.12)

where

$$
\begin{array}{lll}
s^* = R & \text{if} & s = L, \\
s^* = L & \text{if} & s = R, \\
s^* = T & \text{if} & s = B, \\
s^* = B & \text{if} & s = T.
\end{array}
$$

Let us introduce some notation. Let $(H^s(\Omega), \|\cdot\|_s)$ denote the usual Sobolev space for all nonnegative integers s. In particular, $H^0(\Omega) = L^2(\Omega)$ and $\|\ \|_0$ is the usual L^2-norm, with inner product

$$
(v, w) = \int_\Omega v \overline{w} \, dx.
$$

Also, as usual, let $H_0^1(\Omega)$ be the closure of C_0^∞ in the $\|\cdot\|_1$ - norm. Next, if Γ is contained in $\partial\Omega$, let

$$\langle v, w \rangle_\Gamma = \int_\Gamma v\overline{w}\,dS$$

denote the inner product on $L^2(\Gamma)$, with the associated norm denoted by $|\cdot|_{0,\Gamma} = (\langle\cdot,\cdot\rangle_\Gamma)^{1/2}$, dS being the surface measure on Γ.

A weak formulation of (1.12) can be stated as follows: Find (u_{jk}, λ_{jk}) such that

i) $(\nabla u_{jk}, \nabla\varphi)_{\Omega_{jk}} + i(au_{jk}, \varphi)_{\Omega_{jk}} + \displaystyle\sum_{\Gamma_{jk}^s \cap \partial\Omega = \emptyset} \langle\lambda_{jk}^s, \varphi\rangle_{\Gamma_{jk}^s} = (f, \varphi)_{\Omega_{jk}},$

$\varphi \in H^1(\Omega_{jk}) \cap H_0^1(\Omega),$

ii) $-\lambda_{jk}^s + i\beta_{jk}^s u_{jk} = \lambda_{j^*k^*}^{s^*} + i\beta_{jk}^s u_{j^*k^*},$ on Γ_{jk}^s , $\Gamma_{jk}^s \cap \partial\Omega = \emptyset.$
$$(14.13)$$

Since the objective of the domain decomposition procedure is to localize the calculations on each subdomain Ω_{jk}, the hybrid formulation (1.13) suggests the following natural iterative algorithm: Choose $(u_{jk}^0, \lambda_{jk}^{s,0})$ arbitrarily; then compute $(u_{jk}^{n+1}, \lambda_{jk}^{s,n+1})$ by solving

i) $(\nabla u_{jk}^{n+1}, \nabla\varphi)_{\Omega_{jk}} + i(au_{jk}^{n+1}, \varphi)_{\Omega_{jk}} + \displaystyle\sum_{\Gamma_{jk}^s \cap \partial\Omega = \emptyset} \langle\lambda_{jk}^{s,n+1}, \varphi\rangle_{\Gamma_{jk}^s}$

$= (f, \varphi)_{\Omega_{jk}},$ $\varphi \in H^1(\Omega_{jk}) \cap H_0^1(\Omega),$

ii) $\lambda_{jk}^{s,n+1} = i\beta_{jk}^s u_{jk}^{n+1} - \lambda_{j^*k^*}^{s^*,n} - i\beta_{jk}^s u_{j^*k^*}^n,$ on Γ_{jk}^s, $\Gamma_{jk}^s \cap \partial\Omega = \emptyset.$
$$(14.14)$$

1.3   The Iterative Hybrid Finite Element Domain Decomposition Procedure

In this section, we will describe a finite element technique to approximately solve (1.14). To simplify, we treat only the case in which the partition of Ω defining the finite element spaces coincides with the partition $\tau_h^{n_x, n_z}$ used to decompose the differential problem.

Let

$V^h = \{\, \varphi \in C^0(\bar\Omega) : \varphi/\Omega_{jk} \in P_{1,1}(\Omega_{jk}), \varphi \equiv 0 \text{ on } \partial\Omega,\ 1 \le j \le n_x,$
$1 \le k \le n_z\},$

where $P_{1,1}(\Omega_{jk})$ denotes bilinear functions on Ω_{jk}.

We define the local finite element spaces by

$$\begin{aligned}
&\text{i)} \quad V_{jk}^h = V^h/\Omega_{jk},\\
&\text{ii)} \quad W_{jk}^h = \bigcup_s P_1(\Gamma_{jk}^s), \quad \Gamma_{jk}^s \cap \partial\Omega = \emptyset,
\end{aligned}$$

with $P_1(\Gamma_{jk}^s)$ denoting linear functions on Γ_{jk}^s. Note that there are two copies of P_1 assigned to each interior boundary Γ_{jk}^s.

The iterative finite element domain decomposition procedure is defined as a discrete analogue of (1.14): Choose $(u_{jk}^{h,0}, \lambda_{jk}^{h,s,0}) \in V_{jk}^h \times W_{jk}^h$ arbitrarily; then, compute $(u_{jk}^{h,n+1}, \lambda_{jk}^{h,s,n+1})$ as the solution of the equations

$$\begin{aligned}
\text{i)} \quad & (\nabla u_{jk}^{h,n+1}, \nabla\varphi)_{\Omega_{jk}} + i(a u_{jk}^{h,n+1}, \varphi)_{\Omega_{jk}} + \sum_{\substack{s \\ \Gamma_{jk}^s \cap \partial\Omega = \emptyset}} \langle \lambda_{jk}^{h,s,n+1}, \varphi \rangle_{\Gamma_{jk}^s} \\
& = (f, \varphi)_{\Omega_{jk}}, \qquad \varphi \in V_{jk}^h, \\
\text{ii)} \quad & \lambda_{jk}^{h,s,n+1} = i\beta_{jk}^s u_{jk}^{h,n+1} - \lambda_{j^*k^*}^{h,s^*,n} - i\beta_{jk}^s u_{j^*k^*}^{h,n}, \quad \text{on } \Gamma_{jk}^s, \quad \Gamma_{jk}^s \cap \partial\Omega = \emptyset.
\end{aligned}$$
$$(14.15)$$

Next we proceed to analyze the convergence of the iterative procedure (1.15). Let u^h be the solution of the standard finite element Galerkin procedure for (1.7) and let $\lambda^{h,s}$ be the Lagrange multiplier for the flux variable in that formulation. Since the convergence of the solution u^h of the standard Galerkin method to the solution u of (1.7) is well–known, we will show the convergence of $u_{jk}^{h,n}$ to $u_{jk}^h \equiv u^h/\Omega_{jk}$. For simplicity we analyze the special case $\beta_{jk}^s = \beta > 0$.

Set

$$\begin{aligned}
&\text{i)} \quad \epsilon_{jk}^n = u_{jk}^h - u_{jk}^{h,n}, \\
&\text{ii)} \quad \mu_{jk}^{s,n} = \lambda_{jk}^{h,s} - \lambda_{jk}^{h,s,n}.
\end{aligned}$$
$$(14.16)$$

Then, the iteration error equations are

$$\begin{aligned}
\text{i)} \quad & (\nabla\epsilon_{jk}^n, \nabla\varphi)_{\Omega_{jk}} + i(a\epsilon_{jk}^n, \varphi)_{\Omega_{jk}} + \sum_{\substack{s \\ \Gamma_{jk}^s \cap \partial\Omega = \emptyset}} \langle \mu_{jk}^n, \varphi \rangle_{\Gamma_{jk}^s} = 0, \quad \varphi \in V_{jk}^h, \\
\text{ii)} \quad & \mu_{jk}^{n+1} = i\beta\epsilon_{jk}^{n+1} - (\mu_{j^*k^*}^n + i\beta\epsilon_{j^*k^*}^n), \quad \text{on } \Gamma_{jk}^s, \quad \Gamma_{jk}^s \cap \partial\Omega = \emptyset.
\end{aligned}$$
$$(14.17)$$

Choose $\varphi = \epsilon_{jk}^n$ in (1.17.i) to obtain:

$$(\nabla\epsilon_{jk}^n, \nabla\epsilon_{jk}^n)_{\Omega_{jk}} + i(a\epsilon_{jk}^n, \epsilon_{jk}^n)_{\Omega_{jk}} + \sum_{\substack{s \\ \Gamma_{jk}^s \cap \partial\Omega = \emptyset}} \langle \mu_{jk}^n, \epsilon_{jk}^n \rangle_{\Gamma_{jk}^s} = 0.$$
$$(14.18)$$

Thus, from (1.18) we see that

$$(a\epsilon_{jk}^n, \epsilon_{jk}^n)_{\Omega_{jk}} + Imag\left(\sum_{\Gamma_{jk}^s \cap \partial\Omega = \emptyset} \langle \mu_{jk}^n, \epsilon_{jk}^n \rangle \Gamma_{jk}^s \right) = 0. \tag{14.19}$$

Next note that for any pair of complex numbers p and q we have

$$|\pm p + iq|^2 = |p|^2 + |q|^2 \pm 2Imag(p\bar{q}). \tag{14.20}$$

Combining (1.18) and (1.20) we obtain

$$\sum_{\Gamma_{jk}^s \cap \partial\Omega = \emptyset} |\pm \mu_{jk}^n + i\beta\epsilon_{jk}^n|^2_{0,\Gamma_{jk}^s} = \sum_{\Gamma_{jk}^s \cap \partial\Omega = \emptyset} \left(|\mu_{jk}^s|^2_{0,\Gamma_{jk}^s} + \beta^2 |\epsilon_{jk}^n|^2_{0,\Gamma_{jk}^s} \right)$$
$$\mp 2\beta(a\epsilon_{jk}^n, \epsilon_{jk}^n)_{\Omega_{jk}}. \tag{14.21}$$

Set

$$R(\epsilon, \mu) = \sum_{j,k} \sum_{\Gamma_{jk}^s \cap \partial\Omega = \emptyset} |-\mu_{jk} + i\beta\epsilon_{jk}|^2_{0,\Gamma_{jk}^s}$$
$$= \sum_{j,k} \sum_{\Gamma_{jk}^s \cap \partial\Omega = \emptyset} \left(|\mu_{jk}|^2_{0,\Gamma_{jk}^s} \right. \tag{14.22}$$
$$\left. +\beta^2|\epsilon_{jk}|^2_{0,\Gamma_{jk}^s} \right) + 2\beta \sum_{j,k}(a\epsilon_{jk}, \epsilon_{jk})_{\Omega_{jk}},$$

$$R^n = R(\epsilon^n, \mu^n) = \sum_{j,k} \sum_{\Gamma_{jk}^s \cap \partial\Omega = \emptyset} |-\mu_{jk}^n + i\beta\epsilon_{jk}^n|^2_{0,\Gamma_{jk}^s}. \tag{14.23}$$

Since

$$|-\mu_{jk}^n + i\beta\epsilon_{jk}^n|^2_{0,\Gamma_{jk}^s} = |\mu_{jk}^{n-1} + i\beta\epsilon_{jk}^{n-1}|^2_{0,\Gamma_{jk}^s},$$

from (1.22) we see that

$$R^n = \sum_{j,k} \sum_{\Gamma_{jk}^s \cap \partial\Omega = \emptyset} |\mu_{jk}^{n-1} + i\beta\epsilon_{jk}^{n-1}|^2_{0,\Gamma_{jk}^s}$$
$$= \sum_{j,k} \sum_{\Gamma_{jk}^s \cap \partial\Omega = \emptyset} \left(|\mu_{jk}^{n-1}|^2_{0,\Gamma_{jk}^s} + \beta^2|\epsilon_{jk}^{n-1}|^2_{0,\Gamma_{jk}^s} \right)$$
$$-2\beta \sum_{j,k}(a\epsilon_{jk}^{n-1}, \epsilon_{jk}^{n-1})_{\Omega_{jk}}.$$

Thus,

$$R^n = R^{n-1} - 4\beta \sum_{j,k} (a\epsilon_{jk}^{n-1}, \epsilon_{jk}^{n-1})_{\Omega_{jk}}, \qquad (14.24)$$

so that

$$R^n = R^0 - 4\beta \sum_{i=0}^{n-1} \sum_{j,k} (a\epsilon_{jk}^i, \epsilon_{jk}^i)_{\Omega_{jk}}. \qquad (14.25)$$

Since R^n is a decreasing sequence of nonnegative numbers, from (1.25) we conclude that

$$\sum_{n=1}^{\infty} \sum_{jk} (a\epsilon_{jk}^n, \epsilon_{jk}^n)_{\Omega_{jk}} < \infty,$$

and, consequently,

$$\epsilon_{jk}^n \to 0 \ \text{ as } \ n \to \infty \ \text{ in } L^2(\Omega). \qquad (14.26)$$

Thus, $u_{jk}^{h,n} \to u^h/\Omega_{jk}$ as $n \to \infty$ and in the limit the standard Galerkin approximation is obtained. This completes the convergence analysis for (1.15).

1.4   Experimental Calculations

To illustrate the implementation of the iterative domain decomposition (1.9) to obtain approximations to the secondary field u, we used a simple two-layer earth model with an imbedded conductive prism as shown in Figure 3.

The electric conductivity for this model was chosen as follows:

$$\sigma(x, z) = \begin{cases} \sigma_0 = 0.000001 \ (\text{ohm.m})^{-1} & \text{air,} \\ \sigma_1 = 0.01 \ (\text{ohm.m})^{-1} & \text{layer 1,} \\ \sigma_2 = 0.05 \ (\text{ohm.m})^{-1} & \text{layer 2,} \\ \sigma_3 = 0.5 \ (\text{ohm.m})^{-1} & \text{prism.} \end{cases}$$

Approximations to the secondary magnetic field components v_x and v_z were obtained using Lagrange multiplier values to approximate $\frac{\partial u}{\partial x}$ and $\frac{\partial u}{\partial z}$ in (1.9). Thus, the iterative domain decomposition method allows us to approximate simultaneously the three secondary field components u, v_x, and v_z. Figure 4 shows the secondary electric field u in Ω for a frequency of 1 Hz.

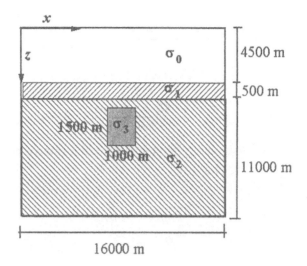

FIGURE 3. Two-layer model with a vertical prism

1.5   Some remarks on the implementation of the algorithm

In order to increase the efficiency of the iterative algorithm (1.15)
we implemented the following alternatives.

Underrelaxation of the solution

The underrelaxation of the solution consists of replacing $u_{jk}^{h,n+1}$ by
$\frac{1}{2}(u_{jk}^{h,n+1} + u_{jk}^{h,n})$ and $\lambda_{jk}^{h,s,n+1}$ by $\frac{1}{2}(\lambda_{jk}^{h,s,n+1} + \lambda_{jk}^{h,s,n})$ after each iteration step. This simple modification reduces significantly the number
of iterations to achieve a prescribed error reduction (see Table 1).

Red-black iterative scheme

A full iteration step consists of computing the solution in all subdomains Ω_{jk}. To compute the solution in the subdomain Ω_{jk} we use
"information" from nearest neighbors given by the Robin boundary
condition on the boundaries Γ_{jk}^s. The idea of the *red-black* scheme is
to use updated "information" to speed-up the convergence.

Let us consider a *red-black* scheme of Ω as in figure 5. Then, each
iteration can be performed as indicated below:

1. Solve the system on the *red* subdomains using the solution
   and Lagrange multipliers for neighboring *black* subdomains as
   boundary conditions.

2. Update Lagrange multipliers for *red* subdomains.

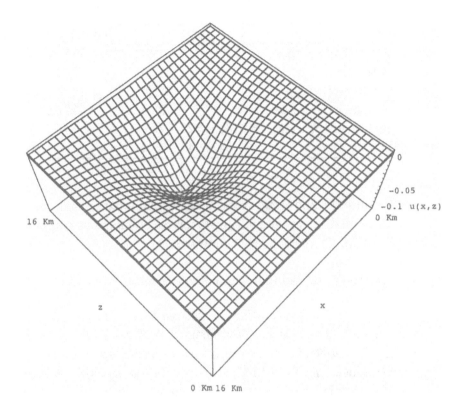

FIGURE 4. Secondary electric field u in Ω for a frequency of 1 Hz.

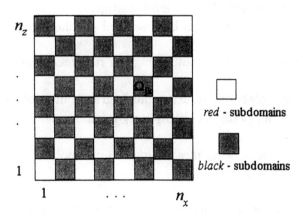

FIGURE 5. *Red-black* scheme.

TABLE 14.1. Number of iterations and CPU time.

Scheme	Iterations		Time (sec)
without relaxation		7919	7398.7
with relaxation		2626	2518.8
red-black (usual)		2066	1576.7
red-black (cache-aware)		2066	1221.6
multigrid	32×32 :	679	341.7
	64×64 :	244	

3. Perform the above two steps on *black* subdomains, using the solution and Lagrange multipliers for *red* subdomains computed in i) and ii) as boundary conditions.

4. underrelax the solution on all subdomains.

In the usual implementation of the *red-black* scheme, the data passes twice through *cache* (fast memory of the processor). Implementing the *red-black* scheme using a *cache-aware* algorithm according to the idea suggested in [7] the data passes only once through *cache*. This implementation reduced the computational cost as shown in Table 1.

Multigrid scheme
We also inplemented a two-level multigrid scheme. The procedure to solve a $n_x \times n_z$ mesh can be described as follows:

1. Solve a $\frac{n_x}{2} \times \frac{n_z}{2}$ mesh using the *red-black* scheme.

2. Interpolate the solution and Lagrange multipliers computed in i) to all points of the $n_x \times n_z$ mesh.

3. Solve a $n_x \times n_z$ mesh using the interpolated values computed in ii) as initial guesses.

The different schemes described above were implemented on a Sun SPARC Station 20 system with Sun4m processor. Table 1 shows the perfomance of the different schemes for a 64×64 uniform grid and a relative error reduction of 0.0001.

2 Conclusions

We presented an iterative domain decomposition method to solve the two-dimensional forward problem in magnetotelluric for the $TE-$ mode which allows us to approximate simultaneously the three secondary field components. As other domain decomposition techniques, this procedure is specially suited for implementation in computers of parallel architecture.

3 References

[1] D. N. Arnold and F. Brezzi Mixed and nonconforming finite element methods: implementation, postprocessing and error estimates In *R.A.I.R.O. Modélisation Mathématique et Analyse Numérique Vol. 19*, pp. 7 - 32, 1985.

[2] J. H. Coggon Electromagnetic and electrical modelling by the finite element method. In *Geophysics, Vol. 36*, pp. 132 - 155, 1971.

[3] L. C. Cowsar and M. F. Wheeler Parallel domain decomposition method for mixed finite elements for elliptic partial differential equations. In *Proceedings of the Fourth International Symposium on Domain Decomposition Methods for Partial Differential Equations.* R. Glowinski, Y. Kuznetsov, G. Meurant, J. Périaux, and O. Widlund (eds.), 1990.

[4] B. Després Domain decomposition method and the Helmholtz problem. In *Mathematical and Numerical Aspects of Wave Propagation Phenomena.* G. Cohen, L. Halpern, and P. Joly (eds.), SIAM, Philadelphia, pp. 44 - 52, 1991.

[5] B. Després Domain decomposition method and the Helmholtz problem (part II). In *Mathematical and Numerical Aspects of Wave Propagation.* R. Kleinman, et al. (eds.), SIAM, Philadelphia, pp. 197 - 206, 1993.

[6] B. Després Méthodes de décomposition de domaines pour les problèmes de propagation d'ondes en régime harmonique. *Thèse.* Université Paris IX Dauphine, UER Mathématiques de la Décision, 1991.

[7] C. C. Douglas Caching in with multigrid algorithms: problems in two dimensions. To appear, 1996.

[8] J. Douglas, Jr., P. J. Paes Leme, F. Pereira, and L. M. Yeh. A massively parallel iterative numerical algorithm for immiscible flow in naturally fractured reservoirs. In *Flow in Porous Media, International Series of Numerical Mathematics, Vol. 114.* Birkhäuser Verlag, Basel, pp. 75 - 94, 1993.

[9] J. Douglas, Jr., P. J. S. Paes Leme, J. E. Roberts, and J. Wang. A parallel iterative procedure applicable to the approximate solution of second order partial differential equations by mixed finite element methods. In *Numerische Mathematik, Vol. 65,* pp. 95 - 108, 1993.

[10] J. Douglas, Jr., F. Pereira, J. E. Santos. A parallelizable approach to the simulation of waves in dispersive media. In *Proceedings of the Third International Conference on Mathematical and Numerical Aspects of Wave Propagation.* G. Cohen (edr.), SIAM, pp. 673 - 682, 1995.

[11] J. Douglas, Jr., F. Pereira and L. M. Yeh. A parallelizable characteristic scheme for two phase flow I: Single porosity models. In *Computational and Applied Mathematics, Vol. 14,* pp. 73 - 96, 1995.

[12] R. E. Ewing and J. Wang. Analysis of the Schwarz algorithm for mixed finite element methods. In *R.A.I.R.O. Modélisation Mathématique et Analyse Numérique, Vol. 26,* pp. 739 - 756, 1992.

[13] B. X. Fraeijs de Veubeke. Stress function approach. In *International Congress on the Finite Element Method in Structural Mechanics.* Bournemouth, 1975.

[14] R. Glowinski, W. Kinton, and M. F. Wheeler. Acceleration of domain decomposition algorithms for mixed finite elements by multi–level methods. In *Third International Symposium on Domain Decomposition Methods for Partial Differential Equations.* R. Glowinski (edr.), SIAM, Philadelphia, pp. 263 - 290, 1990.

[15] R. Glowinski and M. F. Wheeler. Domain decomposition and mixed finite element methods for elliptic problems. In *Domain Decomposition Methods for Partial Differential Equations.* SIAM, Philadelphia, pp. 144 - 172, 1988.

[16] L. Guarracino. Un método iterativo paralelizable de descomposición de dominio para modelado magnetotelúrico. In *Práctica de la especialidad.* Facultad de Cs. Astronómicas y Geofísicas, UNLP, 1995.

[17] K. H. Lee and H. F. Morrison. A numerical solution for the electromagnetic scattering by a two–dimensional inhomogeneity. In *Geophysics, Vol. 50*, pp. 466 - 472, 1985.

[18] P. L. Lions. On the Schwarz alternating method, I. In *Domain Decomposition Methods for Partial Differential Equations.* R. Glowinski, G. Golub, G. Meurant, and J. Periaux (eds.), SIAM, Philadelphia, 1988.

[19] P. L. Lions. On the Schwarz alternating method III: a variant for nonoverlapping subdomains. In *Domain Decomposition Methods for Partial Differential Equations.* T. F. Chan, R. Glowinski, J. Periaux, and O. B. Widlund (eds.), SIAM, Philadelphia, pp. 202 - 223, 1990.

[20] R. L. Mackie, T. R. Madden, and P. E. Wannamaker. Three-dimensional magnetotelluric modeling using difference equations – Theory and comparisons to integral equation solutions. In *Geophysics, Vol. 58*, pp. 215 - 226, 1993.

[21] B. J. Travis and A. D. Chave. A moving finite element method for magnetotelluric modelling. In *Physics of the Earth and Planetary Interiors, Vol. 53*, pp. 432 - 443, 1989.

[22] P. E. Wannamaker and J. A. Stodt. A stable finite element solution for two–dimensional magnetotelluric modelling. In *Geophys. J. R. Astr. Soc., Vol. 88*, pp. 277 - 296, 1987.